Environmental Consulting Fundamentals

Environmental Consulting Fundamentals

Investigation, Remediation, and Brownfields Redevelopment

Second Edition

Benjamin Alter

CRC Press
Taylor & Francis Group
Boca Raton London New York

CRC Press is an imprint of the
Taylor & Francis Group, an **informa** business

CRC Press
Taylor & Francis Group
6000 Broken Sound Parkway NW, Suite 300
Boca Raton, FL 33487-2742

Library of Congress Cataloging-in-Publication Data

Names: Alter, Benjamin, author.
Title: Environmental consulting fundamentals : investigation, remediation, and brownfields redevelopment / authored by Benjamin Alter.
Description: Second edition. | Boca Raton : Taylor & Francis, a CRC title, part of the Taylor & Francis imprint, a member of the Taylor & Francis Group, the academic division of T&F Informa, plc, 2019. | Includes bibliographical references.
Identifiers: LCCN 2019002047| ISBN 9781138613201 (hardback : acid-free paper) | ISBN 9780429464379 (ebook)
Subjects: LCSH: Environmental impact analysis. | Environmental impact consultants.
Classification: LCC TD194.6 .A47 2019 | DDC 333.71/4068--dc23
LC record available at https://lccn.loc.gov/2019002047

Visit the Taylor & Francis Web site at
http://www.taylorandfrancis.com

and the CRC Press Web site at
http://www.crcpress.com

eResource material is available for this title at
https://www.crcpress.com/9781138613201

To my wife, Jean, who encourages me to pursue my dreams, revels in

my successes, and is there for me through my failures.

Contents

Section II   Site Investigations and Remediations

Section III   Land Development and Redevelopment

Section IV   Indoor Environmental Concerns

Preface to Second Edition

My goal, when I wrote the first edition of this book, was to provide the textbook that wasn't available when I was teaching my course at Hunter College in the 2000s. I believed that the lack of a survey book for the environmental consulting profession was a gap that needed to be filled. That supposition was vindicated by all of you who purchased and read the first edition of my book.

This new edition of the book is more of an enhancement than an update. There have been many changes in the field of environmental consulting since I wrote the first book, and I've incorporated many of those changes into the book. More importantly, the scope of the first edition was limited essentially to what I was able to cover in a semester-long course on environmental consulting. As I indicated in the preface to the first book, most of the topics covered in the book's chapters could be books themselves. As I had done in my lectures, I had truncated portions of many chapters and omitted other topics altogether so that the first edition of the book could be read through and taught in a college semester.

In this new edition of the book, I removed those self-imposed shackles. I expanded the scope of many chapters, particularly the chapter on environmental projects. It now includes many of the environmental consulting elements that I want my entry-level staff to know—the parts of an environmental project, how the team is assembled, and how the business side of the consulting practice works. I added discussion on emerging contaminants to the chapter on chemistry, and discussions regarding the Superfund sites and the RCRA (Resource Conservation and Recovery Act) corrective action process. I added a new chapter on brownfields for the same reason I was motivated to write this book in the first place—I could not find in the literature an adequate, readable introduction to brownfields. For the first time, I now understand many of the nuances and terminologies that are woven into the brownfields process.

There are two other major enhancements in the second edition of the book. Most of the chapters of this edition contain descriptions of what an environmental consultant actually does regarding the chapter's subject matter, whereas just a few chapters in the first edition of the book contained such discussions. Many of us veterans of the profession overestimate how difficult the transition can be from academic studies to a job that involves so much cross-disciplinary activity. Lastly, all but the introductory chapters contain problems and exercises to assist the professor in using this book in an academic setting.

CRC Press will establish and maintain a web site that will support this book. On that web site, I periodically post additional problems and exercises, so that the book will be a living textbook, enhancing the educational experience and adapting to changes in the profession. I also will have PowerPoint presentations available to academicians who intend to build a course around this book or use portions of it for their lectures.

I again thank all of you for your support and encouragement. It is a major task to write a book, especially when you have a full-time job and other extracurricular responsibilities and interests. It is particularly gratifying when that book is read and appreciated. I hope that you will find this enhanced second edition even more worthwhile. I hope that it encourages you to become an environmental consultant, a profession that has given me so much satisfaction.

Ben Alter

Preface to First Edition

This book was born of necessity.

In 2000, I began teaching a graduate course, entitled "Environmental Investigations and Remediations" at Hunter College in New York City. I designed the course to be a survey course that would provide someone interested in becoming an environmental consultant upon graduation with the basic building blocks needed to start their career in this dynamic, multi-disciplinary field. There was only one problem—I couldn't find a textbook to support the course. There were textbooks out there that covered only 1/3 of the course material, but none that covered the other 2/3 of the course. I was forced to cover that material with a combination of regulations and guidance documents that were, in general, far too detailed and obtuse for a survey course, or with material written for "citizens" that was far too simplistic.

So, in the course of the ten years that I taught the course, I started to fill in the reading assignments with my own writings, which led me to pursuing the "fully Monty," namely this book.

This is the book that I wish was available when I was designing the course. In fact, it is the book I wish someone had written when I began my career as an environmental consultant! It is designed for the student as well as the beginning practitioner.

Each chapter of this book covers a topic that merits a book itself; however, they are *not* intended to make the reader an expert in any of the topics presented. Nor, for that matter, will the reader be able to go out and perform an environmental investigation or remediation after reading this book. That skill requires not just the knowledge obtained from reading this book, but the field experience, ideally obtained under the apprenticeship of a more experienced hand.

Rather than placing an emphasis of formulas, equations, and regulatory requirements, this book emphasizes the *thought processes* that go into designing an environmental study, interpreting the data obtained from the study, and selecting the next step, be it further investigation or remediation. It also discusses the specific roles played by the environmental consultant, and the roles played by others in the investigation and remediation activities, thereby giving the reader the "big picture" of how these activities actually happen in real life.

The book begins with an overview of the environmental consultant—the typical consultant's educational background, formal training, and on-the-job training. It follows with brief summaries on three of the important building blocks of environmental investigations and remediations—their regulatory structure in the United States, and the scientific underpinnings of the processes that occur in the environment. The next several chapters take the reader through the steps of subsurface investigations and remediations, going from Phase I and Phase II environmental site assessments through to remediation. This sequence of chapters is followed by a chapter on ecological studies and a chapter on environmental impact assessments, a huge sub-field of environmental consulting.

The next set of chapters takes the reader indoors, as they cover environmental issues related to buildings (being from the Northeast and teaching in the fall semester, this transition dove-tailed nicely with the ensuing seasonal change!) The topics covered include asbestos, lead-based paint, radon, mold, and indoor air quality. The book concludes with a short chapter on describing a typical environmental consulting project, including designing the scope of work and developing a prospective budget and project schedule.

Connecting many of the chapters are examples of environmental problems at a fictitious factory and office building. The examples are designed to put flesh on some of the theoretical concepts presented in the book. They also provide a thread through which many of the book's chapters can become connected in the mind of the reader.

Because of its diversity and ever-changing landscape, I've never stopped learning the varied aspects of environmental consulting. Neither will you. Enjoy the journey! I'm pleased you've chosen this book as one of your starting points.

Acknowledgments to Second Edition

As with the first edition of the book, I was fortunate to have many knowledgeable and diligent people review the second edition of the book. First off, I thank Dr. Larry Feldman, LSP, Jamie Bents, AICP, Blaine Rothauser, Dr. Ben Sallemi, CIH, and Deborah Zarta-Gier, CNRP, just a few of my valued colleagues at GZA GeoEnvironmental, Inc. This edition also benefits from the expertise and encouragement of Dr. Howard Apsan of City University; Dr. Jorge Berkowitz, LSRP of JHB, LLC; Colleen Kokas of Environmental Liability Transfer; Prof. Alec Gates at the Earth and Environmental Sciences Department of Rutgers-Newark; Prof. Barry Hersh of New York University's Schack Institute of Real Estate; Will Moody at Provectus Environmental Products, Inc.; and Paul Simms at Alpha Analytical. I also owe a debt of gratitude to Kim Kramer, Shirleen Laubenthal, and David Allen, P.G., with the Global Casualty Group at AIG. Thanks also to Rob Gray of Abscope Environmental, Inc. and Bob Kreilick of Summit Drilling Co. for graciously contributing photographs for use in the book.

A big TY to Marilyn Rose (marilynroseart.com), who did her usual amazing job preparing the illustrations for this book. Last but not least, I thank all of my colleagues at GZA GeoEnvironmental, Inc. who encouraged me to tackle this huge but gratifying project.

Author

 **Benjamin Alter** has been an environmental consultant for over 30 years. He is a Principal and Senior Vice President with GZA GeoEnvironmental, Inc. in its Fairfield, New Jersey office. He also co-chairs GZA's technical practice for environmental investigations. Prior to becoming a consultant, Alter was a geophysicist in the oil industry, tasked with exploring for oil and natural gas along the Gulf Coast. He has a Bachelor of Science degree in geology and mathematics from State University of New York (SUNY)/Albany, a Master of Science degree in geophysics from Cornell University, and a Master of Business Administration degree in finance and management from Columbia University.

As an environmental consultant, he has designed and managed multiple remedial investigations and remediations and has conducted hundreds of site assessments throughout the United States. Alter has provided litigation services for numerous hazardous waste cases and has been a key contributor in the development of New Jersey's Licensed Site Remediation Professional (LSRP) program. He sits on the Board of Trustees of the LSRP Association and co-chairs its College Outreach Committee. He also sits on the Board of Trustees of the LSRPA Foundation, which he founded. The Licensed Site Remediation Professional Association Foundation (LSRPAF) provides scholarships to undergraduate and graduate students who are interested in becoming environmental consultants in the State of New Jersey.

Alter was an adjunct professor at the Hunter College School of Health Sciences in New York City from 2000 to 2009, which was the initial source of inspiration for this book. He has published numerous articles and given numerous presentations on environmental investigations and remediations. He currently teaches a continuing education course entitled, "Environmental Due Diligence in New Jersey" and a course on technical writing.

In his spare time, Alter is a piano accompanist with a variety group, sings in various choirs, and writes lyrics, vocal arrangements, and musical compositions. He also is an enthusiast of baseball history.

Section I

Environmental Consulting
A Perspective

1

What Is Environmental Consulting?

1.1 The Environment and Environmental Hazards

To understand what constitutes environmental consulting, we first must understand the meaning of "the environment." *Webster's Dictionary* defines "environment" as:

> "the complex of physical, chemical, and biotic factors (as climate, soil, and living things) that act upon an organism or an ecological community and ultimately determine its form and survival."

Let's dissect this definition and discuss how it pertains to the contents of this book.

As the definition indicates, physical factors include climate and soil, where climate includes the air and sunlight, and water, one of the fundamental requirements for life on Earth and the topic of discussion in many of the book's chapters. The chemical factors include the interactions between many of these physical factors as well as chemicals that occur naturally and those introduced by mankind. The "biotic factors" indicated in the definition encompass the full range of living things: microbial, plant, and animal life.

An *environmental hazard* is a condition that can affect the ecosystem and living things (generally termed "receptors"). Chemicals that can create environmental hazards are known as *pollutants* or *contaminants*. For an environmental hazard to exist, three conditions must be present (see Figure 1.1). There must be a *source* of the pollution, a *receptor* for the pollution, and a *pathway* connecting the two.

Although it might appear convenient to place receptors into one of the three categories of living things (microorganisms, plants, and animals), in the context of environmental investigations and remediations it is better to place receptors into one of two categories: humans and everything else. Numerous environmental regulations, such as regulations regarding asbestos-containing materials and lead-based paint, implicitly or explicitly recognize humans as the sole potential receptor.

Sources of contamination can be quite varied. They may be natural sources, such as rock formations that contain lead, arsenic, or emit radon. They may be chemicals or petroleum products that are or were manufactured for the benefit of humans, but which are toxic to humans and a wide variety of biota. They may be building materials such as the above-mentioned asbestos and lead-based paint that were manufactured for the benefit of humans but are toxic to humans exposed to these materials.

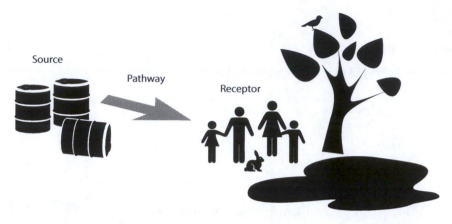

FIGURE 1.1
The three components of an environmental hazard.

Within the category of receptors, there is a subset of receptors, known as *sensitive receptors*, which are so named because of their heightened vulnerability to contaminants. Sensitive human receptors typically include children, the elderly, and disabled people, and locations where these populations typically are found, such as schools, parks, and nursing homes. Depending on the contaminant, this category might also include immune-compromised people, people with allergies, and other such populations. In the environment, sensitive receptors may include wetlands (discussed in Chapter 12), surface water bodies, rainforests, endangered and threatened species, and their habitats.

The most common contaminant pathways leading to receptors in the environment include:

Water: Water can transport contaminants from the source as solutes (chemicals dissolved in water) or as particulates, which are solids suspended in water. In general, water is considered either *"surface water,"* such as rivers, lakes, and salt water bodies, and *"groundwater,"* which is water entrained in geologic formation. Groundwater is discussed in Chapter 5.

Soil and Rock: Contaminants can move through the void spaces of soil and rock. See Chapter 5 for a discussion of this pathway.

Overland Pathways: Chemical spills, especially in liquid form, can migrate on the surface through overland paths, especially if the surface is paved or otherwise impermeable to percolation into the underlying earth.

Air: Air can transport contaminants over great distances. For instance, gases and particulates released from volcanic eruptions have the ability to cross the globe and circle it several times. Air pollution can arise from both stationary sources, such as power plants, and mobile sources, such as motor vehicles. When particulate air contaminants settle on the ground, they become a potential secondary source of pollution to the soils, surface water, and groundwater.

As of 2014, an estimated 25,000 chemicals were in production in the United States. Hundreds of thousands more chemicals, both naturally occurring and man-made, are known to exist. When one considers the billions of humans, the countless number of other biotas, and the various pathways to connect the two groups, one can start to gain awareness of the infinite amount of permutations that can constitute an environmental hazard.

1.2 What Is Environmental Consulting?

To define the second part of the term "environmental consulting," we go back to *Webster's Dictionary*. *Webster's* defines "consulting" as providing professional or expert advice. This advice is provided to people and companies, who, in this context, are known as *clients*. So *environmental consulting* can be defined as providing professional or expert advice regarding the environment.

In practice, however, there is a lot more to consulting than providing advice. To be in a position to provide professional advice, the consultant performs a wide variety of activities, as described in Section 1.4.

1.3 Types of Clients

People whose jobs are associated with the environment may act in various capacities for a wide variety of entities. What distinguishes environmental consultants is their role in providing services to *clients*, who then utilize the consultant's services for their intended needs.

Clients generally fall into one of three categories: private-sector clients, public-sector clients, and not-for-profit companies. Private sector clients include companies or individuals who require assistance in handling environmental issues affecting their business, residence, or place of operations, or need to make decisions involving a particular issue for which environmental expertise is required. They may be companies that have polluted; companies that are considering purchasing a contaminated property; companies that want to invest or avoid investing in a contaminated property, or a company that owns contaminated properties; companies directly involved in the clean-up of a property; and so on.

Public-sector clients include governmental agencies, public utilities, or quasi-governmental agencies, such as development corporations and public-private ventures. Environmental consultants conduct investigations and remediations on publicly-owned properties, such as federally and state-owned lands, properties owned by local authorities (such as counties and municipalities), and properties in which there is a substantial public interest, such as Superfund sites (see Chapter 3). Public entities may retain consultants to assist them in promulgating or implementing regulations, as described below.

Not-for-profit entities include *non-governmental organizations* (NGOs) that have many of the same interests as private sector clients, and environmental advocacy groups. Many NGOs invest in land for various reasons and become involved in construction projects in which environmental remediation plays a key role. Environmental advocacy groups may be policy-oriented, in which case they attempt to influence governmental legislation or practices in support of environmental protection in general or one environmental issue in particular. They seek the expertise of environmental consultants to support their policy positions, perform research which may include the collection of independent data, and provide expert testimony in lawsuits and other court-related activities.

1.4 What Do Environmental Consultants Do?

Because there is an infinite amount of possible environmental problems, the skill sets that environmental consultants bring to the table vary widely. In this book, the term "environmental consultant" is used almost interchangeably with the terms "consultant," "geologist," "chemist," "biologist," "engineer," and "investigator." That is because an environmental consultant may be any one of these things, or all of these things, depending upon the situation and the type of project. They also may be known by the particular role they're playing on a project, such as "air monitor," "field supervisor," or "risk assessor." The role of the environmental consultant can even be more complicated than this, as discussed below.

1.4.1 The Consultant as Consultant

Environmental consultants provide a wide spectrum of consulting services to private and public sector clients, as described below.

One basic type of service provided by environmental consultants involves field investigations, building inspections, scientific and engineering assessments, desktop studies, or computer simulations to identify areas where the client is not in compliance with existing regulations or to understand the scope and severity of the environmental problem. Because of the wide range of activities performed by environmental consultants, providing a comprehensive study of these services could be a book in itself. Some of the broader categories of services provided by environmental consultants are discussed in this and the subsequent subsection.

The United States has a body of environmental regulations that is complex, sometimes conflicting, and often confusing. In printed form, it is many thousands of pages in length, and supplemented by guidance documents, databases, and spreadsheets and formulae that dwarf the regulations upon which they're based. Much of the environmental consultant's job is to guide the client through this labyrinth of overlapping federal, state, and in some cases local laws and regulations. Chapter 3 describes the basic regulations and their applications.

Environmental consultants provide critical support services to the vast infrastructure of commerce throughout the world. Many of these entities are involved in the management of wastes, including the handling, transportation, and processing of wastes. These clients must comply with various environmental regulations, which often require the services of consultants who specialize in providing these services. Consultants assist these facilities in their design and construction; their initial permitting, and subsequent permitting; and ultimately the design and implementation of their closure.

Environmental consultants also will provide independent reviews of environmental compliance programs designed by others, either in-house staff or another consultant. In this role, the environmental consultant will assess whether the program is adequate to comply with existing regulations and whether the program is being implemented as designed. Such a review, known as an *environmental compliance audit*, is described in Chapter 3.

Environmental consultants may perform studies to determine the potential effects of a planned action on the environment. This may include performing some of the analyses required in the preparation of an *environmental impact statement* (EIS), which is discussed in Chapter 13.

Environmental consultants also can assist their clients in going beyond what is required by regulations. Voluntary governmental initiatives in waste minimization and recycling have led to the formation of a so-called "green" economy, which includes various entities that want to minimize their impact on the environment. These entities may be interested in being perceived as "green" for marketing purposes, to save money, to minimize the liability associated with the generation of wastes, or simply to be good global citizens.

In addition, an international initiative started in 1997 by the International Standards Organization, or ISO, encouraged companies to design and implement an *environmental management system* that is designed to minimize impacts on the environment. The guidelines developed by ISO, known as the *ISO 14000 series*, have been adopted by thousands of corporations throughout the industrial world. Environmental consultants assist private entities in formulating and implementing these voluntary measures and auditing the success of these measures. Consultants may also assist corporations in *benchmarking*, a method by which the company's environmental performance is compared to other companies in the same industry, companies of similar size, or companies from the same geographical area.

Emerging sectors in the field of environmental consulting include sustainability, green building design and construction, green product claims, climate change, and renewable energy. As with other voluntary environmental initiatives, environmental consultants assist companies or individuals in these endeavors by designing implementation plans, implementing operational changes at the facility or company level, auditing the effectiveness of these changes, and benchmarking the company's performance.

The subfield of the hazardous waste site investigation and remediation addresses chemicals, petroleum products, and other contaminants that have or may have been released into the environment. Chapters 6 through 9 cover the multi-stage process of investigating and remediating releases of chemicals and petroleum products into the environment. These four chapters are preceded by Chapters 4 and 5, which provide basic building blocks of chemistry and geology needed to understand the investigation and remediation of spills of hazardous wastes and petroleum products. Chapter 10 describes the investigation and remediation of vapor intrusion, an aspect of hazardous waste investigation and mitigation that specifically addresses one pathway into the human body.

Environmental consultants are involved in all phases of hazardous waste investigations and remediations. They may possess almost any background in the science or engineering fields. Complicated hazardous waste investigations and remediations require a multitude of skills, and rarely does one person possess all of the skills needed to see a project through from the initial assessment to the final clean-up. Remediation systems are typically designed by engineers in consultation with scientists. Field engineers generally oversee the construction, start-up, and operation of remediation systems.

Hazardous materials associated with buildings, and naturally occurring substances that present hazards when present within buildings, tend to fall into a different category of environmental consulting sometimes (but somewhat erroneously) called industrial hygiene. The investigation of these hazards is sometimes known as a *hazardous materials inventory*, or HMI. The HMI survey will include a survey of lead and often copper in drinking water (see Chapter 14); potential asbestos-containing materials, (see Chapter 15); a survey for the presence of lead in paint (see Chapter 16), and possibly mold (see Chapter 19). The HMI survey also includes identifying the presence of materials that, if disposed of, would be classified as hazardous wastes or as universal wastes, which are described in Chapter 3. In recent years, HMI surveys have also entailed assessing window caulking for the presence of polychlorinated biphenyls, commonly known as PCBs

(PCBs are described in Chapter 4). While industrial hygiene-related services, asbestos surveys, lead paint surveys, etc. can be cobbled together in the form of an HMI survey or an indoor air survey, most often they are provided separately.

Consultants on the industrial hygiene end of the spectrum also conduct indoor air quality investigations, as described in Chapter 17 and radon gas surveys, as described in Chapter 18.

Environmental consultants with backgrounds in biochemistry or ecology may specialize in studies of natural ecological systems, or the design and implementation of the restoration of degraded ecological systems. Chapter 12 describes some of the major components of natural ecosystems, such as surface water bodies, wetlands, and associated flora and fauna (including endangered or threatened species), and some of the procedures that environmental consultants design and implement to mitigate problems associated with damaged components of ecosystems.

1.4.2 The Consultant as General Contractor

A large portion of the work, and certainly the lion's share of the revenue earned by environmental consultants, comes from the investigating and remediating pollution that has impacted real estate, which is land that can be bought or sold. Contractors will perform many of the field activities to assist the consultant in investigating and remediating real estate.

To conduct a subsurface investigation, as described in Chapter 7, an environmental consultant may retain a drilling contractor to install soil boreholes and monitoring wells. Other contractors used by consultants in environmental investigations include geophysical contractors and contract laboratories. When remediation involves excavation, environmental consultants will retain a contractor, typically one that specializes in environmental remediation. Remediation contractors will have expertise constructing all or some of the remediation systems described in Chapter 9. The consultant's administration of contractors and subcontractors is described in Chapter 2.

1.4.3 The Consultant as Client

Many of the larger waste handling firms and industrial sector clients have in-house professionals whose job it is to help them achieve and maintain regulatory compliance. However, in many cases these "in-house" experts are actually environmental consultants retained by private sector companies on a full- or part-time basis (known as "outsourced employees") to supply their expertise where needed. Their desk may be in the firm's office, but in many cases their primary desk may be in the factory or corporate office of their client. In such circumstances, the outside consultant is indistinguishable from an employee of the client except for the company name on their paycheck.

1.4.4 The Consultant as Regulator

Public agencies that are responsible for environmental regulations and enforcement often need assistance in the drafting, promulgation, and enforcement of those regulations. They also may need outside consultants to develop guidebooks for understanding environmental regulations, or to staff help lines for people with questions regarding regulations. In these situations, the regulatory agency is the client with the consultant as the outsourced employee as described above. Such consultants may also staff help lines for citizens and professionals who have questions in the interpretation and implementation of the regulations.

1.4.5 The Consultant as Expert

Environmental consultants who, due to their education, training, and experience, have developed expertise and specialized knowledge in technical or regulatory matters, may provide support to attorneys who are involved in litigation regarding an environmental matter or a matter containing an environmental component. In this role the consultant becomes an *expert witness*, providing an interpretation of the data that the attorney will use in crafting an argument for the plaintiff (the party alleging damage) or the defendant (the party who the plaintiff claims is responsible for the damage). Consultants will produce expert reports to be used by the attorney for the plaintiff or the defendant in the litigation. As an expert witness, a consultant may need to give a deposition, which is sworn testimony given outside of the courtroom to attorneys who are privy to the litigation. Environmental consultants may also assist the attorney in preparing the various pieces of written and verbal correspondence that occur during the litigation or take part in calculating the assessment of monetary damages. Some litigations go to trial, with the expert witness taking the stand and explaining to the court his or her interpretation of the data.

1.5 Credentials and Certifications of Environmental Consultants

The educational backgrounds of environmental consultants are almost as varied as their work. A bachelor's degree is generally recognized as the minimum education required for an environmental consultant, while many consultants have master's degrees and doctorates. Fields of study are generally in science or engineering, although some consultants have backgrounds in economics or business. Increasingly, environmental consultants are graduating with environmental science or environmental engineering degrees that are directly oriented to the tasks performed by environmental consultants.

Training for environmental consultants is quite varied as well. In general, there is no set level of training required of environmental consultants, although there are regulatory requirements that depend upon the type of consulting services to be provided. For instance, consultants pursuing the field of hazardous waste remediation are required by federal law to take a 40-hour course known as Hazardous Waste Operations and Emergency Response (HAZWOPER). There are a wide variety of federal and in many cases state licenses required for consultants working in the fields of asbestos and lead-based paint, as well as licenses dealing with radon investigation and mitigation.

Environmental consultants often have certifications that allow them to perform certain functions on a project. For instance, the professional engineer (PE) license, granted by states or territories, is held by environmental consultants responsible for the design of remediation systems and other structures. In most cases, the signature of a PE is required on a system design before a regulatory entity will allow its construction. A professional geologist (PG) certification is granted by some states and many independent associations, such as the American Institute of Professional Geologists (AIPG).

A consultant holding a certified hazardous materials manager (CHMM) license, which is issued by the Institute of Hazardous Materials Management (IHMM), certifies individuals with expertise in the management of hazardous materials. A certified industrial hygienist

(CIH) license, which is issued by the American Board of Industrial Hygiene (ABIH), is given to individuals with expertise in worker safety. Since environmental protection often overlaps with health and safety issues, CIHs play many roles in the investigation and remediation of hazardous wastes and other environmental hazards.

Certain states issue licenses to qualified environmental consultants and require the licensed individual to conduct or supervise all or certain aspects of investigations and remediation of hazardous wastes and petroleum. Such individuals may be licensed as registered environmental assessors (REAs) in California; licensed environmental professional (LEPs) in Connecticut; licensed site professionals (LSPs) in Massachusetts; licensed site remediation professionals (LSRPs) in New Jersey; or as certified professionals (CPs) in Ohio. As is the case with PE, PG, and other professional licenses, the individual must first qualify for these licenses based on educational background and a high level of relevant experience in the field, and then pass a comprehensive test to obtain the license.

Continuing education (CE) is a necessary aspect of environmental consulting. Some training courses require annual or biennial updates. For instance, consultants with HAZWOPER certifications are required to take an eight-hour refresher course annually, as are consultants with asbestos inspector licenses and lead paint risk assessor licenses. Environmental consultants holding one of the above-mentioned licenses typically have CE requirements associated with their licenses. In most cases, they are required to take several courses relevant to their certification in a given time period so that they can maintain their licenses. These courses are taught by academics, such as professors, as well as environmental consultants with the requisite expertise.

1.6 Career Pathways in Environmental Consulting

At most environmental consulting firms, there are only a few levels in the pyramid. Employees who typically have less than five years of experience in environmental consulting or a related field are data gatherers, be it in the field, in the catacombs of a data repository, or in the data available through the Internet. They are unlikely to play a role in the design and management of a project except perhaps in small projects, in which they may be given a management role. If the employee's goal is to become an expert at a technical discipline, their work will be focused on just that discipline. Because the field of environmental consulting is so interdisciplinary, most consultants take a generalist career path.

Field-level employees on the technical career path will report to a technical specialist. Field-level employees on a generalist career path report to a mid-level employee known as a project manager. The project manager typically has at least five years of experience and often much more. They manage the staffing, scheduling, and fiscal aspects of the project. Project managers may be data gatherers, but to a lesser degree than field-level employees. They need to have a broader background of experience than field-level employees because many different technical areas may be tapped in the course of a project. The project manager is responsible for the preparation of the written documentation for the project as well as written and verbal correspondence in the course of the project.

At the top of the project pyramid is a senior-level employee known as the project director or principal. The project director typically has at least 15 or 20 years of experience in environmental consulting or a similar field. The project director oversees the work of the project manager (or managers, on larger, more complex projects). The project director is responsible for the overall performance, timeliness, and profitability of the project. The project director rarely collects data or prepares documentation but reviews the data and documentation and is responsible for the interpretation of the data and the subsequent recommendations to the client. Client management is an essential component of the job of a senior-level employee. Most people in that position are called "seller-doers," responsible for finding clients and maintaining and nurturing those relationships.

The senior-level employees in an environmental consulting firm report to senior employees who manage the company. These employees are technical people who have "risen through the ranks" as environmental consultants. They may have spent their entire careers at the same firm or have switched firms on the way up. Many of the senior level employees fit the seller-doer model and still manage their important clients while others are in charge of technical disciplines within the company.

The next chapter discusses the various aspects of the environmental consulting project.

Bibliography

Roundtable on Environmental Health Sciences, Research, and Medicine; Board on Population Health and Public Health Practice; Institute of Medicine, 2014. Identifying and Reducing Environmental Health Risks of Chemicals in Our Society. Washington, DC: National Academies Press (US).

2

Environmental Projects: The Technical Side and the Business Side

2.1 The Technical Side of Environmental Consulting

The environmental consultant, when operating as an environmental consultant (as opposed to a client or regulator, as described in Chapter 1), conducts environmental investigations and remediations, as the title of this book suggests. Environmental investigations and remediations cover an extremely broad band of activities, as described in the subsequent chapters. That said, the investigations and remediations adhere to certain basic principles.

2.1.1 Environmental Investigations and Remediations

The objective of the environmental investigation is to assess whether there is environmental impairment on a certain property or area and, if it exists, its severity and extent. Media impacted include soil, groundwater, surface water, sediment, soil vapor, air, and building materials, as described in subsequent chapters. The environmental consultant must collect various types of data, evaluate and interpret the data, decide what steps to take if additional data are needed, then develop and implement a remedy to mitigate the given environmental impairment.

To perform these duties, the environmental consultant must understand the pros and cons inherent in the data collection and interpretation processes.

2.1.1.1 Data Collection

In general, data collected by a competent environmental consultant have validity because the data were collected and analyzed using the appropriate methodologies. Data collected in the field include what the environmental consultant sees, smells, touches, and is told. Such data is primary data, that is, data obtained through first-hand evidence by the environmental consultant. Other primary sources of data include original documents such as un-doctored photographs, information available on the internet from official and unofficial sources, and personal testimony by a person in a responsible position, such as a factory manager or a homeowner.

A secondary source of data comes from a third-party review of a primary data source. The environmental consultant must assess the reliability of secondary data. Examples of secondary data include summaries or presentations of primary data made by someone else. This may include maps, data tables, narratives, as well as third-person recollections of events. Secondary data is subordinate to primary data, although there are instances in which primary data is wrong or misleading (for instance, people often make observational mistakes).

2.1.1.2 Sampling Objectives

The objective of sampling in an environmental investigation is to provide quantitative data regarding the existence, extent, or severity of an environmental problem. Data can be collected from different areas on or off a property; at different depths in the case of subsurface investigations, and at different times in the course of a day or a year. Data collection can be biased or unbiased, and is subject to various quality assurance and quality control objectives. In addition, data collection is subject to various rules, regulations, and guidelines, as discussed in Chapter 3 of this book. Most importantly, the data must have validity, which is to say that it is factually accurate and provides an appropriate basis for the selected remedy to the environmental problem.

2.1.1.3 Biased versus Unbiased Data

Depending on the goals of the environmental investigation, data collection can be designed as biased or as unbiased. If the objective of the environmental investigation is to identify the worst-case scenario, biased data is preferred, that is, sampling biased towards areas or at times when contamination is most likely to be present. Examples of biased data sampling are: collecting a sample from soils that appear to be stained with petroleum as opposed to nearby soils that lack staining; and collecting noise or air emissions data at a street corner during rush hour rather than at a time of day when there is an average number of vehicles on the road.

Unbiased data is designed to provide an overview of an environmental condition or a set of conditions. The data should not be collected from locations or at times where contamination is more likely to be present than normally would be expected. Examples of unbiased data collection are soil samples collected at a set depth and a set interval spacing regardless of circumstances observed and measured by the environmental consultant, or collecting traffic data over a 24-hour period and calculating a straight average of traffic conditions at the given location (see Figure 2.1).

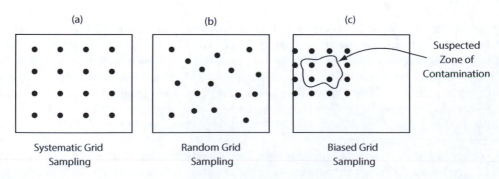

FIGURE 2.1
Examples of (a) systematic grid sampling; (b) random grid sampling; and (c) biased grid sampling.

2.1.1.4 Quality Assurance and Quality Control

For data to be usable, it must be reliable, which is to say, provide an accurate representation of the environmental condition being measured. Environmental investigations are subject to elaborately designed standard operating procedures designed to provide validity to the data being collected. Such processes are collectively known as *quality assurance*. Quality assurance protocols include specifications regarding the equipment being used to collect the data; the qualifications of the people collecting the data; the conditions under which the data should be collected, procedures designed to avoid cross-contaminating the sample with contamination derived from a different source; analytical protocols; methodologies for evaluating the data, and so on. Portions of the quality assurance procedures are prescribed in regulations and included in sampling plans. Others are mandated by regulatory authorities, and still others fall under the banner of best management practices. All are designed to provide valid data for use by the environmental consultant.

Quality control is a set of systems designed to ensure that the product, or in the case of environmental consulting, the service being provided meets the client's requirements. The client's requirements are established in the contract, which is described later in the chapter. By adhering to the client's requirements, and doing so in a manner that assures the quality of the data, the environmental consultant is able to provide sound professional advice to the client.

2.1.1.5 Units of Measurement

In the United States, units from the English system of measurement are routinely mixed in with units from the metric system of measurement. A general rule of thumb is that metric units are used to measure micro-scale items, English or metric units are used to measure meso-scale items, and English units are used to measure macro-scale items. Table 2.1 provides examples of metric concentration units common to environmental investigations.

At the meso scale, the weight of human beings, a parameter used in risk assessment, is given in kilograms; the weight of chemicals or chemical wastes may be given in kilograms or in pounds; the weight of soils or other bulk wastes usually is given in tons. Lengths and distances as measured by a microscope typically are given in millimeters. Distances given on a property, between two places, in depth below the ground surface, or height above the ground surface typically are given in feet or in miles. Temperature in a controlled setting such as a laboratory is given in Centigrade. Temperature in the field, however, usually is given in Fahrenheit. The environmental consultant needs a working knowledge of both systems to perform his or her duties.

TABLE 2.1

Metric Concentration Units Common to Environmental Investigations and Remediations

Generic Concentrations	Metric Concentrations for Solids	Metric Concentrations for Liquids
parts per million (ppm)	milligrams per kilogram (mg/kg) micrograms per gram (µg/g)	milligrams per liter (mg/L)
parts per billion (ppb)	micrograms per kilogram (µg/kg)	micrograms per liter (µg/L)
parts per trillion (ppt)	nanograms per kilogram (ng/kg)	nanograms per liter (ng/L)

2.1.2 Environmental Remediation

Environmental remediation consists of the removal of the source of the environmental hazard, its pathway to the receptor(s), or the receptors themselves. Practically speaking, of these three options, only source removal offers a permanent solution to the environmental hazard. Source removal could entail the physical removal of the pollution source, such as the removal of contaminated soil or asbestos-containing material. Source removal can also entail altering the pollutant so that it becomes less hazardous to human health or the environment, such as adding chemicals to change the pH of a liquid or adding an oxidant to chemically transform the pollutant into a different, more benign molecule.

When environmental remediation addresses either the pathway or the receptors, it is generally a temporary remedy since the environmental hazard can re-establish itself if the pathway is reconstructed or the receptors return. Removal of the pathway entails construction of a physical barrier between the source and the receptor, as with the encapsulation of an asbestos-containing material or painting over a surface containing lead-based paint. It can also take the form of installing a concrete or asphalt surfacing so that people, who are the potential receptors, cannot come into direct contact with the underlying contaminated soils. Permanent remedies that address the pathway involve in-situ stabilization of the contaminant, as described in Chapter 9.

Receptor removal takes the simple form of denying people admittance into the contaminated area. This form of remediation is the least desirable since, because people can be unpredictable, the barriers to entry would have to be highly robust.

Environmental remediation can be partial or comprehensive. In some circumstances, it may be desirable to conduct an *interim remedial measure* (IRM) to reduce the hazard posed by the contamination to human health and the environment while data are being gathered and interpreted so that a permanent remedy can be designed and implemented. Remediation could address just part of the source, perhaps the most severely impacted area or the area that is closest to or most easily accessed by receptors. Examples of this type of remediation would be remediating a portion of a groundwater plume that is being used as a drinking water source, or abating lead-based paint located in the portion of a building where a day care center is operating.

In all cases, the environmental consultant must be able to assess the problem, the potential hazards associated with the problem, and the applicable regulations so that a remedy can be devised that is protective of human health and the environment while providing the appropriate service to the client.

2.1.3 Documenting the Environmental Investigation and Remediation

When the agreed-upon scope of work has been completed, the environmental consultant generally prepares a written document, often referred to as a *project deliverable*. This document, which may be in the form of a report or a letter, indicates that the scope of work has been completed. It documents the methodologies and findings of the scope of work, and usually includes recommendations, if any, for follow-up activities. Some documents must be certified by a licensed individual, such as a professional engineer or an environmental consultant with one of the licenses mentioned in Chapter 1 of this book. Many of these documents must be submitted to the applicable regulatory agency to assess their compliance with applicable regulations and for evaluation whether the proposed remedial actions are appropriately protective.

2.2 The Business Side of Environmental Consulting

Just as environmental consulting is a profession, an environmental consulting firm is a business. As with all businesses, their goal is to realize a profit (unless the environmental consulting firm is a not-for-profit organization). This section describes the environmental project from a business perspective.

2.2.1 Project Contract

Since consultants provide professional services to clients, there must be a contractual relationship between the consultant and client. The consulting contract can take several forms. The most common form in the private sector is the proposal, which is a document in which the consultant describes the scope of work to be provided, the schedule in which the work will be completed, and the budget it will take to complete the scope of work within the stated timeframe. The proposal also states the objective of the scope of work and often describes what types of technical skills will be needed to perform the scope of work adequately.

Contracts that lack defined scopes of work and contracts with an undefined scope of work are known as *indefinite delivery/indefinite quantity* (IDIQ) contracts. These contracts typically are issued by public agencies; they do not commit the issuing agency to a particular scope of work or a particular expenditure of money, but establish the terms and cost structure by which future expenditures under the contract will be governed. IDIQ contracts typically expire after a defined period, after which the consultant would have to enter into a new contract to continue working for the client.

Some consulting contracts take the form of a purchase order, which is a document issued by the client to the consultant with a budget amount. That budget may be the amount to be spent on the contract or a maximum amount that the consultant is allowed to spend. In a purchase order, the scope of work and usually the project schedule have been decided in advance, sometimes through a proposal from the consultant and sometimes through a document known as the request for proposal (RFP) or request for bid (RFB). Clients also will use RFPs and RFBs to solicit work scopes from various consultants in an attempt to identify the consultant best qualified to assist the client on a specific project or projects.

Each project has three components—a scope of work, a schedule, and a budget. Each component should tie into each other so that each piece of the proposal logically reinforces the others.

2.2.2 Project Scope of Services

The project scope of services outlines the basic steps of the environmental investigation or remediation, the quality assurance/quality control procedures to be put into place, and the project deliverables to be provided to the client and/or the applicable regulatory agency. The scope of services should include at a minimum the regulations that apply to the work to be performed, any permits that might be required to perform the work, and the required licensing, if any, of the personnel to perform the work.

When the scope of services changes, the contract must be changed as well, typically through the issuance of a change order to the client. As with the proposal, a change order changes the scope of services, with a commensurate change to the project schedule as well as the project budget.

2.2.3 The Project Team

Most environmental consulting projects are team efforts, involving many people at various levels within the organization. The larger the project, the greater the involvement in terms of number of consultants, their levels within the organization, and the skill sets they bring to the table. At the helm is a senior-level person who directs the project, and usually has the title of principal or equivalent within the organization. That person is responsible for putting together the project team and seeing that the team fulfills its contractual goals. The project manager is the point person on the team, managing the scope of services, the schedule and the project budget. The project manager reports to the person directing the project. Assisting the project manager are various team members who may be junior to and less experienced than the project manager or may be more-experienced technical experts who perform various defined tasks within their areas of expertise on the project. Figure 2.2 shows the organizational structure of a typical project team.

Performance of the field work portion of the scope of services is often aided by outside parties known as contractors (sometimes called "subcontractors") and vendors. *Contractors* generally perform field work that the environmental consulting firm does not routinely do on its own. This work may involve the usage of heavy equipment, such as earth-moving equipment and drilling equipment, or the performance of a specific technical task outside the field of expertise of the environmental consultant, such as asbestos or lead-based paint removal, or the installation of a vapor mitigation system. *Vendors* either provide project services at locations other than the project site, such as a fixed-base laboratory, or provide equipment to the project, either through purchase or rental. As with the client/environmental consultant relationship, the contract, usually known as the *subcontract*, governs the activities performed by subcontractors and vendors on a project.

2.2.4 The Project Schedule

The project schedule defines, at a minimum, the deadline for the completion of the scope of services or the timeframe in which the work will be performed. For all but the most

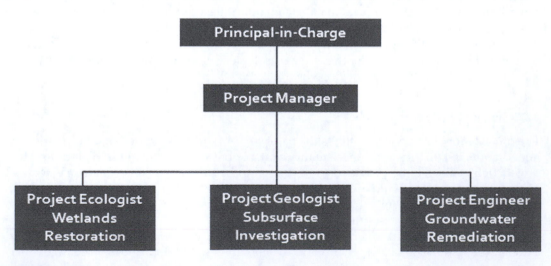

FIGURE 2.2
An organizational chart of a typical environmental project team.

basic projects, the project schedule also will define interim target dates and timeframes for completion of project tasks or sub-tasks. These deadlines typically are tied into the various stages defined in the contract scope of services.

2.2.5 The Project Budget

Contracts are awarded on a variable budget or a fixed-base budget. In a *time-and-materials* (T&M) project, the cost can vary up or down. This type of contract is commonly employed when the scope of services and work conditions are unknown or likely to change. A fixed-base or *lump-sum contract* is common when the scope of services and the work conditions are well-defined and unlikely to change. If the scope of services or work conditions change, or the project budget needs to be increased, the initial contract is amended through the issuance of a *change order*.

The project budget has three types of costs:

- *Labor costs*, which result from time spent by the environmental consultant in performing the scope of services;
- *Pass-through costs*, which result from the consultant's use of outside services (vendors and subcontractors, as described below)
- *Other direct expenses*, which include travel costs, lodging and meals costs, and other incidental costs incurred by the consultant while performing the scope of services.

2.2.5.1 Labor Costs

Labor costs are accrued in one of two ways. The consulting contract may contain a fee schedule which would define the rates (typically by the hour) of the personnel who will be working on the project. In general, the more senior and more technically expert consultants have the highest billing rates and the most junior and least technically skilled consultants have the lowest billing rates. Table 2.2 is a condensed version of a fee schedule.

The other method of accruing labor costs is using a *labor multiplier*. The consulting agreement defines a number that is used to calculate the billing rate of a particular consultant based on that consultant's salary. For example, assume that a field technician of the consulting firm earns \$30/hour, which, when multiplied by 2,080 hours (52 weeks/year × 5 days/week × 8 hours/day) is equivalent to an annual salary of \$62,400. If the consulting agreement sets the labor multiplier at 3.5, then the field engineer will bill to the project at a rate of \$105/hour.

TABLE 2.2

Condensed Billing Rate Schedule

Position	Hourly Rate
Principal	\$210
Chief Hydrogeologist	\$210
Project Manager	\$160
Field Engineer	\$120
Field Technician	\$105
Draftsperson	\$80

2.2.5.2 Pass-Through Costs and Other Direct Expenses

Most consulting firms do not have their own laboratories and instead use contract laboratories to analyze the environmental samples they collect. Most consulting firms do not operate heavy construction equipment or drilling rigs. They may not own a piece of specialty equipment or carry a special license or permit with which they would perform a task, file paperwork, or a myriad of other tasks. For these and a host of other reasons, environmental consultants often must rely on outside expertise and services in the course of a project.

As with the client, the consulting firm will enter into a contract to obtain the services of an outside vendor or contractor (see Chapter 1 for a description of activities performed by contractors). This contract is called a *subcontract* and the contractor retained known as the *subcontractor*. As with the client contract, the subcontract will define the scope of the services to be rendered, the schedule for providing those services, and the cost of the services. As with the client contract, the contract can be secured through the issuance of a purchase order to the subcontractor or vendor, especially in cases where the scope of work and schedule are well-defined.

Services rendered by third-party entities are called pass-through costs because the costs are passed directly through the consultant to the client. For example, if a laboratory provides $3,000 worth of laboratory analyses, then the laboratory will invoice the consulting firm for $3,000. The consulting firm may then invoice the client for the same $3,000 to recoup the cost of the laboratory service. Often, however, the consulting firm will seek to recoup the cost of handling the subcontract and the subcontractor and make a profit on the outside services by applying a mark-up to the subcontractor's and vendor's costs. If, for example, the consulting firm applies a 10% mark-up (a common figure) to the laboratory's invoice, it will invoice the client $3,300 for this outside service. A mark-up may apply to other direct expenses as well, such as photocopying and equipment rentals.

2.2.5.3 Net Revenue

Net revenue is the revenue that a firm accrues minus its labor costs, pass-through costs, and direct expenses. On a lump-sum contract, net revenue is the difference between the contract value and these costs. On a time-and-materials contract, net revenue is the difference between labor charges billed to the client minus the labor costs plus the revenue earned through the mark-up on pass-through costs and direct expenses. For example, a consulting firm will earn $160/hour in revenue from a principal who earns $50/hour and bills out at $210/hour. The $10 earned on the 10% mark-up of $100 in project expenses becomes part of the net revenue as well.

2.2.5.4 Project Invoicing

The environmental consultant charges the client for services rendered through the submittal of an *invoice*. An invoice may cover all or part of the project and, correspondingly, all or part of the project's budget. In a fixed-base contract, the invoice will be expressed either as a set dollar figure or as a percentage of the agreed-upon budget. Figure 2.3 shows a hypothetical invoice on a fixed-base contract. For time-and-materials contracts, the invoice will be based on actual labor hours performed as well as subcontractor costs, vendor costs, and other direct expenses, with or without a mark-up, as described earlier in the chapter. A typical invoice for a time-and-materials project is provided in Figure 2.4.

June 12, 2019
Project No: 6543
Invoice No: 123457

Client Name
Client Company
Client Address
Client City/St/Zip

Project Name:　Phase I Environmental Site Assessment at XYZ Industries
Invoice for Period Ending May 31, 2019

TOTAL AMOUNT DUE:　　　　**$3,000.00**

FIGURE 2.3
Typical invoice on a fixed-fee project.

June 12, 2019
Project No: 6543
Invoice No: 123456

Client Name
Client Company
Client Address
Client City/St/Zip

Project Name:　Soil Investigation at XYZ Industries
Invoice for Period Ending May 31, 2019

Task 1: Permitting

	Hours	Rate	Amount	
Principal	1.00	$210.00	$210.00	
Project Manager	4.50	$160.00	$720.00	
Total Labor	5.50		**$930.00**	
Reproduction (185 pages $0.10/page)			$18.50	
		10% markup	$1.85	
Total Contractors and Expenses			**$20.35**	
			Total for Task 1:	**$950.35**

Task 2: Sampling and Analysis

	Hours	Rate	Amount	
Principal	3.00	$210.00	$630.00	
Project Manager	6.00	$160.00	$4,470.00	
Field Geologist	24.00	$120.00	$2,880.00	
Total Labor	33.00		**$960.00**	
Field equipment and travel			$225.91	
Drilling subcontractor			$2,800.00	
10% markup			$302.59	
Total Contractors and Expenses			**3,328.50**	
			Total for Task 2:	**$7,798.50**
			TOTAL AMOUNT DUE:	**$8,748.85**

FIGURE 2.4
Typical invoice on a time-and-materials project.

Problems and Exercises

1. Because contamination was suspected to be present inside the warehouse portion of a facility, ten samples were collected inside the warehouse. They were spaced ten feet apart from each other. No samples were collected in other portions of the facility. What is this type of sampling called?

2. Convert 10 parts per million (ppm) into milligrams per kilogram, nanograms per liter, and micrograms per liter.

3. Calculate the project multiplier on a $5,000 lump-sum contract if the principal bills 8 hours to the project, the project manager bills 24 hours to the project, and $500 in expenses are incurred. Assume that the principal earns $50/hour and the project manager earns $35/hour.

4. Using the billing rate schedule provided in Table 2.2, calculate the total invoice on a time-and-materials contract if the principal bills 6 hours to the project and the project manager bills 30 hours to the project, and $200 in expenses are incurred, assuming a 10% mark-up on expenses. Calculate the net revenue on this project.

Section II

Site Investigations and Remediations

3

Framework of Environmental Regulations

More than almost any other business, the business of environmental consulting is driven by regulations. A complete description of all federal regulations, not to mention the myriad of state and territorial environmental regulations, would require its own book. In this chapter, we limit our discussion to the legal framework, history, and scope of the major federal environmental regulations that pertain to the topics discussed in the subsequent chapters of this book.

3.1 The Formation of United States Environmental Protection Agency and Occupational Safety and Health Administration

3.1.1 Pre-history of Environmental Regulations

Prior to 1969, environmental laws were geared primarily to the protection of navigable waterways and the conservation of natural resources. The Rivers and Harbors Act of 1899, the oldest federal environmental law in the United States, prohibited the dumping of refuse into a navigable water body or its tributaries without a permit. This act also prohibited the excavation, filling, or altering of any port, harbor, or channel, without a permit.

Beginning with the tenure of Gifford Pinchot in 1898 and the formation of the U.S. Forest Service under the Roosevelt Administration in 1905, the conservation of natural resources became the primary environment-related goal of the federal government. A series of laws aimed towards the conservation of natural resources was passed in the ensuing decades, the most notable of which was the Wilderness Act of 1964, whose mission was to preserve designated national forest lands in their natural condition.

The first federal law aimed at chemical hazards was the *Federal Insecticide, Fungicide, and Rodenticide Act*, known as FIFRA, which was first enacted in 1947 and revised and expanded in subsequent years. The Act, FIFRA, originally was oriented towards consumer protection and required pesticide manufacturers to register their products. The Delaney Clause of FIFRA, which took effect in 1958, required that manufacturers of consumer products demonstrate that their products would not cause cancer in people or animals.

Meanwhile, the first half of the twentieth century saw an explosion of growth in the invention and usage of synthetic chemicals in the industrialized countries of the world. Sanitary engineers who were used to dealing with naturally occurring chemicals that would biodegrade naturally were unaccustomed to dealing with synthetic chemicals, many of which would not biodegrade naturally and therefore would persist in the environment. Many of these chemicals proved to be hazardous to a wide variety of flora and fauna (especially humans) and posed a growing health threat in the world's industrialized countries.

The general public gained awareness of this threat after author Rachel Carson published *Silent Spring* in 1962. In that landmark book, Ms. Carson documented the degradation of the environment caused by these chemicals, and warned that, if we humans are to live among these chemicals, we should be aware of their effects on us. Her wake-up call helped to launch the modern environmental movement.

3.1.2 Establishment of the United States Environmental Protection Agency

The passage of the *National Environmental Policy Act* (NEPA) in 1969 set the stage for modern environmental regulations. NEPA created the *Council on Environmental Quality* (CEQ) within the executive branch of the United States government. For the first time, an environmental law in the United States acknowledged the "profound impact" of man's activity on the environment and the "critical importance of restoring and maintaining environmental quality to the overall welfare and development of man." Chief among the new requirements under NEPA was the requirement to prepare an *environmental impact statement* (EIS) for actions contemplated by the federal government that may have a "significant impact" on the environmental quality in the area. The National Environmental Policy Act is discussed in greater detail in Chapter 13.

Following on the heels of NEPA was the establishment of the *United States Environmental Protection Agency* (USEPA). Soon after its inception, the USEPA began to develop a regulatory framework to protect the environment of existence, the USEPA passed six major laws, with major amendments and reauthorizations in subsequent years. With one exception, major laws implemented by USEPA are forward-looking, that is, designed to prevent future environmental degradation. The one backward-looking law is the Comprehensive Environmental Response, Compensation, and Liability Act (CERCLA), more commonly known as Superfund. These laws are described below.

Table 3.1 below is an incomplete list of major federal environmental laws passed since 1970. Many of these laws, amendments, and reauthorizations are discussed in this chapter.

TABLE 3.1

Major Environmental Laws after Establishment of USEPA

1970	Clean Air Act Amendments
1972	Federal Water Pollution Control Act Amendments (Clean Water Act)
	Federal Environmental Pesticides Control Act of 1972 (amended FIFRA)
1973	Endangered Species Act
1974	Safe Drinking Water Act
1976	Resource Conservation and Recovery Act (RCRA)
	Toxic Substances Control Act (TSCA)
1977	Clean Air Act Amendments
	Clean Water Act Amendments
1980	Comprehensive Environmental Response, Compensation, and Liability Act (CERCLA) (Superfund)
1984	Hazardous and Solid Waste Amendments (RCRA Amendments)
1986	Safe Drinking Water Act Amendments
	Superfund Amendments and Reauthorization Act (SARA) (.a.k.a. Emergency Planning and Community Right-To-Know Act [EPCRA])
1987	Water Quality Act
1990	Clean Air Act Amendments of 1990
	Oil Pollution Act of 1990
	Pollution Prevention Act
2002	Brownfields Act
2005	Energy Policy Act

Source: Kraft, M.E., *Environmental Policy and Politics*, HarperCollins College Publishers, 1996.

3.1.3 Establishment of the Occupational Safety and Health Administration

The Occupational Safety and Health Act of 1970 established the *Occupational Safety and Health Administration*, better known as OSHA. This administration's primary responsibility is to protect the health and safety of workers in the workplace. The primary role of OSHA in environmental investigations and remediations regards the protective measures to be taken by workers at sites with hazardous wastes or some other environmental hazard. Therefore, this book discusses OSHA regulations within the context of environmental investigations and remediations.

3.2 Legal Framework of Federal Environmental Regulations

In the United States, laws are passed by Congress and signed by the president. The president also can issue executive orders, which bypass congressional approval. Once a law or an executive order is in place, it is sent to the appropriate department, which is charged with creating regulations designed to carry out the directives in the law or executive order. Once the regulations have been finalized, the department that issued the regulations is responsible for enforcing them.

At the federal level in the United States, the USEPA is tasked with promulgating and enforcing regulations designed to protect the environment. The USEPA is not a cabinet department, although its chief administrator is usually part of the executive branch's cabinet.

3.2.1 Code of Federal Regulations

Regulations promulgated by the USEPA appear in the *Code of Federal Regulations* (CFR), which can be accessed at www.gpoaccess.gov/cfr/index.html. Each regulation has a numerical title, which refers to the department that administers the regulations. Regulations administered by the USEPA appear in Title 40 of the *Federal Register*. Specific parts of a regulation are ordered by part, section and sub-part. For instance, 40 *CFR* 761.30(p) is immediately recognizable as an environmental regulation by virtue of its title number. The part of the regulation is 761, the section is 30, and the sub-part is (p). The sections of the federal environmental regulations are arranged by statute. Table 3.2 provides the subchapters and sections of the regulations that are discussed in this chapter.

The OSHA regulations appear in 29 *CFR*. The portion of the OSHA regulations with the most bearing on environmental investigations and remediations is 29 *CFR* 1926, which are regulations designed to protect workers at construction sites. Subpart D, Occupational Health and Environmental Controls, includes regulations for lead (29 *CFR* 1926.62), which is the only chemical with its own set of OSHA regulations not directly related to air hazards, and general hazardous waste operations and emergency response (29 *CFR* 1926.65). 29 *CFR* 1910 regulates the health and safety of chemical air hazards and includes the requirement for protection from airborne asbestos (29 *CFR* 1910.1001) and the requirement to communicate chemical hazards in the workplace to workers (29 *CFR* 1910.1200).

The USEPA is divided into ten regions, as shown in Figure 3.1. Each region is responsible for implementing federal environmental regulations within its jurisdiction. They do not have the authority to issue environmental regulations—that authority resides at USEPA headquarters in Washington, DC.

TABLE 3.2

Relationship between Key Environmental Statutes and the Code of Federal Regulations

Sections of 40 CFR	Subchapter	Related Statute
50–99	C	Clean Air Act
100–149	D	Safe Drinking Water Act
150–189	E	Federal Insecticide, Fungicide, and Rodenticide Act
239–299	I	Resource Conservation and Recovery Act
300–399	J	Superfund
400–699	N	Clean Water Act
700–799	R	Toxic Substances Control Act

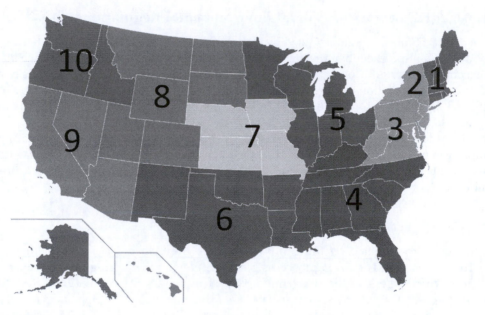

FIGURE 3.1
The ten regions of the USEPA. (Courtesy of U.S. Environmental Protection Agency.)

3.2.2 Legal Framework of State Environmental Regulations

Many of the federal regulations require the states to develop plans, known as *state implementation plans*, for the implementation of the federal regulations. Once a state implementation plan is approved by the USEPA, the state must implement the plan. The USEPA region will monitor the state's implementation of the federal regulation. Appendix B provides a listing of the 50 departments that regulate the environment within their states.

Most states promulgate statutes and regulations in a manner similar to the USEPA. However, no state can promulgate regulations that are less stringent than the existing federal regulation.

States also can pass their own environmental statutes, independently from the USEPA. From these statutes, the state environmental regulatory entity typically issues environmental regulations, environmental guidance, etc. Environmental consultants must familiarize themselves with state as well as federal regulations (and in some cases, local regulations) to perform their jobs properly.

3.3 Major Federal Environmental Laws

3.3.1 Clean Air Act

The *Clean Air Act*, originally passed in 1963, actually predates the USEPA. The original law provided federal support for air pollution research and had little impact on private or public sector entities. However, the Clean Air Act Amendments of 1970 vastly strengthened the Clean Air Act and included enforcement provisions, providing some of the regulatory structure that exists today.

The 1970 amendments established *National Ambient Air Quality Standards* (NAAQS) for certain *criteria pollutants*. These pollutants included sulfur dioxide (SO_2), nitrogen oxides (NO_x), which included nitrous oxide (NO_2) and nitric oxide (NO_3), ozone (O_3), carbon monoxide (CO), and PM_{10}, which is an abbreviation for particulate matter with a diameter of ten microns or more (this threshold for particulates was set as a practical limit of attainment rather than a scientific limit based on health risk). It established non-attainment areas, which are parts of the county with worse air quality than the NAAQS.

The 1977 Clean Air Act Amendments added to the attainment requirements in non-attainment areas by calling for the prevention of significant deterioration (PSD) for attainment areas. The PSD applies to new sources of pollutants or major modifications at existing sources of pollutants. It requires installation of the *best available control technology* (BACT) for the new or modified emissions source and analyses of the source's impacts on air quality and other parameters such as water, vegetation, and soils. These amendments also established three classes of "clean areas":

Class I: National parks and similar areas, where air quality would be protected from any deterioration

Class II: A specified amount of deterioration would be permitted

Class III: Air pollution could continue to deteriorate, but not beyond national standards

In addition, the Clean Air Act in 1977 also contained a standard for lead emissions.

The Clean Air Act Amendments of 1990 introduced the following changes to the Clean Air Act under what is known as Title V:

1. The passage of the *National Emissions Standards for Hazardous Air Pollutants* (HAPs), known as NESHAPs (pronounced knee-shaps), which increased the number of hazardous air pollutants with NAAQS to 187 (including asbestos, see Chapter 15), plus acid rain and chlorofluorocarbons (CFCs). The USEPA identified non-attainment areas for the new HAPs and required the states to devise plans to meet NAAQS criteria for non-attainment areas in their respective jurisdictions.

2. The establishment of emissions standards for mobile and stationary sources.

3. A requirement that new air pollution sources, such as new power plants and new car models, must use *best available technology* (BAT) to comply with the standards set for the criteria pollutants.

4. The establishment of an emissions trading program through which the owners of pollution sources in non-attainment areas could trade pollution "credits" with owners of pollution sources in attainment areas, thereby meeting their regulatory requirements.

5. Requirements for oxygenated fuel, which would result in more efficient fuel combustion and therefore less air pollution during the cold winter months. Methyl tertiary-butyl ether (MTBE), which was a commonly used oxygenate in gasoline, became a significant pollutant of the subsurface and is discussed in Chapter 4.

6. The establishment of the *Title V permitting program*, which applies to major stationary sources of air pollutants. A major stationary source is defined by its air emission rates and depends upon whether the source is in an attainment area, a non-attainment area, or an area of serious non-attainment for the specific HAP in question. A Title V permit describes the air emissions; the theoretical air emissions possible from the facility; the type and performance of existing air emissions control equipment; and an air monitoring plan.

3.3.2 Clean Water Act

The *Clean Water Act*, passed in 1972 and amended in 1977 and 1987, regulates the discharge of pollutants into the navigable waters of the United States. At first, the Clean Water Act focused on *conventional pollutants*, so named because of the decades-long focus on these pollutants by sanitary engineers. Conventional pollutants are discussed in Chapter 4.

The 1977 amendments to the Clean Water Act forced the USEPA to focus on toxic pollutants as well as conventional pollutants. In response, the USEPA created a list of *priority pollutants* to be regulated under the Clean Water Act. Dischargers of conventional and priority pollutants to surface waters were required to obtain a permit from the USEPA under the *National Pollutant Discharge Elimination System* (NPDES). Each NPDES permit defines the *total maximum daily load* (TMDL) allowed for each toxic pollutant. Regular monitoring and self-reporting of wastewater discharges to navigable water bodies is required under a NPDES permit. The 1977 amendments to the Clean Water Act also provided funding for municipal waste water treatment plants.

The 1987 amendments to the Clean Water Act (known as the Water Quality Act) added *nonpoint sources* of water pollution to the discharge regulations. Nonpoint sources are sources of pollution that lack a specific point of origin. One example of a nonpoint source of pollution is *storm water runoff*, which picks up pollutants that are diffuse and lacking in a specific location, such as drips of oil in a parking lot. Storm water is a concern because it has the ability to transport drips of oil, other contaminants, and debris lying about, into a nearby surface water body, thus adding to the loading factor for the contaminants. Facilities with the ability to create nonpoint-source pollution were required to develop storm water pollution prevention plans (SWPPP) as a best management practice.

Section 404 of the Clean Water Act (CWA) is perhaps the most significant portion of the Act. Under Section 404 of the CWA, the USEPA regulates the introduction of dredge or fill materials into the waters of the United States, including *wetlands*, which are discussed in Chapter 12. This portion of the CWA effectively prohibits the infilling of wetlands, which has a significant impact on real estate development. To dredge or fill a wetland, a permit must be obtained from the U.S. Army Corps of Engineers, a branch of the U.S. military that effectively oversees Section 404 compliance, although some state regulatory agencies have permitting authority. Certain farming and forestry activities are exempt from Section 404 regulation.

3.3.3 Endangered Species Act

The *Endangered Species Act* (ESA) is administered by the U.S. Fish and Wildlife Service (USFWS) as well as the National Oceanic and Atmospheric Administration (NOAA).

The ESA defines a species as endangered if it is "in danger of extinction throughout all or a significant portion of its range." It defines a species as threatened if it is "likely to become an endangered species within the foreseeable future."

The ESA's primary goal is to prevent the extinction of imperiled plant and animal life; its secondary goal is to recover and maintain those populations by removing or lessening threats to their survival. Endangered animals are protected even on private land under the ESA, although plants are not. When a proposed action has the ability to impact an endangered species, the ESA requires studies to assess and quantify the potential impacts, and implementation of measures to minimize or eliminate those impacts. A study of endangered species is required for all EISs (see Chapter 13). Such studies have resulted in the delay or even cancellation of major construction projects whose implementation would be detrimental to endangered and threatened species.

3.3.4 Safe Drinking Water Act

The *Safe Drinking Water Act* (SDWA), first passed in 1974 and amended in 1986, regulates some 50,000 public water systems (PWSs) throughout the United States. Only permanent water systems that have at least 15 services connections or service at least 25 people are regulated under the SDWA, to avoid an unnecessary regulatory burden on small businesses and homeowners with private wells. The SDWA sets national drinking water standards, known as maximum contaminant levels (MCLs), for a host of chemical, biological, physical, and radiological agents (see Chapter 4 for further discussion of these standards), including lead (see Chapter 14). The SDWA requires the system operator to utilize best available technology to achieve drinking water standards and establishes secondary drinking water standards for publicly owned treatment works (POTWs), such as sewage treatment plants.

3.3.5 Toxic Substances Control Act

The original, primary role of the *Toxic Substances Control Act* (TSCA), passed by Congress in 1976, was to require the testing of new chemicals, as well as chemicals already in commerce, so that their effects on human health and the environment could be assessed. The original TSCA legislation, which became known as Title I, was expanded soon afterwards, as described below.

3.3.5.1 Polychlorinated Biphenyls under the Toxic Substances Control Act

The manufacture and usage of polychlorinated biphenyls (PCBs) were banned under TSCA in 1979 (see Chapter 4 for a description of PCBs, and Chapter 6 for a description of the history of their usage in the United States). The TSCA banned the manufacture or usage of equipment containing PCBs at a concentration greater than 50 milligrams per kilogram (mg/kg), which is equivalent to 50 parts per million (ppm). This numerical threshold is a cleanup standard that applies not only to equipment but also to soils, water, light ballasts, and a whole host of other media and equipment that the original regulation did not foresee. It crosses statutes as well—the TSCA limit of 50 ppm finds its way into RCRA and CERCLA cleanups (see below).

3.3.5.2 Titles II through VI of TSCA

Congress passed five major amendments to the original TSCA, which became known as Titles II through VI. Brief descriptions of these amendments are provided below.

Title II of TSCA, which became law in 1986, is better known as the Asbestos Hazard Emergency Response Act, or AHERA. This law regulates asbestos-containing building materials in schools. In 1990, it was amended by passage of the Asbestos School Hazard Abatement Reauthorization Act (ASHARA). Asbestos-related regulations are discussed in Chapter 15.

Title III of TSCA regulates the presence of radon gas in indoor air. Radon gas investigation and mitigation are discussed in Chapter 18.

Title IV of TSCA became known as the Residential Lead-Based Paint Hazard Reduction Act of 1992. The objective of Title IV is to eliminate lead-based paint hazards from public housing. Lead-based paint investigation and mitigation are discussed in Chapter 16.

Title V of TSCA authorizes the USEPA to provide technical assistance to schools to comply with environmental regulations. It also provides guidance regarding school siting. Title VI of TSCA regulates the usage of formaldehyde in composite wood products. Titles V and VI are not discussed further in this book.

3.3.6 Resource Conservation and Recovery Act

Resource Conservation and Recovery Act (RCRA), which Congress passed in 1976, is a deceptively-named environmental law. While its name implies a relatively simple mission such as plastic and paper recycling, its scope goes far beyond that.

The objectives of RCRA broadly fall into two categories: (1) preventing hazardous wastes from entering the environment and (2) minimizing the generation of all types of solid wastes, which is defined as a waste can be deposited in a landfill, including gaseous, liquid, and semi-liquid waste. RCRA governs the storage, transportation, and disposal of all solid wastes.

3.3.6.1 Solid Waste and Hazardous Waste

Solid waste, as defined under RCRA, means any garbage or refuse, sludge from a wastewater treatment plant, water supply treatment plant, or air pollution control facility and other discarded material, resulting from industrial, commercial, mining, and agricultural operations, and from community activities. Despite its name, solid wastes are not necessarily solid. The wastes could be liquid, semi-solid, or contain gaseous wastes.

Under RCRA, a hazardous waste is a type of solid waste. A waste is classified as a hazardous waste if one of two criteria apply: (1) the waste is specifically listed under RCRA as a hazardous waste (known as a listed waste), or (2) the waste has the characteristics of a hazardous waste.

A *listed hazardous waste*, as implied by the name, appears on one of several lists developed by the USEPA. The following wastes are listed hazardous wastes under RCRA:

- Process wastes from general industrial processes (the F list);
- Process wastes from specific industrial processes (the K list);
- Unused or off-specification chemicals, container residues and spill cleanup residues of acute hazardous waste chemicals (the P list and the U list); and
- Other chemicals, a category that includes a broad spectrum of waste chemicals deemed to be hazardous. These wastes, known as *universal wastes*, include typical residential and commercial wastes such as batteries, commercial, pesticides, mercury-containing equipment, and bulbs from electric lamps.

A *characteristic hazardous waste* (D listed waste) is not a listed waste but does have physical characteristics that are considered to be hazardous to human health. For a non-listed waste to be considered hazardous, it must have at least one of the following four characteristics:

- *Ignitable* (flammable): Flash point below 140°F (60°C). This means that the material can vaporize and spontaneously catch on fire without the benefit of a spark or flame. Materials with flash points above 140°F (60°C) that will catch on fire when exposed to a spark or flame are labeled as combustible. D001 is the code given to ignitable waste.

- *Corrosive*: pH below 2 or above 12.5. D002 is the code given to corrosive waste.

- *Reactive*: Normally unstable, reacts violently with air or water, or forms potentially explosive mixtures with water; emits toxic fumes when mixed with water or is capable of detonation. D003 is the code given to reactive waste.

- *Toxic*: A waste that tests toxic using the *toxicity characteristic leachate procedure* (TCLP). The TCLP is a laboratory procedure that tests the ability of a chemical to leach out of the solid matrix in which it is present and enter the environment.

The tests that are required under RCRA for hazardous wastes include: flash point (to test the waste's ignitability), pH, and reactivity. In addition, RCRA requires analysis for certain volatile organic compounds (VOCs), semi-volatile organic compounds (SVOCs), pesticides, and metals (these chemical classes are defined in Chapter 4). The 40 RCRA compounds/analytes are listed in Table 3.3. Wastes that test hazardous are given codes D004–D043, one code for each contaminant involved.[1]

TABLE 3.3

Compounds Required to Be Tested to Determine Whether a Solid Waste Exhibits the Characteristic of Toxicity

VOCs	SVOCs	Pesticides	Metals
Benzene	Nitrobenzene	Chlordane	Arsenic
Chlorobenzene	Hexachlorobenzene	2,4-D	Barium
1,4-Dichlorobenzene	Hexachloroethane	Endrin	Cadmium
1,2-Dichloroethane	Hexachlorobutadiene	Heptachlor (and its epoxide)	Chromium
Tetrachloroethylene	Pyridine	Lindane	Lead
Trichloroethylene	m-Cresol	Methoxychlor	Mercury
1,1-Dichloroethylene	o-Cresol	Toxaphene	Selenium
Vinyl chloride	p-Cresol	2,4,5-TP (Silvex)	Silver
Methyl ethyl ketone	Cresol		
Chloroform	2,4-Dinitrotoluene		
Carbon tetrachloride	Pentachlorophenol		
	2,4,5-Trichlorophenol		
	2,4,6-Trichlorophenol		

Source: U.S. Environmental Protection Agency, RCRA Waste Sampling Draft Technical Guidance Planning, Implementation, and Assessment, EPA530-D-02-002, August 2002.

---

[1] The USEPA assigned hazardous waste codes D004–D043 in alphabetical order rather than the order shown on Table 3.3.

Another type of hazardous waste is *universal waste*, which contains commonly found materials such as lead-acid batteries, fluorescent light bulbs, mercury-containing equipment, pesticides, and air refrigerants.

RCRA exempts certain types of waste from hazardous waste characterization, such as household waste (other than those mentioned above), agricultural waste returned to the ground, and utility wastes from coal combustion.

3.3.6.2 "Cradle-to-Grave" Concept of Hazardous Waste Management

RCRA seeks to prevent hazardous wastes from entering the environment at the point of generation, at the point of final disposition, and all points in between. Subtitle C of RCRA contains a litany of regulations that govern hazardous waste generators and receivers. Subtitle C codifies the most famous of RCRA concepts, namely the "cradle-to-grave" concept of hazardous waste management. Under the cradle-to-grave concept, the generator of hazardous waste is legally responsible for the waste from the time of generation and forevermore. Therefore, once the wastes are removed from their place of origin, they remain the responsibility of the owner. It also requires every party that handles or stores hazardous waste to obtain an operating permit from the USEPA, so that proper handling and storage of the waste can be tracked and enforced.

Regulations governing *large quantity generators* (LQGs), facilities that generate more than 1,000 kilograms of hazardous waste per month, dictate how the wastes are stored at their point of origin; where they can be stored; the characteristics of the waste storage area; labeling of the containers; how long they may be stored at their point of origin; and many other aspects of the management of this waste. Less complex regulations apply to *small quantity generators* (SQGs), which are facilities that generate between 100 and 1,000 kilograms of hazardous waste per month, and *conditionally exempt small-quantity generators* (CESQGs), that generate less than 100 kg of hazardous waste per month.

A *uniform hazardous waste manifest* must accompany each shipment of hazardous waste from its point of generation to its point of disposal. Figure 3.2 shows is an example of a hazardous waste manifest. The manifest has a unique number in the right-hand corner for easy identification. It also contains information about the waste, including waste code and quantity. It contains information about the generator, the transporter(s), and the designated receiving facility for the waste. Representatives of the waste generator, transporter, and receiving facility must sign the waste manifest.

Each time a waste changes hands, a representative of the entity taking over the responsibility for the waste signs the manifest. Each entity has a legal responsibility under RCRA for its temporary control of the hazardous waste. This part of the process was designed to eliminate "midnight dumping," in which irresponsible, unlicensed parties would accept waste and then deposit it in an unlicensed facility or even on the side of the road rather than incur the expense of bringing it to a properly-licensed receiving facility.

Subtitle C also establishes the qualifications for receiving facilities, which are known as *treatment, storage, and disposal facilities* (TSDFs). There are elaborate and detailed regulations governing the construction and management of hazardous waste landfills and other TSDFs. The process by which a TSDF obtains its operating license is known as a *RCRA Part B permit*.

A Part B permit is a comprehensive document designed to ensure the proper operation of RCRA facility to prevent pollution of the soil, water, and air. It contains the following information:

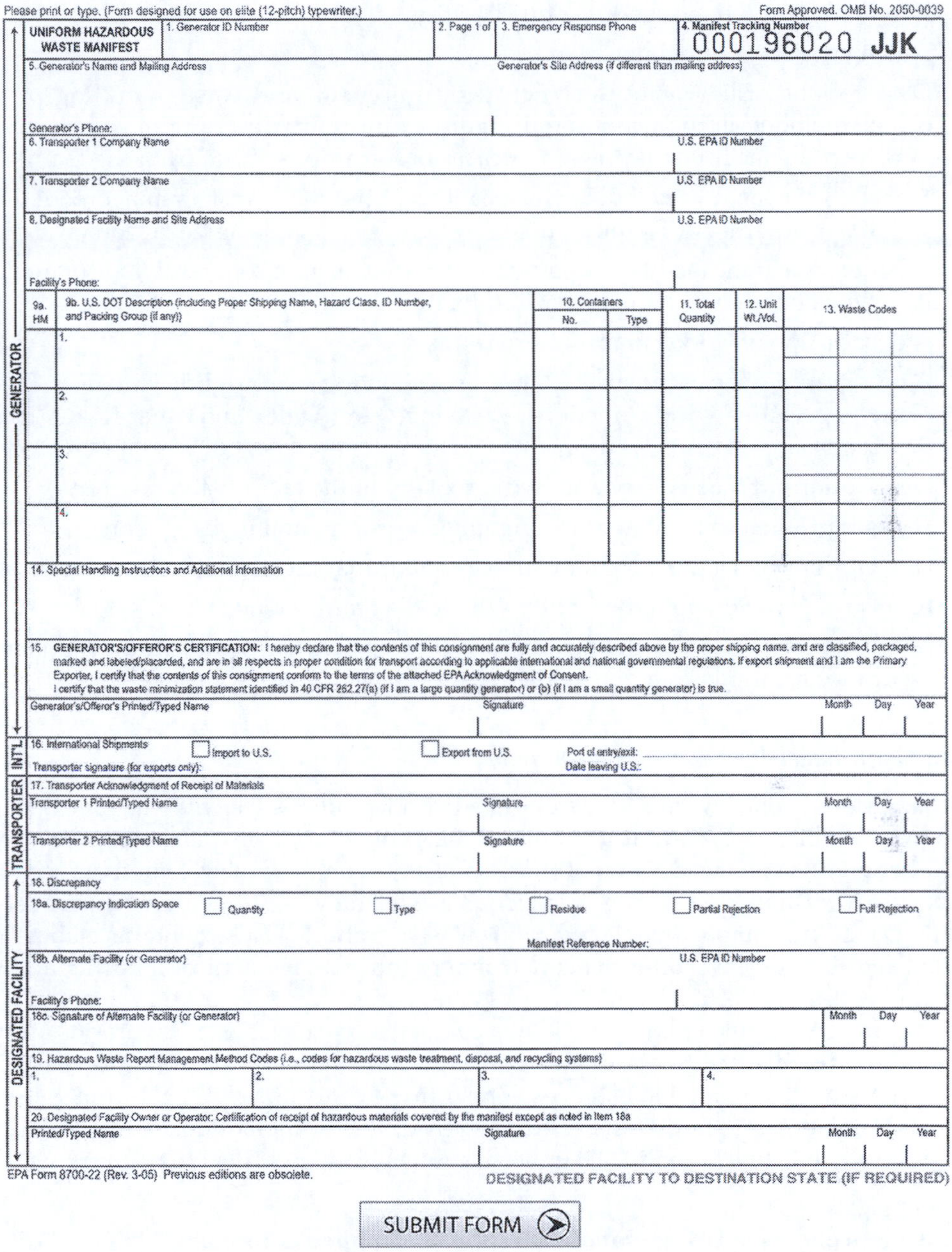
Please print or type. (Form designed for use on elite (12-pitch) typewriter.)
Form Approved. OMB No. 2050-0039

UNIFORM HAZARDOUS WASTE MANIFEST
1. Generator ID Number
2. Page 1 of
3. Emergency Response Phone
4. Manifest Tracking Number 000196020 JJK
5. Generator's Name and Mailing Address
Generator's Site Address (if different than mailing address)
Generator's Phone:
6. Transporter 1 Company Name
U.S. EPA ID Number
7. Transporter 2 Company Name
U.S. EPA ID Number
8. Designated Facility Name and Site Address
U.S. EPA ID Number
Facility's Phone:

9a. HM	9b. U.S. DOT Description (including Proper Shipping Name, Hazard Class, ID Number, and Packing Group (if any))	10. Containers No.	10. Containers Type	11. Total Quantity	12. Unit Wt./Vol.	13. Waste Codes
	1.					
	2.					
	3.					
	4.					

14. Special Handling Instructions and Additional Information
15. GENERATOR'S/OFFEROR'S CERTIFICATION: I hereby declare that the contents of this consignment are fully and accurately described above by the proper shipping name, and are classified, packaged, marked and labeled/placarded, and are in all respects in proper condition for transport according to applicable international and national governmental regulations. If export shipment and I am the Primary Exporter, I certify that the contents of this consignment conform to the terms of the attached EPA Acknowledgment of Consent.
I certify that the waste minimization statement identified in 40 CFR 262.27(a) (if I am a large quantity generator) or (b) (if I am a small quantity generator) is true.
Generator's/Offeror's Printed/Typed Name | Signature | Month Day Year
GENERATOR
INT'L
16. International Shipments ☐ Import to U.S. ☐ Export from U.S. Port of entry/exit:
Transporter signature (for exports only): Date leaving U.S.:
TRANSPORTER
17. Transporter Acknowledgment of Receipt of Materials
Transporter 1 Printed/Typed Name | Signature | Month Day Year
Transporter 2 Printed/Typed Name | Signature | Month Day Year
DESIGNATED FACILITY
18. Discrepancy
18a. Discrepancy Indication Space ☐ Quantity ☐ Type ☐ Residue ☐ Partial Rejection ☐ Full Rejection
Manifest Reference Number:
18b. Alternate Facility (or Generator) U.S. EPA ID Number
Facility's Phone:
18c. Signature of Alternate Facility (or Generator) Month Day Year
19. Hazardous Waste Report Management Method Codes (i.e., codes for hazardous waste treatment, disposal, and recycling systems)
1. 2. 3. 4.
20. Designated Facility Owner or Operator: Certification of receipt of hazardous materials covered by the manifest except as noted in Item 18a
Printed/Typed Name | Signature | Month Day Year
EPA Form 8700-22 (Rev. 3-05) Previous editions are obsolete.
DESIGNATED FACILITY TO DESTINATION STATE (IF REQUIRED)
SUBMIT FORM

FIGURE 3.2
A blank hazardous waste manifest. (Courtesy of U.S. Environmental Protection Agency.)

- A general description of the facility
- The types of hazardous wastes expected to be stored or treated at the facility
- A waste analysis plan, whose purpose is to ensure that the wastes being accepted by the facility meet the standards set forth in the facility's operating permit
- Security procedures at the facility

- A general inspection schedule of the facility
- An emergency contingency plan
- Descriptions of the facility's structures, equipment, and waste handling procedures. Included in the waste handling procedures are descriptions of the means by which the facility will prevent releases of hazardous waste from the facility
- A map of the facility, including the locations of the solid waste management units (SWMUs), and the traffic routes taken by vehicles containing hazardous wastes
- Seismic data, if the facility is located in an active seismic zone, and floodplain data if it is located within the 100-year floodplain
- A description of worker training programs
- A topographic map of the surrounding area showing surrounding land uses, floodplains, prevailing wind directions, drinking water wells, and other RCRA facilities, if any
- Descriptions of the geology and hydrogeology of the facility and environs
- A description of any previous or ongoing environmental investigations
- Descriptions of documented subsurface contamination at the facility, as applicable
- A closure plan for when the facility will be decommissioned
- A cost estimate to close the facility, and documentation demonstrating financial assurance for the costs of the facility closure

3.3.6.3 Non-hazardous Waste Management

Non-hazardous solid wastes are regulated under *RCRA Subtitle D*. Subtitle D is analogous to Subtitle C, in that it regulates the generation, transportation, and disposal of non-hazardous solid waste, except that it is far less comprehensive. For instance, it does not require information relating to hazardous waste analysis, storage, and treatment since Subtitle D facilities cannot accept hazardous wastes. Subtitle D also contains elaborate and detailed regulations governing the construction and management of landfills and other receiving facilities.

In 1984, RCRA Subtitle I (the letter "I") was passed as part of RCRA Hazardous and Solid Waste Amendments. Subtitle I established regulations for underground storage tanks (USTs) that store "regulated substances," including hazardous chemicals and petroleum products. Classes of USTs that are exempted from RCRA include residential heating oil tanks with a capacity less than 1,100 gallons, septic tanks, and fuel tanks on farms.

3.3.7 Comprehensive Environmental Response, Compensation, and Liability Act

The *Comprehensive Environmental Response, Compensation, and Liability Act* of 1980, known officially by its acronym of CERCLA but more popularly as *Superfund*, establishes the procedures and standards used in subsurface investigations and remediations.

3.3.7.1 Origins of Superfund

The impetus for the passage of CERCLA in 1980 was a series of major contamination cases that made the front pages of the newspapers in the late 1970s. Thousands of drums that contained hazardous wastes were discovered in what became known as Valley of the Drums in Kentucky. Times Beach, Missouri, made the headlines due to widespread dioxin contamination. However, no hazardous waste site was more infamous than the case known as Love Canal.

Love Canal was a neighborhood in Niagara Falls, New York, that in 1953 was owned by Hooker Chemical. Love Canal was at one time a canal constructed by a man named William T. Love as a means to bring hydroelectric power to the area from the falls. His plan failed, and by the 1920s, the canal became a dump site for the City of Niagara Falls and a dump site for the U.S. Army during World War II. By the mid-1940s, Hooker Electrochemical Company, later known as Hooker Chemical Company, began dumping chemical wastes into the canal.

After World War II, the expanding City of Niagara Falls asked Hooker Chemical to sell the land so that it could build a school on the property. Hooker refused, citing its concerns regarding locating a school over a toxic dump. Hooker eventually yielded to political pressure, selling the land to the City for $1, provided the following appeared in the contract deed:

> …[Niagara Falls] has been advised by [Hooker Chemical] that the [Love Canal property] has been filled…with waste products resulting from the manufacturing of chemicals…and [Niagara Falls] assumes all risk and liability incident to the use thereof…as a part of the consideration of this Conveyance and as a condition thereof, no claim, suit, action, or demand of any nature whatsoever shall ever be made by [Niagara Falls]…against [Hooker Chemical]…in connection with or by reason of the presence of said industrial wastes…

Despite the dire and explicit warning in the contract deed, a neighborhood and a school were built in and around the chemical dump. By the late 1970s, the buried chemicals were identified as the cause of the sickness and death that afflicted the residents of the Love Canal neighborhood. Widespread publicity and community activism pressured the U.S. government to take action against Hooker Chemical.

Since no law was available at that time under which action could be brought against Hooker Chemical, Congress responded to public pressure by passing such a law in 1980—CERCLA. Superfund gave the president of the United States the right to take "any actions necessary to abate…imminent and substantial danger to the public health or welfare or the environment." Through Superfund, the USEPA has the authority to identify parties responsible for the dumping of chemicals at inactive hazardous waste sites and force them to remediate the sites.

3.3.7.2 Liability under Superfund

There are many aspects of Superfund that make it unique among federal laws. One such aspect of Superfund is *retroactive liability*. Retroactive liability enables Superfund to reach back in time to snare the party responsible for the hazardous waste release with no statute of limitations. Retroactive liability is coupled with *strict liability*, which means that the party responsible for the chemical dumping is liable *even though its actions were permissible when they occurred*. Strict liability is consistent with RCRA's cradle-to-grave concept, in that the generator is held liable for the hazardous waste regardless of what party is in possession of the waste. The concept of strict liability enabled USEPA to force Occidental Petroleum, successor company to Hooker Chemical, to pay for the cleanup at Love Canal.

Lastly, Superfund created *joint and several liability* for the responsible parties. This means that if there were multiple parties that contributed to an inactive hazardous waste site, the USEPA had the ability to recover all of the costs from any one of the parties, leaving it to that party to pay the entire cost of the cleanup or attempt to recover costs from the other responsible parties, as described below.

3.3.7.3 Petroleum Exclusion

The definition of a hazardous substance under CERCLA specifically excludes petroleum. This portion of the definition, known as the *petroleum exclusion*, appears to exclude contamination caused by any type of petroleum product or crude oil. However, the USEPA interprets this clause as excluding only crude oil and fractions of crude oil, including any hazardous substances, such as benzene, that may be part of the crude oil. Any chemicals added to a crude oil in the refining or manufacturing substance are therefore regulated under CERCLA.

3.3.7.4 National Priorities List

In 1981, the USEPA established the *hazard ranking system* (HRS) by which the hazards posed by sites could be assessed to determine whether they should be included in the Superfund program (see Chapter 6 for an explanation of the HRS). This led to the establishment 1 year later of the *National Priorities List* (NPL), which listed all the sites slated for assessment and remediation under CERCLA. Sites under consideration for inclusion on the NPL were placed on the CERCLA Information Systems list, which was known by its acronym: CERCLIS. The information list, CERCLIS, was succeeded by the Superfund Enterprise Management System (SEMS) in 2013. As of 2014, there were over 1,300 sites on the NPL (see Figure 3.3).

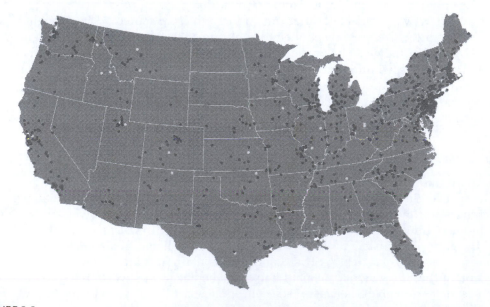

FIGURE 3.3
Superfund sites in the 48 contiguous states as of 2014. The dark dots are NPL sites and the light dots are proposed NPL sites. (Courtesy of Locus Technologies, Mountain View, CA.)

Partial funding for the Superfund program comes from a tax on chemical companies. However, most of the funding comes from USEPA cost recovery actions against *potentially responsible parties* (PRPs) under the "polluter pays principle." The USEPA will pursue identifiable, available, and financially viable PRPs for cleaning up the site or, at a minimum, cost recovery for a USEPA-directed cleanup. These PRPs will generally seek to recoup at least some of their costs by identifying other PRPs and coercing or, oftentimes, suing them for contributory compensation. The original CERCLA law, contained no provision for *de minimis* contributors, which resulted in numerous small businesses bearing a disproportionate burden of the Superfund costs since they lacked the resources to challenge better financed PRPs in court. Cleanup costs typically range on the order of tens of millions of dollars, so determining cost allocation can be a costly and bitter process. The Brownfields Act of 2003 addresses the problem of *de minimis* contributors (see below).

3.3.7.5 National Contingency Plan

Activities under CERCLA are governed by the *National Contingency Plan* (NCP), the body of regulations which establishes the framework—and in many cases specific requirements—for subsurface investigations and remediations. The NCP or variations on the NCP have since been used at Superfund and non-Superfund sites. If a site is placed on the NPL, it undergoes a series of steps that eventually result in the remediation of the contamination originating at the site. The first three steps—the preliminary assessment, the site investigation, and the remedial investigation/feasibility study (RI/FS)—are discussed in Chapters 6 through 8, respectively.

The end of the investigation process is the issuance of a *record of decision* (ROD), in which USEPA selects a final remedy.

3.3.7.6 Superfund Amendments and Reauthorization Act of 1986

The 1986 amendments to Superfund, known as both the *Superfund Amendments and Reauthorization Act* (SARA) and the *Emergency Planning and Community Right-To-Know Act* (EPCRA), had three purposes. Firstly, SARA established technical requirements to guide Superfund cleanups, and required that the selected remedy take into consideration protection of human health and the environment as well as costs and schedule. The statute placed a bias on permanent rather than temporary solutions. On-site treatment of hazardous wastes was preferred, and alternative treatment technologies were encouraged, giving rise to the *Superfund Innovative Technology Evaluation* (SITE) program (see http://www.epa.gov/nrmrl/lrpcd/site/).

Since the passage of SARA, cleanups under the Superfund program have been guided by regulations specific to Superfund and by applicable or relevant and appropriate requirements, commonly known as ARARs (pronounced AY-rarz). The ARARs fall into three categories: (1) chemical-specific criteria, which can include the cleanup standard for PCBs under TSCA and state-mandated standards; (2) action-specific, such as the criteria for landfill design under RCRA for cases where landfilling and capping the wastes at the Superfund site is the selected remedy; and (3) location-specific, such as meeting the criteria for wetlands protection under the Clean Water Act.

The second purpose of SARA was to define an innocent purchaser defense, through which the purchaser of a contaminated property could avoid CERCLA liability. The SARA specified that purchasers who performed "all appropriate inquiries" prior to purchasing the property would qualify for an innocent purchaser defense, although it didn't define the steps needed to obtain this defense.

The third objective of SARA, embodied in Title III of the SARA statute, was a public notification law that was precipitated by the catastrophic chemical release incident in Bhopal, India, in 1984, which killed over 2,000 people. The EPCRA required industrial facilities to report information about the chemicals they manufactured, stored, and released into the air, which became known as the *toxic release inventory* (TRI).

The Community Right-To-Know portion of SARA also established the *local emergency planning committee* (LEPC), which is a committee comprised of first responders, such as fire fighters, police, and medical personnel, who might be exposed to hazardous chemicals when responding to an emergency and would need to understand the nature of a person's chemical exposure in order to provide adequate treatment.

A major part of the information provided to the LEPC is the *safety data sheet* (SDS), formerly known as the material safety data sheet, or MSDS. Safety data sheets are prepared by the manufacturer and contain toxicity and safety information on the chemicals used at a facility. Any facility that stores or uses hazardous materials must have them on hand for anybody who wants to understand what chemicals are present at the facility.

3.3.7.7 Small Business Liability Relief and Brownfields Revitalization Act of 2002

The *Small Business Liability Relief and Brownfields Revitalization Act of 2002*, commonly known as the *Brownfields Act*, addressed some of the shortcomings of the original Superfund law. Title I established a *de minimis* threshold for small business contributors, specified exemptions for *municipal solid waste* (MSW), and provided for expedited settlements with PRPs based on their ability to pay. Title II provided federal grant money for the cleanup of so-called *brownfield* sites, which are discussed in Chapter 11. This legislation also provided a definition for "all appropriate inquiries" (see Section 3.2.7.6), which is described in Chapter 6.

3.3.7.8 Energy Policy Act of 2005

The Energy Policy Act contained a host of provisions relating to energy exploration, consumption, and conservation in the United States. Among those provisions relating directly to the environment are the following:

- It provides incentives for the development of nuclear energy and other energy sources associated with minimal or no emissions of greenhouse gases (GHG);
- It encourages the usage of biofuels such as ethanol;
- It includes funding for "clean coal" initiatives, i.e., technologies to control the amount of pollutants emitted during the combustion of coal;

The act also prohibited the manufacture and importation of mercury-vapor lamp ballasts after January 1, 2008.

3.4 Environmental Regulations and Environmental Consultants

Manufacturing facilities and companies that store, sell, process, or use hazardous materials and petroleum products are subject to a myriad of environmental regulations through the various federal statutes described in this chapter, as well as the various state statutes and local regulations. Many companies employ environmental professionals to assist them in complying with these regulations. Large companies develop systems to manage their regulatory responsibilities, which can include not just the physical handling, storage, and disposal of chemicals, petroleum products, and chemical and petroleum wastes, but properly managing the paperwork involved in these activities. This will include the timely filing of various reports, records, and associated fees to the appropriate regulatory agency.

Rather than hiring full-time employees, many companies retain outside environmental consultants to perform part or all of their environmental compliance duties. The purview of the environmental consultant could range from as specific a task such as handling an annual regulatory filing for a specific facility or an air emissions permit for a new boiler installation (as described below) to directing all corporate environmental compliance activities. The consultant might be working at their usual desk in the company's office, or might be sitting right there in the factory, indistinguishable from the other employees save the origin of their paycheck.

The following discussion presents examples of the types of compliance assistance that can be performed by environmental consultants. The discussion is organized by statute.

3.4.1 Clean Air Act Compliance

Environmental consultants who specialize in air permitting may prepare Title V air permit applications and renewals and design air pollution control systems for regulated air emissions sources. They also measure and quantify air pollutants for the purposes of compliance with the Clean Air Act, establishing the effectiveness of air emissions control technologies, as well as establishing credits to be traded in the emissions trading program.

3.4.2 Clean Water Act Compliance

Environmental consultants will assist companies in obtaining NPDES permits for their operating facilities, and monitoring wastewater discharges for compliance with the Clean Water Act. Consultants also prepare storm water pollution prevention plans and assist in the identification and mitigation of wetlands. Facilities that store petroleum products above certain thresholds are required to have a *spill control and countermeasures* (SPCC) plan. Environmental consultants prepare SPCC plans and can provide training to facility workers responsible for implementing the plan.

3.4.3 Endangered Species Act Compliance

Compliance with the Endangered Species Act (ESA) becomes necessary when development threatens the habitat or potential habitat of a threatened or endangered species, as discussed in Chapter 12. Environmental consultants will assist clients in complying with the requirements of the ESA and perform studies to assess whether a proposed project will violate the tenets of the ESA.

3.4.4 Safe Drinking Water Act Compliance

Environmental consultants may assist public water systems (PWSs) in complying with the SDWA by collecting water samples for laboratory analysis, and then documenting the results of the water testing in a report to the PWS. Compliance with the SDWA extends into subsurface investigations and remediations where groundwater is a source of drinking water.

3.4.5 Toxic Substances Control Act Compliance

Environmental consultants assist regulated entities in compliance with TSCA regulations, including the testing and verification of new chemicals in accordance with TSCA regulations. TSCA plays a key role in the investigation and remediation of PCBs, asbestos, and other TSCA-regulated substances. TSCA regulations governing the remediation of PCBs are especially complex and require an environmental consultant experienced in the implementation of these regulations to do so competently.

3.4.6 Resource Conservation and Recovery Act Compliance

The RCRA generates a variety of opportunities for environmental consultants, including facility compliance with the management and disposal of hazardous wastes and waste oil. Consultants may register, manage, and eventually close regulated USTs, assist TSDFs in the preparation of Part B permit applications for Subtitle C facilities, permit applications for Subtitle D facilities, and the operating of Subtitle C and Subtitle D facilities. Because it regulates USTs and hazardous waste generation, transport, and disposal, RCRA plays an important role in the investigation and remediation of contaminated properties. Environmental consultants also investigate and remediate RCRA corrective action (CORRACTS) sites, which are described in Chapter 6.

3.4.7 Superfund and Superfund Amendments and Reauthorization Act Compliance

Work at Superfund sites employs thousands of consultants nationwide. Consultants are also directly involved in Superfund by providing technical bases for cost allocations. The Phase I environmental site assessment, which are discussed in Chapter 6, essentially was created to address Superfund concerns, and Superfund has had a profound influence on the investigation and remediation of all types of contaminated sites, Superfund or not.

Environmental consultants assist client in preparing toxic release inventories (TRIs) under EPCRA annually to the USEPA and to state-equivalent agencies, as applicable.

3.4.8 Brownfields Act

The Brownfields Act has been a boon for the investigation and remediation of contaminated sites and suspected contaminated sites across the United States. It is discussed in detail in Chapter 11.

3.4.9 Environmental Compliance Audits

Many companies retain environmental consultants to conduct environmental compliance audits. The objective of the *environmental compliance audit* is to verify that the procedures being conducted at the facility, the division, or the entire company are complying with environmental regulations. Although the people managing the environmental compliance activities know those activities best, bringing in a professional from outside the operation can provide a fresh perspective on the operations and see things that people too close to the action may not see. There are four parts to the environmental compliance audit:

- *Review of environmental compliance plans:* Companies are required to develop various plans to comply with federal and state environmental regulations.
- *Inspecting operations:* A physical inspection of the plant operations allows the consultant to verify that the procedures described in the compliance plans are being followed.
- *Review of environmental compliance records:* Companies must retain records of past activities regarding the storage, usage, and disposal of hazardous wastes and petroleum products, as well as emissions to the air and to the water. These records allow the auditor to look back into past practices. Correspondences with and records of inspections by regulatory authorities also provide insight into past compliance practices, with implications for the future.
- *Interviews with personnel:* Speaking with the parties responsible for environmental compliance, from the executive vice president in the corporate headquarters to the clerk in the shipping department responsible for loading and unloading trucks often provides insight on the company's compliance efforts. Sometimes it will reveal past or current problems that aren't revealed in the official paperwork.

The environmental compliance audit may end with a written report, or maybe just a debriefing by the consultant.

Problems and Exercises

1. Define a regulatory process that addresses an environmental regulation.
2. Determine what federal regulations would apply to:
   a. Disposal of lead-acid batteries in landfills
   b. Disposal of PCBs in landfills
   c. Discharge of wastewater with a pH of 13 into a river
   d. Discharge of gases with a flashpoint of 130°F into the atmosphere
3. Determine what regulations would apply in your state or equivalent jurisdiction for:
   a. Disposal of lead-acid batteries in landfills
   b. Disposal of PCBs in landfills
   c. Discharge of wastewater with a pH of 13 into a river
   d. Discharge of gases with a flashpoint of 130°F into the atmosphere
4. Using the compound list provided in Table 3.3, name three compounds that would be classified as hazardous waste under RCRA based solely on their physical properties.

Bibliography

ASTM International, 2006. Standard Practice for Environmental Regulatory Compliance Audits. E2107-06.

Bregman, J.I., 1999. *Environmental Impact Statements*, 2nd ed. Boca Raton, FL: Lewis Publishers.

Cahill, L.B., and Kane, R.W., 1992. *Environmental Audits*, 6th ed. Rockville, MD: Government Institutes, Inc.

Carson, R.L., 1962. *Silent Spring*. Boston, MA: Houghton Mifflin.

Kraft, M.E., 1996. *Environmental Policy and Politics*. New York: HarperCollins College Publishers.

Newton, L.H., and Dillingham, C.K., 1993. *Watersheds: Classic Cases in Environmental Ethics*. Belmont, CA: Wadsworth Publishing Co.

U.S. Environmental Protection Agency, December 1998. Protocol for Conducting Environmental Compliance Audits of Treatment, Storage and Disposal Facilities under the Resource Conservation and Recovery Act. EPA-305-B-98-006.

U.S. Environmental Protection Agency, December 1998. Protocol for Conducting Environmental Compliance Audits of Treatment, Storage and Disposal Facilities under the Resource Conservation and Recovery Act (RCRA). EPA-305-98-006.

U.S. Environmental Protection Agency, December 1998. Protocol for Conducting Environmental Compliance Audits under the Comprehensive Environmental Response, Compensation and Liability Act. EPA-305-B-98-009.

U.S. Environmental Protection Agency, March 2000. Protocol for Conducting Environmental Compliance Audits of Facilities Regulated Under Subtitle D of RCRA. EPA 300-B-00-001.

U.S. Environmental Protection Agency, March 2000. Protocol for Conducting Environmental Compliance Audits of Facilities with PCBs, Asbestos, and Lead-based Paint Regulated under the Toxic Substances Control Act (TSCA). EPA 300-B-00-004.

U.S. Environmental Protection Agency, March 2000. Protocol for Conducting Environmental Compliance Audits of Public Water Systems under the Safe Drinking Water Act. EPA 300-B-00-005.

U.S. Environmental Protection Agency, March 2000. Protocol for Conducting Environmental Compliance Audits of Storage Tanks under the Resource Conversation and Recovery Act. EPA 300-B-00-006.

U.S. Environmental Protection Agency, September 2000. Protocol for Conducting Environmental Compliance Audits under the Federal Insecticide, Fungicide, and Rodenticide Act (FIFRA). EPA 300-B-00-003.

U.S. Environmental Protection Agency, December 2000. Protocol for Conducting Environmental Compliance Audits for Municipal Facilities under US EPA's Wastewater Regulations. EPA 300-B-00-016.

U.S. Environmental Protection Agency, March 2001. Protocol for Conducting Environmental Compliance Audits under the Emergency Planning and Community Right-To-Know Act and CERCLA Section 103. EPA 300-B-01-002.

U.S. Environmental Protection Agency, August 2002. RCRA Waste Sampling Draft Technical Guidance Planning, Implementation, and Assessment. EPA530-D-02-002.

U.S. Environmental Protection Agency, January 2005. Protocol for Conducting Environmental Compliance Audits under the Stormwater Program. EPA 300-B-05-004.

U.S. Environmental Protection Agency, December 2010. National Priority List Site Totals by Status and Milestone. www.epa.gov/superfund/sites/query/queryhtm/npltotal.htm.

Watts, R.J., 1998. *Hazardous Wastes: Sources, Pathways, Receptors*. New York: John Wiley & Sons.

4

Environmental Chemistry

4.1 Introduction

Environmental investigations and remediations are designed to mitigate the effects of contamination of the environment. To understand how to investigate and remediate contamination, contamination itself must be defined. This chapter discusses the principal chemical substances of interest in environmental investigations and remediations and their physical properties.

4.2 Chemical Nomenclature

Chemicals often go by more than one name, either because of their complex molecular structure, ambiguous accepted terminology, or because they are generally known by trade names, abbreviations, or nicknames. The *Chemical Abstract Service* (CAS), a division of the American Chemical Society, manages a CAS Registry™ that has assigned a unique *CAS Registry Number®* to every chemical of which it has knowledge. CAS registry numbers are very useful in identifying chemicals, especially those with more than one names. The numbers contain up to ten digits that are separated by two hyphens. For instance, the CAS number for benzene, a chemical that is associated with gasoline and other common products, is 71-43-2. This book generally employs the chemical names used by the U.S. Environmental Protection Agency (USEPA), although chemical names often vary between various regulatory agencies, even within the USEPA itself. In addition, this book often identifies certain chemicals by their trade name if that is the moniker by which they are more commonly known.

The *National Institute for Occupational Safety and Health* (NIOSH) publishes a guide book that lists hundreds of chemicals, their physical properties, and their various names. Environmental consultants use this book as a reference when they encounter an unfamiliar chemical name. The on-line version of this guide book can be accessed at http://www.cdc.gov/niosh/npg/npgsyn-a.html.

4.3 Chemical Lists

There are thousands of chemicals that are toxic to human health and/or the environment. As indicated in Chapter 3, several different lists of chemicals were developed for different sets of regulations. These lists, however, contain a tiny fraction of the chemicals that exist today.

TABLE 4.1

Chemical Lists Established by the USEPA

Chemical List	No. Chemicals on List	Regulatory Program
Target Compound List/Target Analyte List	175	Comprehensive Environmental Responsibility and Cleanup Liability Act (Superfund)
Priority Pollutant List	126	Clean Water Act
RCRA Hazardous Wastes	463	Resource Conservation and Recovery Act
Hazardous Air Pollutants	187	Clean Air Act

The lists make investigations and remediations manageable, but the vast majority of chemicals are rarely subject to environmental investigations or remediations. Periodically, the USEPA and State environmental protection agencies will add chemicals to the list of "targeted" chemicals as scientific knowledge of their toxicity becomes known or better understood or their usage becomes more prevalent. The USEPA produces fact sheets regarding so-called emerging contaminants (see https://www.epa.gov/fedfac/emerging-contaminants-and-federal-facility-contaminants-concern). The prevalent emerging contaminants as of the writing of this book are discussed later in this chapter.

Just as there are inconsistencies within the USEPA and the State environmental regulatory agencies regarding the naming of various chemicals, there are also various lists of chemicals that are used for equivalent purposes. There are numerous lists of chemicals employed at the USEPA, one for almost every regulatory program that involves chemicals. However, only a few of these lists are used in environmental investigations and remediations. Table 4.1 identifies the lists that are most often used in environmental investigations, and the regulatory program with which they are associated. Each of these lists are discussed in this chapter.

4.4 Chemical Classifications

Chemicals fall into two broad categories: organic and inorganic. Organic compounds contain carbon and are generally (but not necessarily) biological in origin. Inorganic analytes (existing usually as compounds, less often in elemental form) are not biological in origin, and either don't contain carbon, or contain carbon in a subordinate role. Lists of targeted chemicals generally include organic and inorganic chemicals, except for the hazardous air pollutants (HAPs) list, which includes only organic chemicals. Organic chemicals are further subdivided into sub-classes, as described later in this chapter. The lists of primary and secondary drinking water contaminants described in this chapter include non-chemicals such as biological agents and radionuclides.

Chemicals also can be classified by their propensity to cause cancer in humans. *Carcinogens* cause cancer or are believed to cause cancer in humans. Depending on the reliability of the toxicological research, a chemical can be classified as a known carcinogen, a probable carcinogen or a possible carcinogen. *Non-carcinogens* are not known or suspected to be directly involving in causing cancer in humans. The implication of carcinogenicity to environmental investigations and remediations is discussed in Chapter 9.

4.5 The Target Compound List/Target Analyte List

In addition to being the basis of environmental investigations in the Superfund program, many states mandate or recommend using the target compound list/target analyte list (TCL/TAL) as the basis for environmental investigations in their jurisdiction. This part of the chapter uses the TCL/TAL as the basis of discussion regarding chemicals of concern in environmental investigations. It defines the categories and sub-categories into which these chemicals commonly are placed, and describes the basic chemical compositions of these categories and sub-categories. Chemicals of special interest are noted in the text.

4.5.1 Inorganic Analytes

Inorganic analytes found in the environment may be naturally-occurring or manufactured by humans. Table 4.2 lists the inorganic analytes on the TAL, with the RCRA 8 metals, discussed in Chapter 3, in italics and underlined. With the exception of cyanide, the inorganic analytes listed in Table 4.2 are elements, although they more often are found in nature as

TABLE 4.2

Target Analyte Metals

Analyte	CAS No.
Aluminum	7429-90-5
Antimony	7440-36-0
Arsenic[a]	7440-38-2
Barium[a]	7440-39-3
Beryllium	7440-41-7
Cadmium[a]	7440-43-9
Calcium	7440-70-2
Chromium[a]	7440-47-3
Cobalt	7440-48-4
Copper	7440-50-8
Iron	7439-89-6
Lead[a]	7439-92-1
Magnesium	7439-95-4
Manganese	7439-96-5
Mercury[a]	7439-97-6
Nickel	7440-02-0
Potassium	7440-09-7
Selenium[a]	7782-49-2
Silver[a]	7440-22-4
Sodium	7440-23-5
Thallium	7440-28-0
Vanadium	7440-62-2
Zinc	7440-66-6
Cyanide	57-12-5

Source: U.S. Environmental Protection Agency. www.epa.gov/clp/epa-contract-laboratory-program-statement-work-inorganic-superfund-methods-multi-media-multi-1.

[a] RCRA 8 metals.

compounds. A standard laboratory analysis does not distinguish between elemental and compound forms of these analytes, nor does it distinguish between an element's various valences, that is, the number of electrons in an atom's structure that are available for combining with other atoms. In environmental investigations, the total concentration of these analytes is usually what is available to the investigator and what the investigator needs to comply with the regulations.

The different valences of one particular metal deserve special notice. *Hexavalent chromium* refers to the element chromium in the +6 oxidation state (trivalent chromium [+3] is the more common oxidation state for the element). Hexavalent chromium, which can have a number of industrial uses, especially in electroplating, is a recognized carcinogen and is highly toxic in other respects. On the other hand, there is little evidence that trivalent chromium is toxic to humans. To assess the presence of hexavalent chromium in the environment requires *speciation*, a process that can be performed in a laboratory.

Because of their toxicity and their special regulatory status, the so-called *RCRA 8 metals* tend to be among the major drivers of environmental investigations.

4.5.2 Organic Compounds

The TCL contains numerous organic compounds with a wide variety of physical properties. Organic compounds on this and other lists are separated into the following categories for reasons explained below:

- Volatile organic compounds
- Semi-volatile organic compounds
- Pesticides
- Polychlorinated biphenyls

4.5.2.1 Volatile Organic Compounds

Volatility, which is a compound's ability to change from liquid to vapor phase, is a primary defining parameter of organic compounds. *Volatile organic compounds*, or VOCs, are the class of compounds which attract attention throughout the spectrum of environmental settings. When released into the environment, a portion of an organic compound will evaporate into the gaseous phase. VOCs readily go into the gaseous phase at temperatures typically found in the environment.

The measure of a compound's volatility is its *vapor pressure*, which is the pressure exerted by vapor on a liquid at equilibrium. Vapor pressure varies with temperature, so published vapor pressures must indicate the temperature associated with that vapor pressure.

Vapor pressure is governed by *Henry's Law*, which states that under equilibrium conditions, the partial pressure of a volatile chemical above a liquid is proportional to its concentration in the liquid. In mathematical terms:

$$p = K_H c$$

where p is the partial pressure of the solute in the gas above the solution, c is the concentration of the solute and K_H is a constant with the dimensions of pressure divided by concentration.

Because of their ability to go into gaseous phase, VOCs are the most mobile of compounds, frequently leaving their source and evaporating, only to condense elsewhere, sometimes far away from their point of origin. Their ability to go into gaseous phase also makes them significant indoor air contaminants, as discussed in Chapter 17.

Table 4.3 lists the VOCs on the TCL, organized by their molecular structures. Forty-nine of the 52 listed VOCs are volatile *hydrocarbons*, which are organic molecules that contain hydrogen as well as carbon atoms. In volatile hydrocarbon molecules, carbon atoms are arranged either in chains containing varying numbers of carbon atoms, or in rings, with six carbon atoms forming a hexagon.

TABLE 4.3

Volatile Organic Compounds on the Target Compound List

Chemical Name	CAS No.
Aliphatics	
Alkanes	
Methanes	
Chloromethane	74-87-3
Bromomethane	74-83-9
Dibromochloromethane	124-48-1
Methylene chloride	75-09-2
Chloroform	67-66-3
Bromoform	75-25-2
Trichlorofluoromethane	75-69-4
Ethanes	
1,1,2,2-Tetrachloroethane	79-34-5
1,1,2-Trichloroethane	79-00-5
1,1,1-Trichloroethane	71-55-6
1,2-Dichloroethane	107-06-2
1,1-Dichloroethane	75-34-3
Chloroethane	75-00-3
1,2-Dibromoethane	106-93-4
1,1,2-Trichloro-1,2,2-trifluoroethane	76-13-1
Propanes	
1,2-Dichloropropane	78-87-5
1,2-Dibromo-3-chloropropane	96-12-8
Hexanes	
Cyclohexane	110-82-7
Methylcyclohexane	108-87-2
Alkenes	
Ethenes	
Vinyl chloride	75-01-4
1,1-Dichloroethene	75-35-4
trans-1,2-Dichloroethene	156-60-5
cis-1,2-Dichloroethene	156-59-2
Trichloroethene	79-01-6
Tetrachloroethene	127-18-4
Propenes	
cis-1,3-Dichloropropene	10061-01-5
trans-1,3-Dichloropropene	10061-02-6

(*Continued*)

TABLE 4.3 (*Continued*)

Volatile Organic Compounds on the Target Compound List

Chemical Name	CAS No.
Aromatics	
Benzene	71-43-2
Toluene	108-88-3
Ethylbenzene	100-41-4
M, p-Xylene	179601-23-1
o-Xylene	95-47-6
Isopropylbenzene	98-82-8
Styrene	100-42-5
Chlorobenzene	108-90-7
1,2-Dichlorobenzene	95-50-1
1,3-Dichlorobenzene	541-73-1
1,4-Dichlorobenzene	106-46-7
1,2,3-Trichlorobenzene	87-61-6
1,2,4-Trichlorobenzene	120-82-1
Ketones	
Acetone	67-64-1
2-Butanone	78-93-3
4-Methyl-2-pentanone	108-10-1
2-Hexanone	591-78-6
Other VOCs (volatile organic compounds)	
Methyl tertiary-butyl ether	1634-04-4
Methyl acetate	79-20-9
Carbon tetrachloride	56-23-5
Carbon disulfide	75-15-0

Source: U.S. Environmental Protection Agency. www.epa.gov/clp/ epa-contract-laboratory-program-statement-work-organic-superfund-methods-multi-media-multi-1.

Molecular variations occur primarily by the addition of carbon atoms or by the substitution of hydrogen with an ion or a *halogen*, which are a group of elements on the periodic table. The halogens include, in order of increasing atomic number, fluorine, chlorine, bromine, iodine, and astatine. Chlorine is the most common halogen found in compounds on the TCL, followed by fluorine and bromine. Iodine and astatine do not appear in compounds on the TCL.

Most of the molecules shown in Table 4.3 have unique molecular formulas. Some molecules, however, have the same atoms in the same quantity but in a different configuration. These molecules are known as *isomers*. Examples of isomers are shown in Figure 4.1. This figure shows the three dichlorobenzene isomers, all of which appear on the

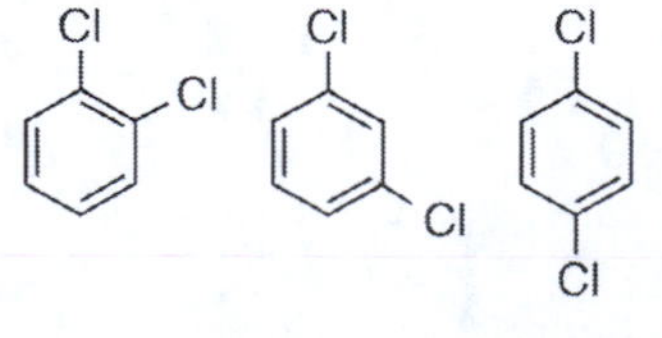

FIGURE 4.1

The three dichlorobenzene isomers. From left to right: 1,2-dichlorobenzene, 1,3-dichlorobenzene, and 1,4-dichlorobenzene.

TCL. The numbers preceding the chemical name refer to the "slots" in which the chlorine atoms appear. The "1" slot is defined arbitrarily, and the 2, 3, and 4 slots are defined in relation to the 1 slot. Note that the 5 and 6 slots don't exist; by rotating the molecule, the "1,5" arrangement would appear identical to a rotated "1,3" arrangement, and the "1,6" arrangement would look identical to the "1,2" arrangement.

Chained hydrocarbon molecules are known as *aliphatics* and ringed hydrocarbon molecules are known as *aromatics*. Aliphatics are divided into three groups, depending on the nature of their carbon bonds: alkanes, alkenes, and alkynes.

> *Alkanes* are volatile hydrocarbons that have only single bonds in their molecular structure. There are numerous categories of alkanes, four of which are represented on the TCL.

> *Methane,* the simplest alkane, is the basic building block for all alkanes. It is comprised of one carbon atom and four hydrogen atoms (see Figure 4.2). Alkanes in which one or more hydrogen atom has been replaced usually will contain "methane" in its name. For instance, replacement of one of the hydrogen atoms in the methane molecule with a chlorine atom results in the formation of chloromethane, a VOC on the TCL. Replacing all four hydrogen atoms with two chlorine atoms and two fluorine atoms results in the formation of dichlorodifluoromethane, also known as Freon 12, which is also on the TCL.[1]

> *Ethanes* are alkanes that contain two carbon atoms linked by a single bond. Ethane has six available slots that are occupied by six hydrogen atoms (see Figure 4.3). As with methane, replacing one or more hydrogen atoms with one or more halogens creates several compounds that are on the TCL. For instance, in 1,2-dichloroethane, two chlorine atoms have replaced two of the hydrogen atoms (the numbers in front of the chemical name refer to the slots that the chlorine atoms occupy in the molecule.). In 1,1,1-trichloroethane, commonly

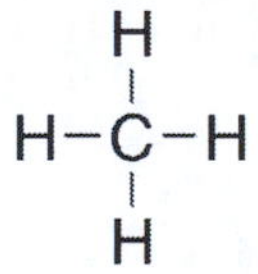

$$
\begin{array}{c}
\text{H} \\
| \\
\text{H}-\text{C}-\text{H} \\
| \\
\text{H}
\end{array}
$$

FIGURE 4.2
A methane molecule.

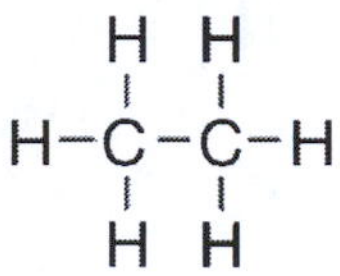

$$
\begin{array}{c}
\text{H}\quad\text{H} \\
|\quad\ | \\
\text{H}-\text{C}-\text{C}-\text{H} \\
|\quad\ | \\
\text{H}\quad\text{H}
\end{array}
$$

FIGURE 4.3
An ethane molecule.

---

[1] Dichlorodifluoromethane, because of its extremely low vapor pressure, is an excellent refrigerant. Unfortunately, as its name implies, it is a chlorofluorocarbon, or CFC, known for its imperilment of the earth's ozone layer. This compound (commonly known as Freon-12), as well as its sister CFCs, were placed under a world-wide production ban in the 1990s.

known as TCA, three chlorine atoms have replaced three hydrogen atoms, and
so on.

Propanes are alkanes that contain three carbon atoms linked by a single bond,
and *hexanes* are alkanes that contain six carbon atoms linked by a single bond
(see Figure 4.4).

Alkenes are volatile hydrocarbons that have a double bond linking its carbon
atoms. The alkenes represented on the Target Compound List fall into one of
two categories: ethenes and propenes.

Ethylene (also known as *ethene*), is the basic building block of all alkenes. It is comprised of two carbon atoms and four hydrogen atoms (see Figure 4.5).

There are six chlorinated ethenes on the TCL. They are related compounds, differing from
each other only by the number and placement of the chlorine atoms attached to the carbon
atoms. The molecule in which all four hydrogen atoms have been replaced with four chlorine atoms is known by several names: *perchloroethene*, perchloroethylene, tetrachloroeth-
ene, tetrachloroethylene, PCE, or "perc" by the dry cleaning industry. The ethene in which
three of the four hydrogen atoms have been replaced by chlorine atoms also is an effective
cleaner with several names: *trichloroethene*, trichloroethylene, "trichlor," or TCE. There are
three isomers of *dichloroethene*, which has two chlorine atoms and two hydrogen atoms.
Lastly, the ethene with one chlorine atom and three hydrogen atoms in its atomic structure
is chloroethene, but is more commonly known as *vinyl chloride*.

Propenes contain three carbon atoms, with one double bond and one single bond
(see Figure 4.6). This leaves six slots for halogens or an ion with the appropriate valence. The two propenes on the TCL, cis-1,3-dichloropropene (also known as

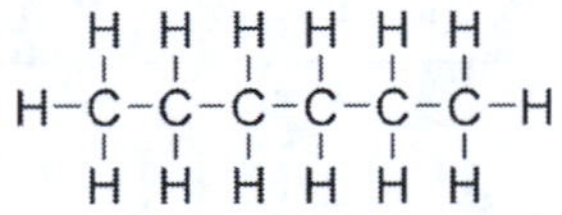

FIGURE 4.4
A hexane molecule.

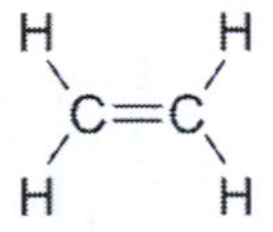

FIGURE 4.5
An ethylene molecule.

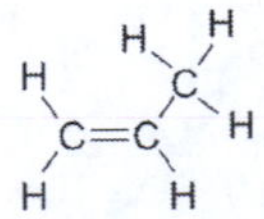

FIGURE 4.6
A propene molecule.

R−C≡C−R'

FIGURE 4.7
An alkyne molecule, where "R" represents any molecule or atom with the appropriate valence.

cis-1,3-dichloropropylene) and trans-1,3-dichloropropene (also known as trans-1,3-dichloropropylene), are isomers, each containing two chlorine atoms in place of hydrogen atoms.

Alkynes are compounds that possess a carbon-carbon triple bond. An example of an alkyne is shown in Figure 4.7. There are no alkynes on the TCL.

As mentioned above, *aromatics* are molecules comprised of six carbon atoms that form a ring. *Benzene* is the simplest aromatic compound (see Figure 4.8a). It has hydrogen atoms filling all of six available slots around the carbon ring. Toluene (Figure 4.8b) and ethylbenzene (Figure 4.8c) are aromatic compounds that have replacements for one of the hydrogen atoms in the benzene ring, and the two xylene isomers (Figure 4.8d and e) on the TCL have two methyl groups (CH_3) replacing two of the hydrogen atoms in the benzene ring. Together, benzene, toluene, ethylbenzene, and the two xylene isomers form a group of chemicals commonly referred to as *BTEX*. The BTEX compounds are related chemically as well as functionally, as discussed later in this chapter. Benzene-related compounds on the TCL with chlorine atoms in their atomic structure include chlorobenzene (one chlorine atom), 1,2-dichlorobenzene, 1,3-dichlorobenzene, and 1,4-dichlorobenzene (two chlorine atoms), and 1,2,3-trichlorobenzene and 1,2,4-trichlorobenzene (three chlorine atoms).

Benzene is a known carcinogen. Along with PCE and TCE, benzene is considered to be one of the most potentially harmful VOCs on the TCL. These three compounds often are the primary contaminants of concern in environmental investigations and remediations.

Ketones contain a carbonyl group, which is an oxygen atom double-bonded to a carbon atom, which is in turn bonded to two other carbon-based molecules. Acetone, which is shown in Figure 4.9, is the simplest ketone. It is a common industrial cleaner and is often used to clean laboratory equipment.

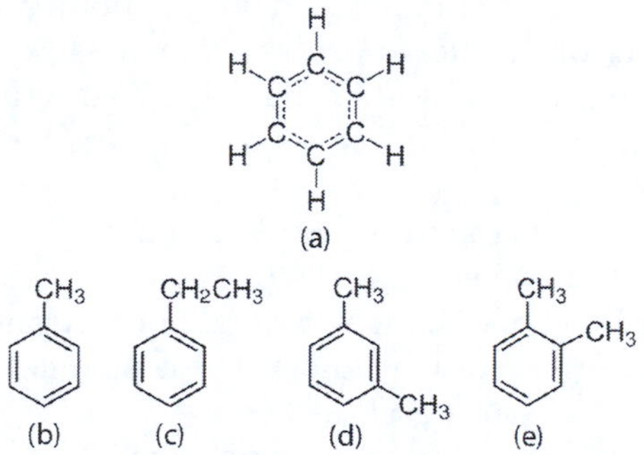

FIGURE 4.8
(a) A benzene molecule; (b) a toluene molecule, with implied carbon atoms in the benzene ring; (c) an ethylbenzene molecule; and (d,e) m-xylene and o-xylene molecules, two of the three xylene isomers.

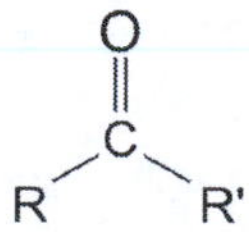

FIGURE 4.9
A ketone molecule.

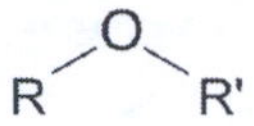

FIGURE 4.10
An ether molecule.

Ethers (see Figure 4.10) contain an ether group, which consists of an oxygen atom attached to a combination of two of the following: one or two alkyl groups (which are alkanes minus one hydrogen atom) and one or two aryl groups (which are benzene rings minus one hydrogen atom). There are two ethers on the TCL. One ether, 1,4-dioxane, is discussed in Section 4.6.7.1, below. The other ether, methyl tertiary-butyl ether (MTBE), is an additive to gasoline that figures prominently in numerous remediations. Unlike most VOCs on the TCL, MTBE is highly miscible in water and therefore can attain very high concentrations in groundwater and surface water bodies. Due to its effect on drinking water supplies, many states have complete or partial bans on MTBE usage in gasoline.

Carbon tetrachloride and *carbon disulfide* are the only two VOCs on the TCL that are not hydrocarbons and don't derive from hydrocarbons (PCE, because it does not have hydrogen atoms in its molecular structure, technically is not a hydrocarbon, either). As their names imply, they are comprised of one carbon atoms with four chlorine atoms (carbon tetrachloride) or two sulfur atoms.

4.5.2.2 Semi-volatile Organic Compounds

Semi-volatile organic compounds, or SVOCs, are generally larger molecules than VOCs and, as their name implies, are less likely to go into the gaseous phase at temperatures typically encountered in the environment. All organic compounds, to some degree, display volatility—there is no such thing as a non-volatile organic compound. However, the analytical methods which are used for detecting organic contaminants have an upper limit on the boiling point used. That maximum boiling point places an upper limit on the chemical compounds that, for practical purposes, are classified as SVOCs. Table 4.4 lists the SVOCs on the TCL, organized by their molecular structure. Please note that there are groups of SVOCs that are known by their groupings rather than as SVOCs per se, as described later in this chapter.

Some SVOCs on the TCL have the same core molecular structures as the VOCs on the TCL, namely methanes, ethanes, benzenes, ketones, and ethers. These compounds differ from their corresponding VOCs in the greater weight of their molecular structures, which generally correlates with higher boiling points. The major categories with different core molecular structures than VOCs are discussed below.

Phthalates (see Figure 4.11) are esters commonly used as plasticizers, which are substances used in the manufacture of plastics. Of particular note among the phthalates listed in Table 4.4 is bis(2-ethylhexyl) phthalate, also known as diethylhexyl phthalate, or DEHP. It is a common laboratory contaminant, since PVC (polyvinyl chloride), a common plastic, can emit DEHP as an off-gas, especially when the equipment is new.

The largest family of SVOCs on the TCL are the *polycyclic aromatic hydrocarbons*, alternately abbreviated as PAHs (polyaromatic hydrocarbons) or PNAs (*polynuclear aromatics*). As their name implies, the compounds are in the aromatics family. Unlike the volatile aromatics, PAHs are comprised of at least two interconnected benzene rings.

TABLE 4.4

Semi-volatile Organic Compounds on the Target
Compound List

Chemical Name	CAS No.
Methanes/Ethanes	
Hexachloroethane	67-72-1
Bis(2-chloroethoxy) methane	111-91-1
2,2'-Oxybis(1-chloropropane)	108-60-1
Benzenes	
2,6-Dinitrotoluene	606-20-2
2,4-Dinitrotoluene	121-14-2
Nitrobenzene	98-95-3
1,2,4,5-Tetrachlorobenzene	95-94-3
Benzaldehyde	100-52-7
Hexachlorobenzene	118-74-1
Ketones	
Isophorone	78-59-1
Acetophenone	98-86-2
Ethers	
Bis(2-chloroethyl) ether	111-44-4
4-Chlorophenyl-phenyl ether	7005-72-3
4-Bromophenyl-phenyl ether	101-55-3
1,4-Dioxane	123-91-1
Phthalates	
Bis(2-ethylhexyl) phthalate	117-81-7
Dimethyl phthalate	131-11-3
Diethyl phthalate	84-66-2
Di-n-butyl phthalate	84-74-2
Butyl benzyl phthalate	85-68-7
Di-n-octyl phthalate	117-84-0
Polycyclic Aromatic Hydrocarbons	
Naphthalene	91-20-3
2-Methylnaphthalene	91-57-6
2-Chloronaphthalene	91-58-7
Acenaphthene	83-32-9
Acenaphthylene	206-96-8
Anthracene	120-12-7
Fluorene	86-73-7
Fluoranthene	206-44-0
Pyrene	129-00-0
Phenanthrene	85-01-8
Benzo(a)anthracene	56-55-3
Benzo(b)fluoranthene	205-99-2
Benzo(k)fluoranthene	207-08-9
Benzo(a)pyrene	50-32-8
Benzo(g, h,i)perylene	191-24-2
Chrysene	218-01-9
Dibenzo(a, h)anthracene	53-70-3

(Continued)

TABLE 4.4 (*Continued*)

Semi-volatile Organic Compounds on the Target
Compound List

Chemical Name	CAS No.
Indeno(1,2,3-cd)pyrene	193-39-5
Phenyls	
2-Nitroaniline	88-74-4
3-Nitroaniline	99-09-2
4-Nitroaniline	100-01-6
4-Chloroaniline	106-47-8
1,1'-Biphenyl	92-52-4
Phenols	
Phenol	108-95-2
4-Chloro-3-methylphenol	59-50-7
2,4-Dichlorophenol	120-83-2
2-Nitrophenol	88-75-5
2,4-Dimethylphenol	105-67-9
2-Chlorophenol	95-57-8
2-Methylphenol	95-48-7
2,4-Dinitrophenol	51-28-5
2,3,4,6-Tetrachlorophenol	58-90-2
Pentachlorophenol	87-86-5
2,4,6-Trichlorophenol	88-06-2
2,4,5-Trichlorophenol	95-95-4
4-Methylphenol	106-44-5
4,6-Dinitro-2-methylphenol	534-52-1
Other Compounds	
Atrazine	1912-24-9
Caprolactam	105-60-2
N-nitrosodiphenylamine	86-30-6
N-nitrosodi-n-propylamine	621-64-7
Carbazole	86-74-8
Dibenzofuran	132-64-9
3,3'-dichlorobenzidine	91-94-1
Hexachlorocyclopentadiene	87-68-3
Hexchlorobutadiene	87-68-3

Source: U.S. Environmental Protection Agency. www.epa.gov/clp/
epa-contract-laboratory-program-statement-work-organic-
superfund-methods-multi-media-multi-1.

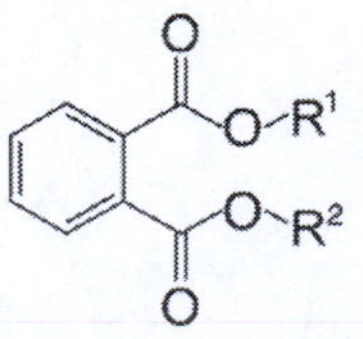

FIGURE 4.11
A phthalate molecule.

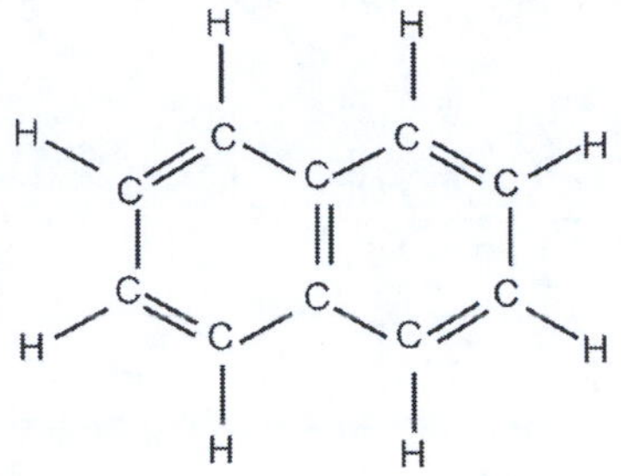

FIGURE 4.12
A naphthalene molecule.

Naphthalene, shown in Figure 4.12, is the simplest of the PAHs. It consists of two fused benzene rings and no other atoms. Naphthalene and a closely-related compound 2-methylnaphalene, are often produced from the distillation of coal tar and the refining of petroleum, and are often found in a variety of petroleum products.

Five PAHs on the TCL, namely acenaphthene, acenaphthylene, anthracene, fluorene, and phenanthrene, have molecular structures based on three fused benzene rings. Two PAHs on the TCL, fluoranthene and pyrene, consist of four fused benzene rings. These seven compounds are derived either from the distillation of coal tar, or from incomplete combustion of organic material, especially fossil fuels. When these compounds become fused with another benzene ring, their toxicity greatly increases. The other nine PAHs listed in Table 4.4, beginning with benzo(a)anthracene (an anthracene molecule fused with a benzene ring), are known as the *carcinogenic PAHs* (CaPAHs). These compounds are known carcinogens, which elevate them to a high level of concern. Their toxicity combined with their ubiquitous appearances in the environment, especially in urban areas, make them the subject of numerous environmental remediations. Benzo(a)pyrene, shown in Figure 4.13, is considered the most toxic of the CaPAHs.

Phenyls are comprised of a phenyl ring, which is a benzene ring with a molecule replacing one of the ring's hydrogen atoms (see Figure 4.14). 1,1'-Biphenyl and n-nitrosodiphenylamine are comprised of two attached phenyl rings. When multiple phenyl rings have multiple chlorine atoms attached to them, they fall into a special category of SVOCs known as *polychlorinated biphenyls*, or *PCBs*, which are discussed below.

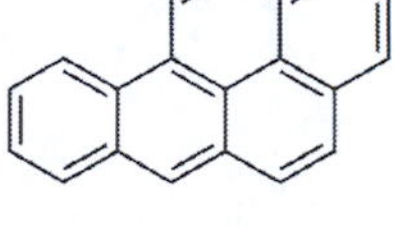

FIGURE 4.13
A benzo(a)pyrene molecule.

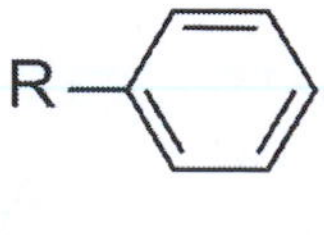

FIGURE 4.14
A phenyl molecule.

FIGURE 4.15
A phenol molecule.

Phenols (see Figure 4.15) consist of a hydroxyl molecule (OH⁻) attached to a benzene ring. This group of compounds is sometimes referred to as *acid extractables*, or AEs, due to the method by which samples are prepared in the laboratory for analysis. In contrast, the other SVOCs on the TCL are known as *base-neutral compounds*, or BNs, again due to the methods by which the samples are prepared for analysis.

4.5.2.3 Pesticides

The TCL pesticides list (see Table 4.5) contains SVOCs with a specific function of killing undesirable animals such as insects and stray rodents. These pesticides, while having

TABLE 4.5

Organic Pesticides on the Target Compound List

Chemical Name	CAS No.
alpha-BHC	319-84-6
beta-BHC	319-85-7
gamma-BHC	58-89-9
delta-BHC	319-86-8
Heptachlor	76-44-8
Aldrin	309-00-2
Heptachlor epoxide	1024-57-3
Endosulfan I	959-98-8
Dieldrin	60-57-1
4,4'-DDE	72-55-9
Endrin	72-20-8
Endosulfan II	33213-65-9
4,4'-DDD	72-54-8
Endosulfan sulfate	1031-07-8
4,4'-DDT	50-29-3
Methoxychlor	72-43-5
Endrin aldehyde	7421-93-4
Endrin ketone	53494-70-5
alpha-Chlordane	5103-71-9
gamma-Chlordane	5103-74-2
Toxaphene	8001-35-2

Source: U.S. Environmental Protection Agency. www.
epa.gov/clp/epa-contract-laboratory-program-
statement-work-organic-superfund-methods-
multi-media-multi-1.

different chemical compositions, all have chlorine atoms in their molecular structures, hence their family name: *organochlorine pesticides*. They form their own chemical category and for practical purposes are considered separately from the family of SVOCs.

The best-known pesticide, dichlorodiphenyltrichloroethane, is known by its initials: *DDT*. It actually is a family of isomers—the TCL contains just one isomer, namely 4,4′-DDT. As its name implies, it is comprised of two phenyl molecules attached to a trichloroethane molecule (that is, an ethane molecule with three chlorine atoms), with two chlorine atoms substituting for two hydrogen atoms (their locations in the molecular structure determine the full chemical name). The pesticide DDT and many of the other pesticides on the TCL are banned in the United States. However, because of their persistence in the environment, they still are often encountered in the locations where they once were applied.

Prior to the rise in the usage of organochlorine pesticides during and after World War II, the pesticide of choice was *lead arsenate*, an inorganic pesticide (chemical formula $PbHAsO_4$). The USEPA banned this pesticide in 1988.

4.5.2.4 Polychlorinated Biphenyls

Prior to their ban in 1979, *polychlorinated biphenyls*, commonly known as PCBs, had widespread usage in electrical equipment due to their stability under high temperature and pressure conditions. They were also commonly used in hydraulic oils and other oils for which stability was a priority. Like organochlorine pesticides, for practical purposes they form their own category, separate from SVOCs.

Polychlorinated biphenyls (PCBs) are composed of two connected phenyl rings, which are similar to benzene rings. Attached to the two phenyl rings (biphenyl) are a varying number of chlorine atoms (see Figure 4.16).

There are 209 types of PCBs known as *congeners*. However, PCBs are commonly grouped as Aroclors, which was a trade name under which Monsanto Corporation manufactured PCBs, but which, over time, has become the generic name for the various PCBs. Other less common names for PCBs include Pyranol, the trade name used by General Electric, and Askerel, origins unknown. The last two digits in the name indicate the chlorine content of the molecule. For instance, Aroclor-1254 is 54% chlorine atoms by weight. The nine Aroclors on the TCL are shown in Table 4.6.

PCBs were designed to be stable at high temperatures and pressures. Not surprisingly, they remain very stable when released to the environment. The USEPA classifies PCBs as *persistent organic pollutants*, or POPs. The USEPA also classifies several pesticides on the TCL as POPs.

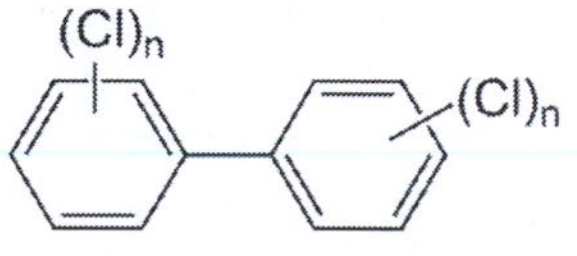

FIGURE 4.16
A PCB molecule. The number of chlorine atoms in its molecular structure varies based on congener.

TABLE 4.6

Targeted PCBs (Aroclors).

Aroclor No.	CAS No.
Aroclor-1016	12674-11-2
Aroclor-1221	11104-28-2
Aroclor-1232	11141-16-5
Aroclor-1242	53469-21-9
Aroclor-1248	12672-29-6
Aroclor-1254	11097-69-5
Aroclor-1260	11096-82-5
Aroclor-1262	37324-23-5
Aroclor-1268	11100-14-4

Source: U.S. Environmental Protection Agency. www.epa.gov/clp/epa-contract-laboratory-program-statement-work-organic-super-fund-methods-multi-media-multi-1.

4.6 Contaminants in Drinking Water

The Safe Drinking Water Act (SDWA) established primary *maximum contaminant levels* (MCLs) and maximum contaminant level goals (MCLGs) for various chemicals, as well as various biological and radiological agents, and MCLGs for various chemical properties. The SDWA also established secondary MCLs for various chemicals and chemicals properties that, while not health-based, are considered nuisances for aesthetic or cosmetic reasons. These chemicals, agents, and properties are important to environmental investigations and remediations when drinking water or an aquifer with the potential of being used as a source of drinking water is affected.

4.6.1 Organic Chemicals with Primary Drinking Water Standards

Table 4.7 lists the organic chemicals that are regulated under the *national primary drinking water regulation* (NPDWR) of the SDWA. Most of the organic chemicals with primary drinking water standards are on the TCL. These chemicals include VOCs, SVOCs, pesticides, and PCBs.

Also included on this list are 11 *herbicides*, which are chemicals designed to kill unwanted vegetation. Because they aren't designed to harm fauna, they tend to have little or no toxicity to humans. The herbicides that are regulated under the SDWA typically are complex molecules with a wide variety of molecular structures. All but two of these regulated herbicides are still in use in the United States, although the European Union has banned most of them because of their ability to contaminate drinking water.

Among the chemicals with primary drinking water standards is one of special note— 2,3,7,8-tetrachlorodibenzo-*p*-dioxin (TCDD). Although 2,3,7,8-TCDD is commonly referred to as dioxin, *dioxin* actually refers to a family of chemical compounds that are generally considered to be among the most hazardous to human health and the environment. Dioxins are unwanted byproducts in the manufacture of organic pesticides or incomplete combustion. They became infamous in United States and across the world as a contaminant in Agent Orange, a defoliant used in the Vietnam War. Generally comprised of two benzene rings connected by oxygen "bridges," with two chlorine atoms at each end of its

TABLE 4.7

Organic Chemicals Regulated by the National
Primary Drinking Water Regulations under
the Safe Drinking Water Act

Volatile organic compounds (VOCs)
Dichloromethane
1,2-Dichloroethane
1,1,1-Trichloroethane
1,1,2-Trichloroethane
1,2-Dichloropropane
Tetrachloroethylene
Trichloroethylene
1,1-Dichloroethylene
cis-1,2-Dichloroethylene
trans-1,2-Dichloroethylene
Vinyl chloride
Benzene
Ethylbenzene
Toluene
Xylenes (total)
Chlorobenzene
o-Dichlorobenzene
p-Dichlorobenzene
1,2,4-Trichlorobenzene
Styrene
Carbon tetrachloride
Semi-volatile organic compounds (SVOCs)
Atrazine
Benzo(a)pyrene (PAHs)
Di(2-ethylhexyl)phthalate
Hexachlorobenzene
Hexachlorocyclopentadiene
Di(2-ethylhexyl)adipate
Ethylene dibromide
Herbicides
Alachlor
Dinoseb
Diquat
Endothall
Glyphosate
Picloram
Simazine
Pesticides
Carbofuran
Chlordane
2,4-D
Dalapon
1,2-Dibromo-3-chloropropare (DBCP)

(Continued)

TABLE 4.7 (*Continued*)

Organic Chemicals Regulated by the National
Primary Drinking Water Regulations under
the Safe Drinking Water Act

Endrin
Heptachlor
Heptachlor epoxide
Lindane
Methoxychlor
Oxamyl (Vydate)
Pentachlorophenol
Toxaphene
2,4,5-Silvex
Others
Polychlorinated Biphenyls
Dioxin (2,3,7,8-TCDD)
Acrylamide
Epichlorohydrin

Source: U.S. Environmental Protection Agency. www.
epa.gov/ground-water-and-drinking-water/
national-primary-drinking-water-regulations.

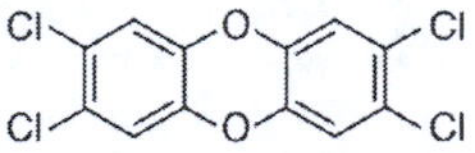

FIGURE 4.17
Diagram of a 2,3,7,8-TCDD (dioxin) molecule.

atomic structure. The chlorine atoms contribute greatly to the toxicity of the molecule.
Figure 4.17 shows the molecular structure of a dioxin molecule.

4.6.2 Inorganic Chemicals with Primary Drinking Water Standards

Table 4.8 provides the inorganic chemicals that are listed in the NPDWR. The metals and
cyanide listed here also appear on the Target Analyte List. The other inorganic chemicals
with primary drinking water standards fall into three categories:

- Disinfectants
- Disinfection byproducts
- Nitrates and nitrites.

4.6.2.1 Disinfectants and Disinfectant Byproducts

Chloramine, chlorine, and chlorine dioxide are commonly added to surface waters that
are sources of drinking water to control microbes. Ingestion of these disinfectants at
concentrations above their MCLs can lead to gastrointestinal distress, anemia, and eye or
nose irritation.

Adding disinfection chemicals to the drinking water supplies can create unwanted byproducts, including bromates, chlorites, haloacetic acids (HAASs), and *trihalomethanes* (TTHMs).

TABLE 4.8

Inorganic Chemicals and Other Parameters Regulated by the National Primary Drinking Water Regulations under the Safe Drinking Water Act

---

Microorganisms

Cryptosporidium

Total coliforms (including fecal coliform and *E. Coli*)

Giardia lamblia

Heterotropic plate count

Legionella

Viruses (eteric)

Metals and Cyanide

Antimony

Arsenic

Barium

Beryllium

Cadmium

Chromium (total)

Copper

Lead

Mercury (inorganic)

Selenium

Thallium

Cyanide (as free cyanide)

Disinfection Byproducts

Bromate

Chloramines (as Cl2)

Chlorine (as Cl2)

Chlorite

Haloacetic acids (HAAS)

Total trihalomethanes

Radionuclides

Alpha particles

Beta particles

Radium 226 and Radium 228 (combined)

Others

Asbestos (fiber<10 micrometers)

Nitrate (measured as nitrogen)

Nitrite (measured as nitrogen)

Physical Property

Turbidity

Source: U.S. Environmental Protection Agency. www.epa.gov/ground-water-and-drinking-water/national-primary-drinking-water-regulations.

Whereas the SDWA regulates total TTHMs, some TTHMs, such as bromoform and chloroform, also are regulated as individual chemicals.

4.6.2.2 Nitrates and Nitrites

Nitrates (NO_3^-) *and nitrites* (NO_2^-) are compounds containing the nitrate (NO_3) or nitrite (NO_2) anion. Both anions contain nitrogen, an essential element for plant and animal growth. Nitrates and nitrites typically derive from runoff from fertilizers and animal wastes. High levels of nitrates and nitrites can lead to eutrophication of the surface water body (eutrophication is discussed in Chapter 12 of this book). In addition, too much nitrogen in drinking water can be toxic to humans.

4.6.3 Radionuclides

Many geological formations contain trace amounts of uranium and radium that undergo radioactive decay, as described in Chapter 18. The erosion of these formations can lead to the freeing of radioactive soil grains that will emit *alpha particles* or *beta particles*. These radioactive emissions can damage the gastrointestinal system when ingested, thus increasing the risk of cancer. Alpha particles are an even larger threat when inhaled. Alpha particles, beta particles, uranium, and radium all are listed under the NPDWR of the Safe Drinking Water Act.

4.6.4 Biological Agents

Six biological agents have primary drinking water standards. *Cryptosporidium*, *Giardia lamblia*, and bacteria coliform are pathogens that can cause gastrointestinal distress when ingested. They typically are associated with inadequately-treated drinking water that is more prevalent in developing countries than in rich countries such as the United States. However, these microorganisms can breed in surface water that has received animal feces upon which the pathogens can breed.

Heterotrophic plate count (HPC) isn't a microorganism but rather a parameter which, if in excess of its MCL, suggests a problem in the drinking water system.

The primary drinking water list also includes viruses, and one particular pathogen, *Legionella*. Legionella, which is a bacterium that attacks respiratory systems, is discussed in Chapter 17.

4.6.5 Turbidity

The NPDWSR lists one water property—turbidity. *Turbidity* measures the opacity of water, which usually is caused by the presence of suspended particles in the water. It has no direct health effects but can interfere with the water treatment process, and the suspended particles can provide a platform for microbial growth. It is recorded as nephelometric turbidity units (NTUs). Turbidity also is a physical parameter measured during groundwater sampling, which is discussed in Chapter 7 of this book.

4.6.6 Secondary Drinking Water Regulations Contaminants

Table 4.9 lists the parameters covered by the national secondary drinking water regulations (NSDWRs), which are chemicals or chemical properties that may affect the taste,

TABLE 4.9

National Secondary Drinking Water Regulations Parameters

Metals	Others
Aluminum	Chloride
Copper	Color
Iron	Corrosivity
Manganese	Fluoride
Silver	
Zinc	Foaming Agents
	Odor
	pH
	Sulfate
	Total Dissolved Solids (TDS)

Source: U.S. Environmental Protection Agency. www. epa.gov/dwstandardsregulations/secondary-drinking-water-standards-guidance-nuisance-chemicals.

odor, or color of the drinking water, or cause cosmetic effects, such as skin or tooth discoloration. The USEPA recommends but does not require regulation of these parameters, except for publicly owned treatment works (see Chapter 3).

The NSDWR list includes six metals: aluminum, copper, iron, manganese, silver, and zinc. The USEPA does not consider ingestion of these six metals to be deleterious to human health. However, their presence will affect the taste and/or appearance of the drinking water, which is why they have secondary MCLs.

Included in this list are chloride, fluoride, and sulfate, which are anions. Chloride, the anion in sodium chloride, or common salt, makes water taste salty, and therefore is undesirable in large concentrations. The presence of too much sulfates also lead to bad tasting water. Fluoride is added to water systems to prevent tooth decay. The World Health Organization has set the target concentration for fluoride in drinking water at 0.5–1.5 milligrams per kilogram (mg/kg). At higher concentrations, fluoride can cause tooth discoloration, which is undesirable but not threatening to human health.

The other parameters regulated under the NSDWR are physical characteristics that are undesirable in drinking water: color (people like their water clear); corrosivity (which is related to pH, also on the list); foaming agents (an oily or "fishy" taste is undesirable in drinking water), odor, and total dissolved solids (TDS), which also is associated with the taste of the drinking water.

4.6.7 Emerging Contaminants in Drinking Water

The lists of chemicals of concern is not static. As information becomes available, the USEPA and state-equivalent regulatory authorities will add chemicals to their respective lists. Typically, these chemicals have caught the public eye and are believed to be prevalent and toxic. They could be in or out of production or could be an unwanted byproduct of a manufacturing process, combustion, or chemical or biological degradation. Emerging contaminants present a special problem to properties that have already been remediated for the chemicals of concern at the time of the remediation. Some of the current, more prominent emerging contaminants are discussed below.

4.6.7.1 1,4-Dioxane

1,4-Dioxane is an ether that in recent years has gained heightened concern at federal and state regulatory agencies due to new studies regarding the compound's physical and toxicological properties. Figure 4.18 shows the structure of a 1,4-dioxane molecule. 1,4-Dioxane was used as a stabilizer in the manufacture of various solvents, especially 1,1,1-TCA. In 2013, the USEPA moved this compound from the TCL VOC list to the TCL SVOC list, and in 2015 lowered its MCL when studies indicated that it was more toxic than originally believed. This reclassification has led to changes in various parameters used to assess human health risk, and a lowering of the threshold concentration considered by the USEPA to be non-toxic to human health. It is highly miscible in water, meaning that it can travel great distances once dissolved in groundwater or surface water. In addition, 1,4-dioxane is *hygroscopic,* meaning it attracts and holds onto water molecules. Although it has a boiling point indicative of a VOC, its high water solubility makes it difficult to analyze as a VOC, which is why USEPA recommends treating it as an SVOC instead. 1,4-dioxane does not readily biodegrade in the environment. Because of its unusual physical characteristics, it is resistant to various remedial technologies (see Chapter 9).

FIGURE 4.18
A 1,4-dioxane molecule.

4.6.7.2 Perchlorate

Perchlorate is an anion consisting of a chlorine atom bonded to four oxygen atoms (ClO_4^-). It can occur naturally, but more often is a manufactured product used in solid rocket propellants, munitions, fireworks, airbag initiators for vehicles, matches, and signal flares. It commonly is found as the anion in perchloric acid and salts such as ammonium perchlorate, sodium perchlorate and potassium perchlorate. Like 1,4-dioxane, it is highly soluble in water, enabling it to travel great distances in groundwater or surface water.

4.6.7.3 Per- and Polyfluoroalkyl Substances (PFASs)

Per- and polyfluoroalkyl substances (PFASs) represent a broad group of more than 6,000 highly fluorinated aliphatic molecules that is emerging as a game changer for site remediation. They were used extensively in a variety of personal, commercial, and industrial products due to their unique ability to repel oil and water.

Perfluorooctane sulfonate (PFOS) and perfluorooctanoic acid (PFOA) were the most produced PFAS chemicals, and most widely studied. They are part of a subset of PFASs known as perfluorinated alkyl acids (PFAAs). Figure 4.19 shows the structure of a PFOS molecule and Figure 4.20 shows the structure of a PFOA molecule. These chemicals are highly toxic. In May 2016, the USEPA issued a drinking water health advisory limit of 70 nanograms per liter (ng/L), which is

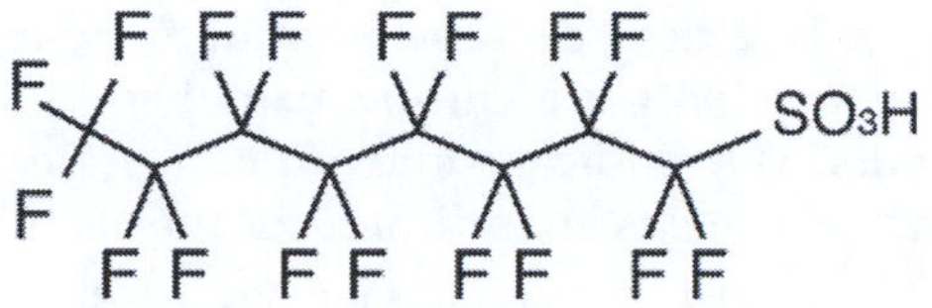

FIGURE 4.19
A PFOS molecule.

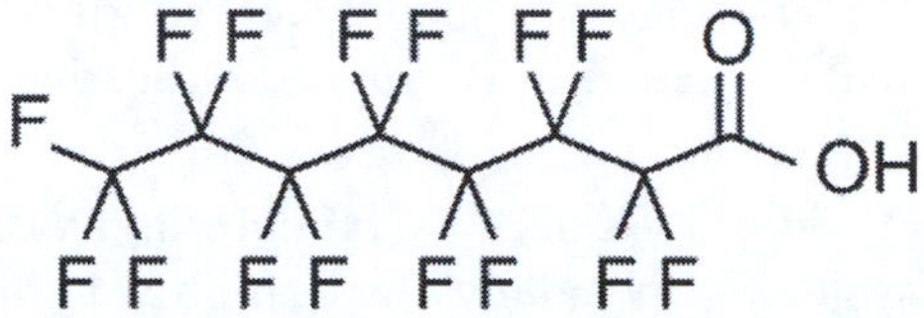

FIGURE 4.20
A PFOA molecule.

a lower threshold than all other chemicals regulated under the Safe Drinking Water Act except for dioxin, which has an MCL of 30 ng/L. The PFASs are persistent in the environment and resistant to typical environmental degradation processes. Because of their unique properties, laboratories must employ special analytical procedures to detect PFASs.

4.7 Chemistry of Surface Waters

Many of the parameters used to study the chemistry of surface waters are also used in the study of drinking water, as discussed above. Many parameters, however, are used only for surface water bodies. Some of the more common parameters are discussed below.

Biochemical oxygen demand (BOD): BOD measures the amount of dissolved oxygen needed by aerobic organisms to stabilize the organic material present in a body of water. Soluble organic matter introduced into a surface water body will increase its BOD, resulting in oxygen depletion in the water and the suffocation of aquatic animals. Animal wastes emanating from humans, livestock, and wild animals, are the most common organic matter that increases BOD in a surface water body. Human wastes commonly originate from a malfunctioning or improperly maintained septic system or sanitary sewer system, whereas livestock (cattle, pigs, sheep, etc.) are the most common sources of non-human animal wastes. BOD is measured in milligrams per liter (mg/L) or in the equivalent unit of parts per million (ppm).

Chemical oxygen demand (COD): COD measures the oxygen demanded by chemicals present in the water body. These chemicals, in some cases domestic and industrial wastes introduced into the water body, compete with aquatic animal life for available oxygen. High COD, like high BOD, can result in the suffocation of aquatic animals. Like BOD, COD is measured in mg/L.

Dissolved oxygen (DO): DO is the amount of oxygen available in water to aquatic animals. DO values of less than 5 ppm typically are considered detrimental to fish. Dissolved oxygen also is a physical parameter measured during groundwater sampling, which is discussed in Chapter 7 of this book.

Total organic carbon (TOC): TOC measures the amount of organic carbon in the aquatic system. It does not include inorganic carbon, usually in the form of carbon dioxide and carbonic acid salts. TOC indicates the amount of decayed organic matter in the water body, which is used as an indicator for the cleanliness of the water.

Turbidity: Turbidity is discussed in Section 4.6.5, above. High turbidity could block sunlight for aquatic plants, or can silt up the water body, reducing its capacity.

Total solids: Total solids is the sum of suspended solids (particulates in the water) and dissolved solids, which are described above. Total solids in a surface water body is usually caused by suspended solids, which are indicative of soil erosion or other infiltration of materials.

pH: pH is described in Section 4.6.6, above. The pH range of 5–9 generally is indicative of a healthy ecosystem. Abnormally low or high pH not only affects organisms directly, but it also affects the solubility of various metals, which can also affect their chemical state, mobility, biological/toxicological effects, and the entire ecosystem.

Salinity: Salinity is the concentration of chlorides [Cl⁻] in the water, as described in Section 4.6.6, above. Water with a salinity reading of less than 0.5% is considered fresh water. Excessive salinity in a fresh water body can impact fresh water flora and fauna, and can increase the solubility of certain undesirable compounds, such as metals (see below) that may impact surface water quality, flora, and fauna. Conversely, a decrease in salinity can cause these same metals to precipitate out of water and fall to the bottom of the water body, thereby impacting the sediments at the bottom of the water body.

Conductivity: Conductivity is the ability of the water body to conduct electricity, which is a function of the concentration of ions and particulates within the water. It is related to salinity, and usually is measured in microsiemens (also known as micromohs) per centimeter.

Hardness: Also known as alkalinity, hardness is measured by the concentration of calcium carbonate ($CaCO_3$) in the water. Measuring hardness is important, especially if metals analyses are to be performed, since metals are more toxic in "soft" water (low calcium concentration) than in "hard" water (high calcium concentration).

Various metals, including iron and manganese: The presence of various metals can raise the COD of a water body.

Heavy metals, such as lead, chromium, copper, and zinc: Heavy metals, naturally present at low concentrations in most aquatic environments, typically enter the environment as constituents of industrial wastes. If allowed to reach toxic concentrations, heavy metals can kill flora and fauna and inhibit living flora and fauna from reproducing.

Sulfides (S^{2-}): Sulfide, an anion of sulfur, is highly basic, and can raise the pH of a water body. Sulfides are formed in oxygen-deficient environments by anaerobic microbial respiration, using sulfates (SO_4^{2-}), which are described above, as an electron receptor instead of oxygen. Sulfides are often seen in natural anaerobic environments such as wetlands (see Chapter 12) or anoxic water bodies.

Phosphorus (P), nitrates and nitrites: Nitrates and nitrites are discussed in Section 4.6.2.2, above. Like nitrates and nitrites, phosphorus is a nutrient necessary for plant and animal life. Also like nitrates and nitrites, phosphorus typically derives from run-off from fertilizers and animal wastes and high levels of phosphorus in a water body could result in eutrophication.

4.8 Petroleum and Petroleum-Related Compounds

There are numerous other chemicals and classes of chemicals that present contamination when released to the environment. The most important class of chemicals doesn't have a specific formula at all—petroleum.

Petroleum is a catch-all term that represents a complex mixture of chemicals with common chemical compositions. Petroleum occurs naturally in the form of crude oil, and is not by nature toxic to humans. It is, however, hazardous to flora and fauna, especially birds and fish, when spilled. Certain petroleum additives can make it toxic to humans, as described below.

4.8.1 Chemical Composition of Petroleum

All types of petroleum have carbon atoms and hydrogen atoms, hence the name hydrocarbons. The simplest type of petroleum is methane, which contains one carbon atom, as described above. Most types of petroleum, however, involve multiple carbon atoms, arranged in chains which consist of two or more carbon atoms bonded together. The hexane molecule shown in Figure 4.4 is a straight-chain hydrocarbon containing six carbon atoms. Adding more carbon atoms to the chain would create ever-larger hydrocarbon molecules (see Figure 4.21). The notation commonly used to describe hydrocarbon chains is C_n, where n is equal to the number of carbon atoms in the chain.

Crude oil goes through a refining process to become a substance that is useful to humans. In the refining process, impurities and other undesirable chemicals are removed from the crude, and desirable chemicals are added to the crude, to form a petroleum product designed with specific uses in mind. Some of the more common types of refined petroleum products are described below.

4.8.2 Gasoline

The most common use for petroleum is in the manufacture of fuels designed for the rapid delivery of power. These usually contain flammable components whose bonds will break apart and yield energy instantly when injected into the internal combustion engine of a vehicle or a similar device. These volatile components also render the fuel flammable, thus hazardous to work with and store, as well as toxic to human beings.

The most common flammable components added to gasoline are the BTEX compounds, discussed earlier in the chapter. The BTEX compounds, when ignited, give the internal combustion engine the punch that is so familiar to those of us who have pressed down on the accelerator. Until the mid-1970s, lead, typically in the form of tetra-ethyl lead, was added to the gasoline to provide that power. The practice of adding lead to gasoline was discontinued under the Clean Air Act to reduce the amount of lead in the air.

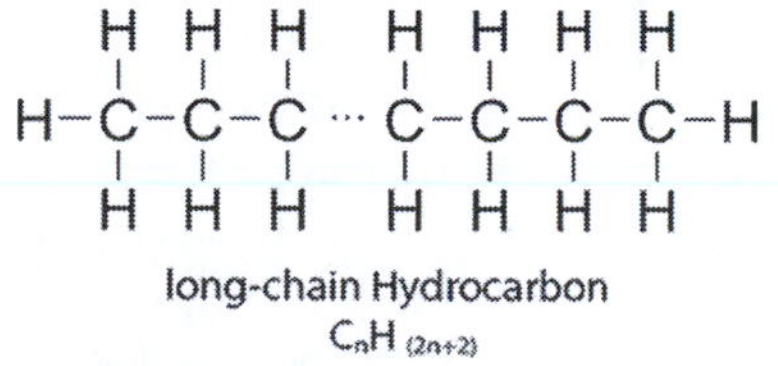

long-chain Hydrocarbon
$C_nH_{(2n+2)}$

FIGURE 4.21
General formula for a straight-chain hydrocarbon.

The use of oxygenates in gasoline began in the 1980s and accelerated into the 1990s after the enactment of the Clean Air Act Amendments of 1990. The role of oxygenates is to enable gasoline to burn more cleanly and create less carbon monoxide emissions, thereby reducing smog. One oxygenate, MTBE, is discussed in Section 4.5.2.1.

4.8.3 Non-volatile Fuels

Non-volatile fuel oils are designed for delivery of power as well as stability in the environment. They tend to be numbered by their degree of volatility.

#1 fuel oil, more commonly known as *kerosene*, is a less refined petroleum product than gasoline. It is the basis for jet fuel; however, most kerosene products are not volatile, and therefore are more stable than gasoline or jet fuel and safer to use and store.

Diesel fuel, or *#2 fuel oil*, is widely used in internal combustion engines of larger motor vehicles, such as large cars, trucks, buses, and trains. These vehicles do not need the acceleration desired in a standard automobile, although turbo-charging automotive diesel engines are becoming a popular way of enhancing the engine's power. This fuel is very stable in the environment, which is why it is widely used to heat buildings, where it is commonly known as #2 fuel oil.

Of the commonly used fuel oils, the heaviest and most stable is known as *#6 fuel oil*. It is often called "bunker oil," although that phrase is generally meant to describe the oil used on ships. It is so viscous (thick) that it usually has to be heated to flow, which has important implications for spills of #6 fuel oil.

#4 Fuel oil and #5 fuel oil are mixtures of #2 fuel oil and #6 fuel oil. #3 Fuel oil technically exists but is rare.

4.8.4 Engineered Oils

In addition to the fuel oils, there are literally thousands of petroleum products that are engineered with other uses in mind. Many petroleum products are designed for stability under extreme conditions of heat and/or pressure, or as lubricants. Some of the more common forms of engineered oils are:

- Mineral spirits, or mineral oil
- Lubricating oil
- Hydraulic oil
- Cutting oil
- Crankcase
- Grease

The composition of the literally thousands of petroleum products cannot be generalized. Each product contains a unique mix of various types of petroleum and petroleum additives. The only ways to understand the chemical composition of a petroleum product, even a fuel oil, is to either analyze the product in a laboratory or read its Safety Data Sheet (SDS, see Chapter 3).

4.9 Synthetic Organic Contaminants

The other thousands of chemicals known to be toxic to humans and the environment and are not on a defined list do not go unnoticed in a laboratory analysis since analyses performed to detect the organic chemicals on the various lists also pick up other chemicals with similar physical properties, especially boiling point. A laboratory can attempt to identify a non-targeted compound based on the physical properties that they exhibit during the analysis. Such compounds are known as *tentatively identified compounds*, or TICs. No standards exist for individual TICs; however, many jurisdictions have standards for total organic compounds in the aggregate. The term used, which includes "targeted" compounds as well as TICs, is total *synthetic organic contaminants* (SOCs).

Problems and Exercises

1. Explain why different analytical methods are needed for volatile organic compounds than other compounds on the Target Compound List.

2. How does the difference in the number of chlorine atoms in the ethanes on the Target Compound List affect their boiling point, molecular weight, and density?

3. Explain how the substitution of bromine atoms for hydrogen atoms in the ethane molecule affects the molecule's boiling point, molecular weight, and density as opposed to the substitution of chlorine atoms for hydrogen atoms in the ethane molecule.

4. Explain the differences between the molecular structures of benzo(a)anthracene, benzo(b)anthracene, and benzo(k)anthracene.

5. Research five different types of manufactured petroleum products on the internet. Determine their utilities and evaluate why their physical properties are suited to the desired utility.

Bibliography

Alpha Analytical, 2018. Technical Bulletin: Analysis of 1,4-Dioxane in Water.

Budavari, S. et al., 1996. *The Merck Index*, 12th ed. Merck & Co., Inc, Whitehouse Station, NJ.

International Union of Pure and Applied Chemistry, 1993. *Quantities, Units, and Symbols in Physical Chemistry*, 2nd ed. Oxford, UK: Blackwell Science.

California Air Resource Board, 2015. California Reformulated Gasoline Phase 3 (CaRGP3).

Lagrega, M.D., Buckingham, P.L., and Evans, J.E., 2001. *Hazardous Waste Management*, 2nd ed. McGraw-Hill, Waveland Press, Long Grove, IL.

Meyer, E., 1989. *Chemistry of Hazardous Materials*, 2nd ed. Prentice-Hall, Inc, Upper Saddle River, NJ.

Schecter, A., Birnbaum, L., Ryan, J.J., and Constable, J.D., 2006. Dioxins: An overview. *Environmental Resource* 101 (3): 419–428.

Schwarzenbach, R.P., Gschwend, P.M., and Imboden, D.M., 2005. *Environmental Organic Chemistry*. Hoboken, NJ: John Wiley & Sons, Inc, Hoboken, NJ.

Suthersan, S., Quinnan, J., Horst, J., Ross, I., Kalve, E., Bell, C., and Pancras, T., 2016. Making strides in the management of "Emerging Contaminants." *Groundwater Monitoring & Remediation*, 36 (1): 15–25.

U.S. Environmental Protection Agency, www.epa.gov/clp/epa-contract-laboratory-program-statement-work-inorganic-superfund-methods-multi-media-multi-1.

U.S. Environmental Protection Agency, www.epa.gov/clp/epa-contract-laboratory-program-statement-work-organic-superfund-methods-multi-media-multi-1.

U.S. Environmental Protection Agency, www.epa.gov/dwstandardsregulations/secondary-drinking-water-standards-guidance-nuisance-chemicals.

U.S. Environmental Protection Agency, www.epa.gov/ground-water-and-drinking-water/national-primary-drinking-water-regulations.

U.S. Environmental Protection Agency, 2007. State Actions Banning MTBE. EPA-420-B-07-013.

U.S. Environmental Protection Agency, 2016. Primary Drinking Water Standards: Guidance for Nuisance Chemicals. https://www.epa.gov/sites/production/files/2016-06/documents/npwdr_complete_table.pdf.

U.S. Environmental Protection Agency, Secondary Drinking Water Standards: Guidance for Nuisance Chemicals. https://www.epa.gov/dwstandardsregulations/secondary-drinking water-standards-guidance-nuisance-chemicals.

U.S. Environmental Protection Agency, 2017. Technical Fact Sheet: Perchlorate. https://www.epa.gov/sites/production/files/2017-10/documents/perchlorate_factsheet_9-15-17_508.pdf.

U.S. Environmental Protection Agency, 2017. Technical Fact Sheet: Perfluorooctane Sulfonate (PFOS) and Perfluorooctanoic Acid (PFOA). https://www.epa.gov/sites/production/files/2017-12/documents/ffrrofactsheet_contaminants_pfos_pfoa_11-20-17_508_0.pdf.

Watts, R.J., 1997. *Hazardous Wastes: Sources, Pathways, Receptors*. John Wiley & Sons, Inc, Hoboken, NJ.

World Health Organization, 2017. *Guidelines for Drinking-water Quality*, 4th ed, Geneva, Switzerland.

5

Fate and Transport in the Subsurface

In a laboratory, chemicals behave in predictable ways, due to the controlled nature of a laboratory. In the uncontrolled environment, chemical behavior is far more complex. Physical, chemical, and biological forces all influence the occurrence and movement of a chemical in the environment. This chapter describes the role of the physical framework of the environment—its geology, hydrogeology, and geochemistry—in determining the fate and transport of chemicals.

5.1 Surface Transport of Chemicals

At the surface, chemicals are transported either through the air or by surface water. When transported in the air, the chemicals may be in solid, liquid, or gaseous phase. When in solid phase, a chemical is typically attached to a particulate, be it a speck of dust or a grain of soil. Airborne, the chemical will travel while attached with the particulate until the particulate settles on the ground or in a surface water body. Particulates can travel great distances before settling, as is evidenced by large volcanic eruptions, which have resulted in the dispersion of fine particles around the world.

If transported in surface water, such as a river, a particle moves in the direction of river flow, i.e., downstream towards its end point, usually a lake, sea, or other larger water body. Suspended particles can travel significant distances in moving water. In the Hudson River PCB Superfund case, PCBs (polychlorinated biphenyls) adhering to suspended particles traveled over 100 miles from their point of origin. Such cases, however, are unusual, and long-distance migration via surface water is more the exception than the rule when considering the migration of contaminants from their point of origin to other properties.

A flowing water body will transport chemicals in one of three ways:

- Soluble constituents will travel as *solutes* dissolved in water at generally the same rate as the water.
- Non-aqueous phase liquids, such as many petroleum products, travel downstream as liquids do, but not necessarily at the same rate as the water.
- Chemicals that are sorbed to silt, sand, or organic particles suspended in the water also travel downstream, but generally at a slower rate than the flowing water.

Dissolved chemicals and chemicals adhering to particles interact with other chemicals in the water, as well as with chemicals adhered to other suspended particles or the underlying sediments. Chemicals in the gaseous phase can travel much greater distances in air than in water.

5.2 Subsurface Formations

Before discussing chemical fate and transport in the subsurface, several geologic and physical principles are identified and described. First is a brief description of the geology that chemicals typically encounter in the subsurface.

5.2.1 Bedrock

The materials in the subsurface generally fall into two categories: bedrock and soils. Bedrock consists of either igneous, sedimentary, or metamorphic rock. *Igneous rocks* originate from cooling magma that has welled up from deep within the earth's crust, or from cooling lava from volcanoes at the surface of the earth. *Sedimentary rocks* are formed by the consolidation of sediments by the removal of water from their pore structure, usually due to the pressure of overlying sediments or dessication at or near the surface, or through chemical or biological activity, as in the case of limestone. *Metamorphic rocks* are created by changing the molecular structure its component minerals by the application of heat and pressure to igneous or sedimentary rocks.

5.2.2 Soils

Soils are unconsolidated sedimentary deposits, and they are classified primarily by grain size. There are several systems of soil classification, but the basic classifications include the following termology. Particles in *clay* are too small to be observed with the naked eye. *Silt* is comprised of very small grains that can be sensed by touch and observed without use of a microscope, despite their small size. *Sand* is comprised of grains that are easily observed with the naked eye. Sand usually carries secondary descriptors based on grain size: very fine, fine, medium, coarse, and very coarse. The next grain size up from sands is *gravel*, which includes granules and pebbles. Larger particles are known as *boulders*.

For a more rigorous delineation of the boundaries between these soil types, geologists usually use either the *Wentworth scale*, the *Burmister System*, or the *United Soil Classification System* (USCS), which was developed by the U.S. Department of Agriculture. The Wentworth scale is a logarithmic scale in that each grain size is twice as large as the next smaller grain size (see Table 5.1 for a summary of the Wentworth scale). The Burmister

TABLE 5.1

Wentworth's Clastic Scale

Major Clasts	Specific Clasts	Sizes (mm)
	Boulder	>256
Gravel	Cobble	256–64
	Pebble	64–4
	Granule	4–2
Mud	Sand	2–1/16
	Silt	1/16–1/256
	Clay	<1/256 mm

TABLE 5.2

Burmister System of Soil Classification

Description	Sieve Size	Visual Description
Boulders	>6 inches	
Cobbles	3–6 inches	Baseball to large grapefruit
Coarse Gravel	¾–3 inches	Grape to baseball
Fine Gravel	¼–¾ inch	Pea to grape
Coarse Sand	1/16–¼ inch	Rock salt
Medium Sand	1/64–1/16 inch	Granular sugar
Fine Sand	<1/64 inch	Finest visible particle
Fines: Silt and Clay	Not visible	Talcum

TABLE 5.3

Unified Soil Classification System (USCS) (Condensed)

Soil Type	Description
Gravels	More than 50% coarse fraction on No. 4 sieve
Sands	50% or more of coarse fraction passes No. 4 sieve
Fine-Grained Soils	50% or more passes the No. 200 sieve

System uses similar textural size ranges as the Wentworth scale. In addition, it can be used to describe the soil's texture, color, and other physical properties (see Table 5.2 for a summary of the Burmister System of soil types). The USCS identifies soil types based on the results of sieve tests (see below). The USCS also can be used to describe certain physical properties as well as particle size (see Table 5.3 for a summary of the USCS soil types).

Describing soils comprised of more than one soil type can be subjective. The primary classification should reflect the dominant soil type, but this can be difficult to discern in certain circumstances. For instance, a soil with both sand- and silt-sized particles can be classified as "silty sand" or "sandy silt." Often, a quantitative approach is needed to determine soil classification, especially in geotechnical surveys, if the grain size, plasticity, and liquid content will be assessed to determine the soil's suitability for construction. The most reliable way to determine the proper classification for soil with more than one soil type is by performing a *grain size analysis*. The simplest and most common grain size analysis is known as a sieve test (see Figure 5.1). In a sieve test, the soil is sent through a series of screens of progressively smaller aperture. The screens are numbered;

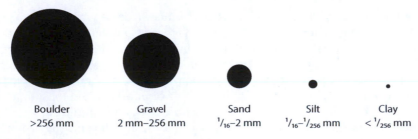

Boulder	Gravel	Sand	Silt	Clay
>256 mm	2 mm–256 mm	$^1/_{16}$–2 mm	$^1/_{16}$–$^1/_{256}$ mm	< $^1/_{256}$ mm

FIGURE 5.1
Comparative grain sizes of major soil types.

the higher the number, the narrower the aperture. The soils retained at each level of screens are then weighed. The type of soil is then assessed using one of the above described classification methods, except for the fine particles.

Other soil descriptors rely on visual classifications, the most obvious being color. The most common method of color identification is the usage of the Munsell Color System, which utilizes a booklet containing color charts, comprising literally hundreds of different shades of colors. An example of a page from a Munsell Color Chart is provided as Figure 5.2.

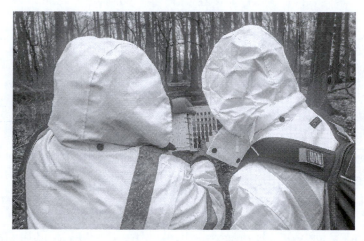

FIGURE 5.2
Using a Munsell Color Chart to classify soil color. (Courtesy of GZA GeoEnvironmental.)

Sands and gravels are also described by the angularity of their grains. These secondary classifications identify the sand or gravel grains as angular, subangular, or rounded (see Figure 5.3). Soils are also described by the uniformity of their grains. Soils with uniform grain size are classified as *well sorted*, and soils with a hodgepodge of soil grain sizes are classified as *poorly sorted*.

The other important visual soil descriptor is its moisture content. *Saturated soils* have only liquids and no gases in their pore spaces. *Unsaturated soils* must contain some gas in their pore spaces, but also may contain liquids. Unsaturated soils that contain some liquids usually are described as "moist," or wet to the touch.

5.2.3 Fill Materials

Environmental investigations inevitably are performed in areas where humans have influenced the environment—there are few untouched areas of land left on the planet. Therefore, the areas may no longer be in their natural state. In some areas, fill materials were imported to raise the elevation of the property above the flood plain, perhaps to cover over wetlands (before this practice was banned), or to level the property. The common objective of land filling is to take a property that was risky, unsuitable or poorly suited for human usage and turn it into a property that would be viable for construction of a building.

Sometimes the fill is soil from a quarry. In such cases, the fill typically is a recognizable soil type and, while not native, is recognizable as a coherent layer of soil. In other cases, the fill consists of detritus from a cheap available source, such as debris from a demolished

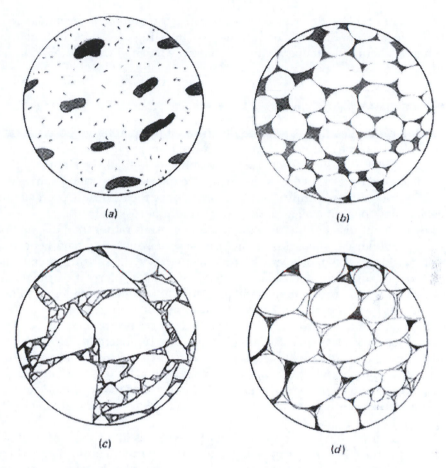

FIGURE 5.3
Variations in grain size and angularity. (a) Poorly sorted sand grains; (b) poorly sorted, well-rounded sand grains; (c) poorly-sorted, angular sand grains; and (d) well-sorted, rounded sand grains.

building. Such fill, generally known as *urban fill*, may consist of concrete, asphalt, bricks, and other man-made, non-natural objects.

Often, urban land has been contaminated due to long-term exposure to contaminants common to urban areas. These contaminants often originate as unwanted combustion byproducts and include polycyclic aromatic hydrocarbons (see Chapter 4), lead, arsenic, and other metals. The presence of urban fill is a crucial consideration when performing environmental investigations in urban areas, especially areas with long histories of urbanization.

5.2.4 Organic Matter

Not all layers in the subsurface are geologic or man-made in origin. Layers of organic matter form when decayed biomass builds up and becomes compressed and dewatered over time. While this is the basic process by which sedimentary rocks are formed, layers of organic matter are far less compressed and dewatered, and still maintain their sponginess and chemical activity. In the United States, such layers form when wetlands and bogs are covered sediments through natural geological processes or man-made processes such as the emplacement of fill, as described above. Because the organic layer may be compressible, it has important implications for the design of buildings. In addition, because organic matter remains chemically active, it can influence the fate and transport of contaminants in the subsurface.

5.2.5 Porosity and Permeability

Two basic soil properties that cannot be determined by qualitative observation but are nonetheless critical to the understanding of chemical transport in the subsurface are porosity and permeability. *Porosity* is the capacity of a soil to hold liquid or air. It is a dimensionless ratio, measured by dividing the volume of pore spaces by the total volume. As shown in Figure 5.3, sand with uniformly sized grains will have the maximum porosity. A rule of thumb establishes "porous" soil as having a minimum porosity of 30%.

Since they are comprised of interlocking mineral crystals rather than discrete particles, igneous and metamorphic rocks do not have *primary porosity*, that is, having pore spaces at the time of their formation. However, differential stresses in the subsurface can create fractures in these rocks, which is referred to as *secondary porosity*. Intersecting fractures in bedrock can contain and transport significant quantities of water and solutes (see Figure 5.4).

Limestone, a type of sedimentary rock, can dissolve over time, creating large dissolution cavities that can form caverns, a dramatic form of secondary porosity, and underground streams. This is known as *karst topography*. The underground streams eventually may undermine the overlying surface, causing it to collapse into *sinkholes,* which are a major hazard. Contaminants enter water in karst terrains easily via streams, sinkholes, or through open fractures in the carbonate rock. Metals tend to precipitate out when they interact chemically with the carbonate rich water in the karst formations. They then will be transported as suspended solids in the water.

Permeability describes the interconnectivity of pore spaces in soil. Soil with good permeability allows for the efficient transport of liquids, in a manner similar to good roadways allowing for the smooth transport of motor vehicles. The wider the pore space, the greater the capacity to carry liquids. Soil with good permeability has many interconnected pore spaces whereas soil with poor permeability has few, poorly interconnected pore spaces. Just as porosity defines the quantity of water that can be stored in a soil or rock formation, permeability dictates the ease by which liquids and their solutes can flow through the soil formation.

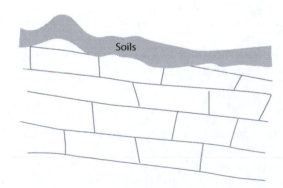

FIGURE 5.4
Complex bedrock fracture pattern.

5.3 Chemical Fate and Transport in the Subsurface

Fate and transport describe what happens to a chemical once it is introduced into the subsurface. Fate refers to the biological interactions with the indigenous microorganisms and the chemical interactions with the myriad of chemicals that the chemical encounters. Transport refers to the physical movement of the chemical in the subsurface. These actions are described in detail below.

5.3.1 Physical State of Chemicals

There are many parameters that are used to measure the physical state of chemicals in the environment. Some of the more common parameters are described below.

5.3.1.1 Vapor-Liquid Partition Coefficient

The ability of a compound to exist in two phases, such as liquid and gas, at the same time is known as its *partition coefficient*. A compound's *vapor-liquid partition coefficient* is the ratio of the concentration of a compound in air to the concentration of the compound in liquid at equilibrium. Other partition coefficients are discussed later in this chapter.

A chemical in the gaseous phase that is released at the surface is unlikely to impact the subsurface; it later condenses and dissolves into the liquid phase or enters a surface water body. A chemical released in a solid phase at the surface will not impact the subsurface unless it dissolves in water or a co-solvent (see Section 5.3.5.1). Regardless of the phase of the chemical, liquids are its primary transport mechanisms into the subsurface.

5.3.1.2 Solubility

Solubility is the ability of a compound, known as the solute, to dissolve in another compound, known as the solvent. It is usually expressed as a simple ratio of the molecules that will dissolve into solution divided by the molecules of solvent.

Chemical literature often categorizes compounds as "soluble" or "insoluble." However, it is misleading to call a chemical "insoluble" within the context of environmental investigations and remediations. The term "insoluble" is convenient for chemical engineering and manufacturing, where the chemicals of interest are present in concentrations that can be expressed in percentages or in tenths of percentages. Environmental investigations and remediations often address chemicals that are present in the environment in far smaller concentrations, expressed in parts per million, parts per billion, or even parts per trillion for particularly toxic chemicals.

Solubility can also be expressed in terms of a chemical's *octanol-water partition coefficient*, which describes whether a contaminant is hydrophilic (attracted to water) or hydrophobic (repelled in water). It is expressed by the ratio

$$K_{ow} = C_o/C_w$$

where C_o is the concentration of the chemical in octanol (a petroleum with eight carbon atoms) and C_w is the concentration of the chemical in aqueous phase (dissolved in water).

5.3.2 The Hydrogeologic Cycle

Water enters the subsurface as part of the *hydrogeologic cycle*. A schematic diagram of the hydrogeologic cycle is shown in Figure 5.5. In the hydrogeologic cycle, water is deposited on the earth's surface as precipitation or condensation, evaporating or transpiring (which is similar to evaporation, but from flora) back into the atmosphere only to fall back to earth as precipitation or condensation, restarting the whole cycle. When water falls to earth, it either becomes part of a surface water body, such as an ocean or a lake, or it lands on an unsaturated surface and infiltrates into the earth. The water that does not return to the atmosphere percolates through the soils, first manifesting itself as soil moisture in the unsaturated zone, and eventually becoming groundwater in the saturated zone. The groundwater eventually discharges to a surface water body, once again becoming a part of the never-ending hydrogeologic cycle.

5.3.3 Vadose Zone

The unsaturated portion of the subsurface through which water moves is known as the *vadose zone*. The vadose zone is a complex area where a myriad of forces affect the fate and transport of a chemical.

Biological processes in the subsurface include those performed by microorganisms living in the soil, which can ingest the chemical and process it as food. This process can result in *biodegradation*, which transforms the chemical into one or more different chemicals.

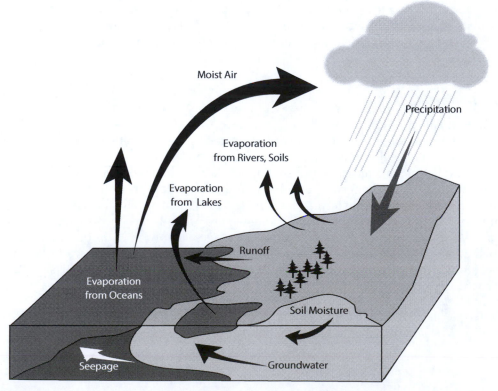

FIGURE 5.5
The hydrogeologic cycle.

Plant roots present in the top few feet of soil may capture the chemical and take it up into the plant as part of its water uptake, or take up a chemical that can then react with the contaminant, creating one or more new compounds.

Various other chemical and physical processes are at work in the subsurface. Carbon tends to act as a sponge, trapping and retaining many chemicals (this ability of carbon is quite useful when remediating organic compounds, as discussed in Chapter 9). Other chemicals present in the soils, or in gaseous phase in the voids between the soil grains, may react with and alter the chemical. The chemical also may spontaneously go into the vapor phase, especially if the chemical is a volatile organic compound (see Chapter 4) and is near the surface. Once in the gaseous phase, the chemical may move upwards through the vadose zone and creating indoor air problems if it finds its way into a building (see Chapter 10).

If the chemical in the vadose zone remains in the liquid phase, in either a pure or dissolved phase, it behaves like a liquid in the subsurface. Liquids obey two basic principles that govern fluid flow:

- Liquids flow downhill, unless under pressure.
- Liquids tend to flow in the path of least resistance.

The chemical generally move downward through the vadose zone until it encounters what is known as the *capillary fringe* (see Figure 5.6). Situated at the boundary of the unsaturated and saturated zones, the capillary fringe is a portion of the soils that technically is in the vadose zone. In the capillary fringe, a liquid can rise above the saturated zone due to its surface tension, the same process that allows liquids to rise above the rim of a cup without spilling.

5.3.4 Saturated Zone

Once a liquid reaches the saturated zone, it merges with the water already present in the saturated zone. Water that is present in the saturated zone is known as *groundwater* (sometimes spelled as two words: "ground water"). Once part of the groundwater system, the chemical is subject to a new and often more complex fate and transport regime.

In ideal conditions, the interface between the saturated and unsaturated portions of the soils is a surface whose configuration mimics the overlying topography. It does so because,

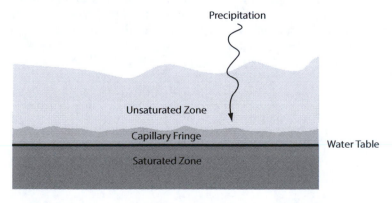

FIGURE 5.6
The capillary fringe is located between the vadose zone and the zone of saturation.

if the soils in the vadose zone are homogeneous, that is, having the same physical properties throughout, then rain water and other liquids should percolate downward in a uniform distribution, saturating the underlying soils in a uniform manner. The *water table* is aptly named, for diagrammatically (and under ideal conditions) it has the configuration of a tilted table.

The capillary fringe should not be confused with the *smear zone*, which is an area that is saturated on a seasonal basis. In the temperate and moist parts of the world, the elevation of the water table can vary several feet or more over the course of the four seasons, creating a portion of the soils that at times is part of the vadose zone and other times lies below the water table.

5.3.4.1 Hydraulic Gradient

When the water table is tilted, the water has a potential energy that becomes kinetic energy as it flows downward. That potential energy is known as the *hydraulic head. Darcy's Law* expresses in mathematical terms the relationship between the factors which govern the flow of groundwater. In Darcy's equation, the velocity of the water depends on two parameters, as follows:

$$V = ki$$

where V is the specific discharge (also known as the flow rate, or Darcy velocity), i is the *hydraulic gradient,* and k is the *hydraulic conductivity.*

Hydraulic gradient is essentially the slope of the water table. It is the rate of change in which energy, or hydraulic head, is lost as water flows through the porous materials. The steeper the water table, the greater the hydraulic gradient. Figure 5.7 shows that hydraulic gradient is calculated by the slope of a line in a right triangle, as follows:

$$i = \Delta h/l,$$

where l is the flow distance. "Up-gradient" refers to a physical location that has a greater hydraulic head than the subject location, while "down-gradient" refers to a physical location that has a lesser hydraulic head than the subject location. A location with the same hydraulic head as the subject location is designated as "cross-gradient."

Hydraulic conductivity is the geologic formation's ability to transmit water. It is not dissimilar in concept to the more familiar electric conductivity, which is a material's ability to transmit an electric current. The greater the hydraulic conductivity, the greater the flow rate of the liquid through the formation.

FIGURE 5.7
Calculating hydraulic gradient. (Courtesy of Prof. Steven A. Nelson, Tulane University, New Orleans, LA.)

5.3.4.2 *Groundwater Flow*

In unconsolidated soils, groundwater, because it flows downhill, generally follows topography. It should be noted, however, that if surface topography is altered, the groundwater flow direction may not change. Therefore, especially in flat, gently sloping, or developed areas, the topographic slope should be treated as a first approximation of groundwater flow direction. Confirmation of groundwater flow direction only can be achieved with data collected at water gauging points (see Chapter 8).

Groundwater does not flow downward forever. Eventually, it reaches a point, known as a *groundwater divide*, where its flow changes direction. In many cases, that divide is indicated by the presence of a water body. Figure 5.8 is a schematic diagram of a *gaining stream*, also known as a *spring-fed stream*. Groundwater on both sides of the stream flow towards the stream and provide it with additional water. Water flows in opposite directions towards the stream on both sides of the stream, with important implications for contaminant migration (see Chapter 8). The opposite of a gaining stream is a *losing stream* (see Figure 5.9), common in arid regions, in which some of the water in the stream migrates downward from the stream, seeking a saturated zone.

Water movement in fractured bedrock can be extremely complicated. Figure 5.10 provides an example of the complexities involved in water and chemical movement in bedrock. In this figure, a chemical, in this case a chemical with a specific gravity greater

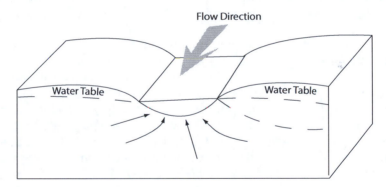

FIGURE 5.8
A schematic diagram of a gaining stream.

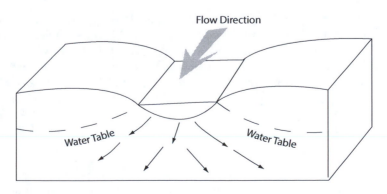

FIGURE 5.9
A schematic diagram of a losing stream.

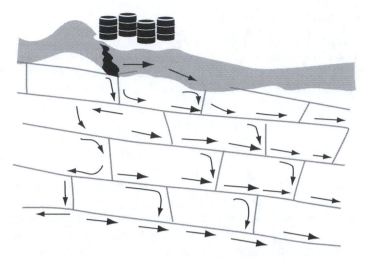

FIGURE 5.10
Complex groundwater flow patterns in fractured bedrock.

than water, is released from a source zone. It generally will travel downward through the vadose zone and then downward and generally down-gradient through the saturated soils.

Once it encounters bedrock, however, the contaminant is forced to navigate a set of complicated pathways that can lead to a migration pattern that may not bear a resemblance to the expected flow pattern or even the bedrock fracture pattern itself. Keep in mind that Figure 5.10 is in two dimensions rather than three dimensions. Fractures that appear to be connected in two dimensions may not actually intersect in three dimensions. The three-dimensional fracture pattern can create migration pathways that could result in contaminants being present in groundwater in certain fractures but not in fractures above, below, up-gradient and down-gradient of the contaminated zone. Therefore, the transport of a contaminant in fractured bedrock can be extremely difficult if not impossible to fully understand.

5.3.4.3 Aquifers

Saturated soil that can transmit a "significant" quantity of liquid is known as an *aquifer*.[1] For an aquifer to exist in a soil formation, that soil must have sufficient porosity to hold an appreciable quantity of water, and sufficient permeability to allow the water to move through the formation. Aquifers can exist in the fractures within a rock formation or between the layers of a sedimentary rock, provided that the fractures or the layers are significantly wide and interconnected as to allow for the transmission of a significant quantity of liquids.

---

[1] This is one of many definitions of what constitutes an aquifer. The environmental consultant should know what the governing regulatory body considers an aquifer before undertaking an environmental investigation within that jurisdiction.

5.3.4.4 Aquitards, Aquicludes, and Confining Layers

Well-sorted sands and gravel formations, with their typical high porosity and permeability, are ideally suited to host aquifers. Soil formations that have a low porosity or permeability, and thus inhibit the flow of liquids in the subsurface, are known as *aquitards*. While there is no fixed rule defining an aquitard, it qualitatively can be described as a soil or rock formation that cannot yield a "significant" quantity of water. Aquitards are typically composed at least partly of silts or clays. Soil or rock formations that are effectively impermeable are known as *aquicludes*.

Aquifers, unless they are overlain by aquitards or aquicludes, can receive water from the surface which has percolated down through the vadose zone. This process is known as *recharge*. A soil formation that acts as an aquitard or an aquiclude in preventing or inhibiting water from recharging the aquifer is known as a *confining layer*. Note that a confining layer works both ways, also preventing upward movement of groundwater, as discussed below.

Confining layers greatly impact the configuration of the water table. In Figure 5.11, water percolates into an aquifer that is overlain by an aquitard. In this case, the water table would rise above the aquitard were it not in the way. The aquitard therefore depresses the water table, keeping the water in the aquifer under pressure. When the pressure is relieved, as happens when a drinking water well penetrates the aquitard and reaches the aquifer, the groundwater rises to a level commensurate with its stored potential energy. The level to which the water rises is known as the *potentiometric surface* (also called *the piezometric surface*), so named because it is the surface at which groundwater reaches its greatest potential energy, or head.

As shown in Figure 5.11, when the potentiometric surface of the groundwater is above the ground surface, groundwater will flow upwards to its potentiometric surface and onto the ground. This is known as an *artesian condition*. (It should be noted that some people consider an artesian condition to exist in any situation where the water table rises once the confining layer has been breached.) It is important to remember that even under artesian

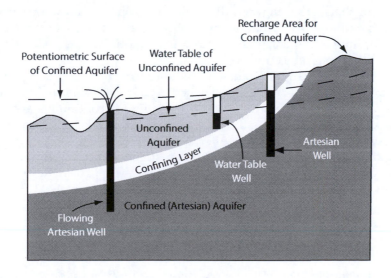

FIGURE 5.11
Diagram showing flowing artesian well conditions caused by the potentiometric surface being at a higher elevation than the land surface. (Courtesy of Minnesota Department of Health, Saint Paul, MN.)

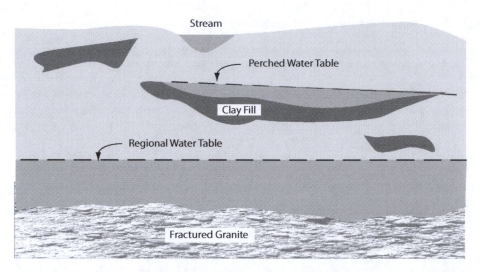

FIGURE 5.12
A perched water table aquifer. Only impermeable layers with the proper configuration can hold water and create a perched water table aquifer. (Adapted from Sterett, R.J., ed., *Groundwater & Wells*, 3rd ed., Johnson Screens, New Brighton, MN, 2007.)

conditions the groundwater still obeys the two principles of liquid flow, since the groundwater beneath the confining layer was under pressure until the pressure was relieved by the penetration of the overlying confining layer.

5.3.4.5 Perched Aquifers

A perched aquifer is formed when a high-permeability formation such as a sand layer is present within a lower permeability formation such as a clay layer (see Figure 5.12). This causes water percolating downward to become trapped by the low-permeability formation and pool there, resulting in a discontinuous zone of saturation. Some scientists debate even the name of this formation, preferring to call it a *perched water table*, since calling it an aquifer implies the potential to transmit a quantity of liquid that the discontinuous zone is unlikely to produce.

It is often difficult to distinguish a perched aquifer from the underlying saturated zone, since both are saturated. In theory, the perched aquifer is discontinuous and underlain by an unsaturated zone whereas the underlying aquifer is continuous and underlain by additional saturated formations. In the field, however, a perched aquifer with significant lateral extent can resemble a regular aquifer to the field observer.

Perched aquifers are important to subsurface investigations because the low-permeability formation not only stops the downward movement of water but also the downward movement of chemicals that are being transported in the water.

5.3.5 Chemical Transport in the Saturated Zone

5.3.5.1 Aqueous Phase Liquids

A chemical in the liquid phase will either be dissolved in a solvent or will be a pure-phase chemical existing in liquid phase at ambient temperatures. Chemicals that are dissolved in water are said to be in the *aqueous* phase. Solvents other than water are known as *co-solvents*, the most important of which is petroleum. A solute can have very different solubility if a co-solvent is present. Some chemicals, such as PCBs, are more soluble in petroleum than in water, which has important implications for their transport in the subsurface.

5.3.5.2 Non-aqueous Phase Liquids

When the chemical released is in its pure rather than dissolved phase, it is known as a *non-aqueous phase liquid*, or NAPL. The chemical's *specific gravity*, which is the density as compared to water, determines the type of NAPL it forms. A chemical with a specific gravity less than 1.00 grams per cubic centimeter (g/cm^3) is lighter than water. The portion of that chemical that does not dissolve in water and is in the liquid phase is called a *light non-aqueous phase liquid*, or *LNAPL*. LNAPL floats on top of water, be it above-ground surface water or below-surface groundwater. A chemical with a specific gravity greater than 1.00 g/cm^3, known as *dense non-aqueous phase liquid*, or *DNAPL*, will sink through the water column, with important implications for environmental investigations and remediations.

5.3.5.3 Advection

Advection, as used in hydrogeology, is the mechanism by which groundwater and its solutes move down-gradient. Under advective forces, when a mass of chemicals released to the subsurface encounters the water table it then moves in the direction of groundwater flow. Such a mass of chemicals is generally known as a *contaminant plume*. The rate of advection in an aquifer is calculated using Darcy's Law, as described above.

5.3.5.4 Diffusion

If advection was the only force affecting water flow in the saturated zone, then it would be fairly easy to understand chemical transport in the saturated zone. However, there are two forces that to some degree counterbalance the force of advection: diffusion and dispersion.

Diffusion is the movement of a chemical from higher concentration to lower concentration. Diffusion causes chemicals to spread out in directions other than the direction of water flow and will occur in the absence of water flow. As such, it enables a contaminant plume to spread cross-gradient and even up-gradient, to a limited extent. A chemical's *diffusion coefficient* describes its ability to move from an area of higher concentration to an area of lower concentration. The quantity of chemical that will flow through a small area during a short time interval (J, also known as *flux*) is determined using Fick's First Law, which states:

$$J = -D \, (dC/dx)$$

where D is the diffusion coefficient (for a given temperature), C is the concentration of the contaminant in solution, and x is the length in the direction of movement.

Diffusion coefficients are calculated in the laboratory and are available in published scientific literature. Therefore, for a given chemical, if the chemical concentration in the center of a chemical plume is known, and the edge of the plume is known, the concentration gradient can be calculated, enabling the geologist to estimate how the plume will spread over a period of time through the process of diffusion.

5.3.5.5 Dispersion

Other than karst terrains, where water may flow in the subsurface via underground streams, subsurface water must flow through rather than over geological formations, enabling the geology to influence or control the flow of the water. Mechanical dispersion, usually known simply as *dispersion*, has the same effect as diffusion in causing chemicals to spread out in directions other than the primary direction of groundwater flow. However, dispersion is a physical rather than a chemical force, caused by the differing fluid velocities within the pores and pathways taken by the fluid.

Since dispersion is a mechanical process, its effect can only be measured once it has occurred rather than by using theoretical formulas. The dispersion coefficient for a soil or rock formation is approximated by

$$D = \alpha \times v$$

where α is the dispersivity of the aquifer and v is the velocity of the ground water or surface water. The dispersion coefficient is measured in distance over time.

Figure 5.13 shows the effects of dispersion on groundwater flow. Figure 5.13a shows a liquid moving along a smooth, even pathway through well-sorted sand. Figure 5.13b shows the liquid moving through a serpentine path in an effort to get through poorly-sorted sand which has some silt and possibly some clay particles. The circuitous path translates into slower transport through the sand and more opportunities to interact with biological agents or chemicals that can impact the fate and transport of the chemical.

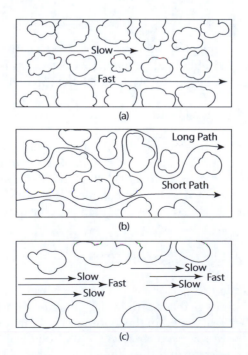

FIGURE 5.13
(a–c) The effect of grain spacing on the permeability of a soil formation. (From Fetter, C.W., *Applied Hydrogeology*, 4th ed., Prentice Hall, New York, 2000.)

Dispersion in various types of soils can be calculated in the laboratory. However, estimating the degree of dispersion in a soil based on laboratory models has inherent flaws and can be even more flawed in the field. Therefore, estimating dispersion effects is generally done using field data Even when there is a clear pathway for the liquid, it may move at different rates depending on its proximity to soil particles, as shown in Figure 5.13c.

5.3.5.6 Attenuation and Retardation

There are also forces in the subsurface that work at preventing dissolved chemicals from moving at the same velocity as the groundwater. When a dissolved chemical sorbs onto a solid surface such as a grain of soil, it ceases to move with the groundwater. This process is referred to as *retardation*. The process in which a dissolved chemical ceases to move with the groundwater because it has been transformed through a biological and/or a chemical process is known as *attenuation*. Figure 5.11c shows the effect that friction plays on the movement of the liquid. When retardation occurs, the differential velocities of groundwater flow result will spread out the chemical. The sum total of the retardation and attenuation of a contaminant plume is known as the *retardation factor*.

The surface that attracts the chemical is known as the *sorbent*, and the chemical that is attracted to the sorbent is known as the *sorbate*. The amount of mass sorbed per unit of chemical can be calculated using K_d, the partition coefficient, as follows:

$$S = K_d \times C$$

where S is the mass sorbed per mass of chemical and C is the chemical's concentration in groundwater.

The tendency of a chemical to be adsorbed by soils or sediment is known as its *soil-water partition coefficient*. A particular type of soil-water partition coefficient is the *organic carbon partition coefficient*, which describes the ability of organic carbon present in the otherwise inorganic, mineral-rich soils to inhibit a chemical's movement. Organic carbon's ability to affect a chemical's mobility plays a role in certain remedial technologies, especially those involving granular activated carbon (GAC), an especially effective sorbent.

Having established the regulatory and scientific framework under which environmental investigations and remediations are conducted, the next several chapters discuss the methodologies by which they are conducted.

Problems and Exercises

1. Under what conditions does the water table differ from the potentiometric surface?

2. In general, how does an increase in temperature affect a chemical's diffusion coefficient?

3. Which ethanes on the Target Compound List would behave as LNAPL if present in non-aqueous form in an aquifer?

4. What role does the organic carbon partition coefficient play in the calculation of the retardation factor in an aquifer?

Bibliography

ASTM International, 2009. ASTM D2488—09a: Standard Practice for Description and Identification of Soils (Visual-Manual Procedure).

ASTM International, 2010. ASTM D2487—10: Standard Practice for Classification of Soils for Engineering Purposes (Unified Soil Classification System).

Fetter, C.W., 2000. *Applied Hydrogeology*, 4th ed. Prentice-Hall, New York.

Freeze, R.A., and Cherry, J.A., 1979. *Groundwater*. Prentice-Hall,

Hemond, H.F., and Fechner-Levy, E.J., 2000. *Chemical Fate and Transport in the Environmental*, 2nd ed. Academic Press.

Lagrega, M.D., Buckingham, P.L., and Evans, J.E., 2001. *Hazardous Waste Management*, 2nd ed. McGraw-Hill.

New Jersey Department of Environmental Protection, 2005. Field Sampling Procedures Manual.

Nyer, E.K. and Gearhart, M.J., Winter, 1997. Plumes Don't Move. *Groundwater Monitoring & Remediation*, pp. 52–55.

Sterett, R.J., ed., 2007. *Groundwater & Wells*, 3rd ed. Johnson Screens.

Swanson, W.R., and Lamie, P., January 2010. *Proceedings of the Annual International Conference on Soils, Sediments, Water and Energy Volume 12 Article 9—Urban Fill Characterization and Risk-Based Management Decisions—A Practical Guide*.

U.S. Department of Agriculture, 1999. *Soil Taxonomy: A Basic System of Soil Classification for Making and Interpreting Soil Surveys*, 2nd ed. Natural Resources Conservation Service, Washington, DC.

Vesper, D.J., Loop, C.M. and White, W.B., 2001. Contaminant transport in karst aquifers. *Theoretical and Applied Karstology* 13 (14): 101–111.

6

Environmental Due Diligence

Due diligence is a legal term that refers to the steps to be taken to satisfy a legal requirement. It especially applies when something is being bought or sold. This chapter discusses the steps commonly performed when undertaking environmental due diligence in the United States.

The first step in the due diligence process typically is a Phase I environmental site assessment (ESA). There are many varieties of Phase I ESAs and they go by a variety of names, such as preliminary assessments, preliminary site assessments, preliminary reviews, and Phase I audits. But in most places and to most people, they're simply known as a "Phase I."

This chapter also discusses the first steps of environmental due diligence for properties regulated under the Comprehensive Environmental Response, Compensation, and Liability Act (CERCLA), commonly known as Superfund, and the Resource Conservation and Recovery Act, commonly known by its acronym, RCRA (see Chapter 3 for an introduction to CERCLA and RCRA). Due diligence specifically focused on the vapor intrusion pathway is discussed in Chapter 10.

6.1 History of the Phase I Environmental Site Assessment

As discussed in Chapter 3, the 1986 Superfund Amendments and Reauthorization Act (SARA) provided for an *innocent purchaser defense* from Superfund liability. To qualify for the innocent purchaser defense, the prospective purchaser had to have acquired the property after the pollution had occurred and had to demonstrate that before the acquisition, it had no knowledge and no reason to know of the contamination. To demonstrate that the person had no reason to know of the contamination at the time the property was purchased, the person was required to have made "*all appropriate inquiries* [italics mine] into the previous ownership and uses of the facility in accordance with generally accepted good commercial and customary standards and practice."

Although SARA had established a benchmark by which a party could obtain innocent purchaser defense under Superfund, it didn't indicate what constituted "all appropriate inquiries." In the meantime, the influx of environmental regulations at the federal and state levels heightened the sensitivity of the business community to the potentially detrimental effects of environmental liabilities on their business. By the 1980s, prospective purchasers of commercial properties had begun hiring environmental consultants to perform basic due diligence activities in support of commercial real estate transactions. These activities had much in common with today's Phase I, but could not be formally represented as constituting an innocent purchaser defense under Superfund since no such protocol had yet been established.

A not-for-profit organization, then known as the American Society for Testing and Materials (now known as *ASTM International*, or commonly by its acronym, ASTM), stepped into the breach. In March 1990, ASTM, which sets standards for products and services in the United States, formed a committee to develop a standard practice for environmental due diligence in property transactions. That committee issued its first Standard for Environmental Assessments in Real Estate Transactions in 1993, known as E1527-93. The ASTM E1527 standard quickly became the standard for the entire real estate industry. Other major organizations, such as insurance companies, had developed their own guidelines for environmental due diligence, but after 1993 virtually all of the guidelines were variations on the ASTM E1527 standard.

The E1527 standard was revised in 1994, 1997, and 2000 to reflect advances in technology and the evolution of the usage of the Phase I ESA as a tool to evaluate business risk in a property transaction posed by environmental conditions. The next revision came in 2005 in response to the 2004 issuance of the U.S. Environmental Protection Agency's (USEPA) *All Appropriate Inquiries (AAI) Rule* as per SARA's requirement regarding innocent purchaser defense. The 2005 version of the E1527 standard was designed to comply with AAI requirements.

One significant aspect of the 2005 standard was the expansion of liability protections to two groups other than the innocent purchaser. They are:

Bona fide prospective purchaser—One who buys property after conducting all appropriate inquiries into the current and historical uses of the property knowing that there are environmental conditions present on the property.

Contiguous property owner—One who owns property that is contiguous to, and may be impacted by, hazardous substances migrating from property they do not own. The required qualifications to obtain this liability protection are discussed later in this chapter.

At the time of the publication of this book, ASTM's latest revision to the Phase I ESA standard was in 2013 and is known in shorthand as the E1527-13 standard. Changes in the 2013 version of the standard are discussed below.

6.2 The ASTM Phase I Standard—Recognized Environmental Conditions, Controlled Recognized Environmental Conditions, and Historical Recognized Environmental Conditions

All environmental due diligence constitutes a procedure that assesses whether a piece of real estate has a potential environmental problem. In the case of the E1527 standard, the environmental problem to be identified is referred to as a "recognized environmental condition." The standard defines a *recognized environmental condition*, or REC (commonly pronounced "Reck"), as follows:

> The presence or likely presence of any *hazardous substances* or *petroleum products* in, on, or at a *property:* (1) due to any *release* to the *environment;* (2) under conditions indicative of a *release* to the *environment;* or (3) under conditions that pose a *material threat* of a future *release* to the *environment. De minimis* conditions are not recognized environmental conditions.

This definition contains three terms that merit discussion.

Hazardous substances or petroleum products—Figure 6.1 shows a vacant lot strewn with debris. While unattractive, the mere presence of solid wastes on a property does not constitute a REC unless there has been a release of hazardous substances or petroleum (as defined in Chapters 3 and 4).

Property—The Phase I ESA only deals with commercial real estate, which is property that can be bought or sold. It does not apply to mobile sources of pollution, such as vehicles, boats, or airplanes, nor does it apply to properties that aren't generally considered sellable, such as public water bodies, public parks, and public roadways. Also excluded from the E1527 definition of "commercial real estate" are residential properties with buildings in which there are fewer than four dwelling units. The assumption is that if a dwelling has four or more units, its principal function is revenue generation and is therefore commercial real estate.

Material threat of a release—A REC does not require the existence of a release or a past release. The material threat of a release is sufficient to establish a REC. A material threat is a physically observable or obvious threat that is reasonably likely to lead to a release that might result in impact to public health or the environment. "Obvious" is the operative term here. If one has to work hard to develop a scenario in which an observed or otherwise identified condition would result in a release to the environment, then it is not a material threat and doesn't qualify as a REC.

FIGURE 6.1
The presence of refuse and other inert wastes by themselves does not constitute a REC. (Courtesy of GZA GeoEnvironmental, Inc.)

ASTM E1527 goes on to state that a REC:

> Is not intended to include *de minimis* conditions that generally…would not be the subject of an enforcement action if brought to the attention of appropriate governmental agencies. Often, it is left to the judgment of the inspector whether a spill is of sufficient size to be considered a REC.

Drips of motor oil underneath a car in a parking lot would therefore not be classified as a REC, whereas the more significant petroleum staining shown in Figure 6.2 might be a REC. State regulations also weigh in on what constitutes a reportable spill.

The ASTM E1527 standard also defines an *historical recognized environmental condition*, or HREC. An HREC is defined as a release of a hazardous substance or petroleum product that has been cleaned up to the satisfaction of the applicable regulatory authority without restrictions on the property usage, as described below. For instance, an oil spill that was reported to the applicable regulatory authority would be a REC. However, once it has been cleaned up to the satisfaction of the relevant regulatory agency it would be reclassified as an HREC under the ASTM E1527 standard.

This is very different from a REC that is due to an historical condition. For instance, assume that the historical records for a property indicate that it once contained a gasoline service station (as described in Section 6.4.4, below). This type of former property usage would present a material threat of a release (unless, of course, a release had been discovered on the property and reported to the regulatory authority). Therefore, it would be a REC, not an HREC, even though the term "Historical REC" sounds like it should apply to this situation.

FIGURE 6.2
Significant oil staining near an above-ground storage tank. (Courtesy of GZA GeoEnvironmental, Inc.)

The 2013 revision to the E1527 standard added a new term to the Phase I lexicon: *controlled recognized environmental condition*, or CREC. Like an HREC, it exists as a result of a past release of hazardous substances or petroleum products. Unlike an HREC, the release has not been cleaned up to the satisfaction of the regulatory authority. Its effects, however, are under control to the satisfaction of the regulatory authority, implying that the contaminants are not able to migrate beyond their current locations. The controls, known as *activity and use limitations*, or AULs, are discussed in Chapter 9 of this book.

6.2.1 Exclusions from the Standard

Given its origins as a CERCLA defense, Standard Practice E1527 addresses only conditions that have the potential to impact the environment. There are numerous environmental issues that are excluded from this definition. The ASTM E1527 standard explicitly excludes the following environmental issues from the Phase I ESA standard:

- Asbestos-containing materials (ACM)
- Lead-based paint (LBP)
- Radon
- Lead in drinking water
- Wetlands
- Regulatory compliance
- Cultural and historic resources
- Industrial hygiene
- Health and safety
- Ecological resources
- Endangered species
- Indoor air quality
- High voltage power lines

The standard does not include these environmental issues because they are excluded from CERCLA. Their exclusion does not minimize their potential impact on a property or on human health. Because they can affect the value of a property without causing any implications regarding CERCLA liability, under the term *business environmental risk*. Business environmental risk is defined as a risk that can have a material environmental impact or environmentally driven impact on the current or planned use of a parcel of commercial real estate. Many of the above-listed environmental concerns that are excluded from the E1527 standard are discussed in subsequent chapters to this book.

6.3 Who Can Perform the Phase I ESA

The E1527 standard does not specify who can perform the various tasks entailed in a Phase I ESA. It does, however, specify who is responsible for those tasks—the *environmental professional*. As defined in the E1527 standard, the environmental professional, or "EP," is a person who is experienced at conducting Phase I ESAs and who has the necessary education, training, and experience to be qualified for the job. The E1527 standard specifies the minimum requirements for an EP. The EP is responsible for the validity of the findings of the ESA and is identified in the final Phase I ESA report as the person responsible for its contents and conclusions.

6.4 Components of the ASTM E1527 Standard

Determining if (and where) pollution has occurred on a property, such as the abandoned factory shown in Figure 6.3, can challenge the most experienced site assessor. The E1527 standard provides a methodology for rooting out most of the salient issues present on such a property.

The research portion of the E1527 standard consists of six steps:

1. Site and vicinity reconnaissance
2. Site interviews
3. User responsibilities
4. Site history review
5. Local agency review
6. Regulatory review

Each of these steps is described below.

FIGURE 6.3
An abandoned factory. (Courtesy of GZA GeoEnvironmental, Inc.)

6.4.1 Site Reconnaissance

The site and vicinity reconnaissance are the most obvious and visible portions of the Phase I ESA. The basic building blocks of the site and vicinity reconnaissance are *areas of concern* (AOCs). AOCs are potential current or former sources or pathways of hazardous substances or petroleum products. There may be many AOCs on a property, yet there may not be any RECs. The EP is responsible for identifying AOCs and offering a professional opinion as to whether they are RECs. Table 6.1 lists some common AOCs that can be encountered at a site.

The most obvious RECs are visually observable releases. The inspector (either the EP or the EP's designate) will look for staining on paved and unpaved surfaces, chemicals or petroleum products leaking out of containers, chemical or petroleum odors, etc. It is important to be able to distinguish between staining, leaks, or odors related to hazardous substances and petroleum products, and staining, leaks, or odors caused by water and other substances that are not covered under the E1527 standard.

The presence of dead or sickly-looking vegetation, as observed around the oil spill shown on Figure 6.4, can be indicative of a spill of petroleum or hazardous substances. The inspector must judge whether the stressed vegetation is due to a spill, or to other reasons, such as lack of water, lack of sunlight, foot traffic, etc. Although the spilled chemicals may have disappeared, the lingering effects on the flora in the area can be direct evidence of contamination.

A container whose contents are hazardous substances or petroleum products has the potential to leak, spill, or otherwise empty its contents into the environment. However, containers are not by themselves RECs; otherwise every hardware store, auto parts store, and home improvement center would be filled with RECs. There must be a release or a

TABLE 6.1

Typical Areas of Concern

Aboveground storage tank (AST) systems
Underground storage tank (UST) systems
Chemical or petroleum storage or handling areas
Chemical waste or petroleum waste storage or handling areas
Dumpsters
Floor drains, trenches, sumps and associated piping
Oil/water separators
Storm water drains, grates and associated piping
Drainage swales, culverts, impoundments, and surface water bodies
Septic systems, leach fields, seepage pits, and dry wells
Open pipe discharges
Landfills and solid waste dumping
Historical fill or other fill material
Staining or stressed vegetation
Electrical transformers or capacitors
Hydraulic equipment, including lifts, elevators, and compactors
Active or inactive production wells
Monitoring wells, former boreholes, or other evidence of environmental investigations
Other observations potentially indicative of the presence of RECs

Source: Adapted from New Jersey DEP.

FIGURE 6.4
Dead, brown vegetation around an oil spill in an abandoned lot. (Courtesy of GZA GeoEnvironmental, Inc.)

material threat of a release for a REC to exist. In addition, there must be a pathway to the environment, either directly to the ground or via a pathway, such as a pipe or a through-going crack in an otherwise impermeable surface. The E1527 standard implies that the pathway to the environment must be obvious—in other words, the EP shouldn't have to perform an elaborate construct for the chemical or petroleum product in question to enter the environment.

Chapter 1 described the three components of an environmental hazard, namely a source, a pathway, and a receptor. A REC needs only a source and a pathway since the entire environment is the receptor, and therefore is always present where real estate is concerned.

The following sections discuss some of the more important and more common potential environmental concerns, and some of the evidence to be sought by the investigator in deciding whether an AOC is in fact a REC.

6.4.1.1 Underground Storage Tanks

Although any vessel that stores or has stored hazardous substances or petroleum merits a concern, certain vessels, by merit of their size or their location, are of greater concern than others. Of very high concern are *underground storage tanks* (USTs), which the USEPA defines as a bulk storage vessel with at least 10% of it and its piping in contact with the ground. USTs can contain hazardous substances or petroleum. However, the overwhelming number of USTs contain petroleum.

Because the UST is in direct contact with the ground, any leak from that portion in contact with the ground will result in a release into the environment—the pathway is a given. USTs also store large quantities of liquids, typically from 500 gallons up to 20,000 gallons or more. There are hundreds of thousands of USTs in the United States, making them a significant environmental issue in a Phase I ESA.

One place where USTs are guaranteed to exist is a filling station. Most filling stations have multiple USTs containing various grades of gasoline and sometimes diesel fuel. Visual evidence of their existence is limited to a fill port, which is used to deliver the liquid to the UST for storage and eventual usage; a vent pipe, which enables air to escape from a UST to make room for the liquid when the UST is being filled; and a dispenser.

As shown on Figure 6.5, USTs are not the only areas of concern at a filling station. Also of concern is the underground piping that allows the gasoline or diesel fuel to flow to the dispensers. The joints and elbows in the underground piping are particularly vulnerable to leakage. The dispensers themselves are areas of concern as is the potential for spills during the filling of vehicular gas tanks and the USTs themselves. Filling stations that also provide vehicle repair services may contain motor oil USTs and waste oil USTs as well.

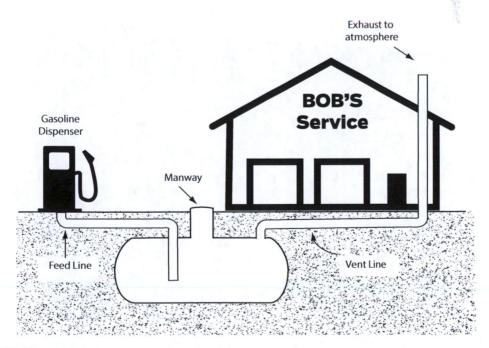

FIGURE 6.5
Conceptual diagram of a gasoline filling station.

Most USTs, however, store heating oil. They can be found at multi-tenant apartment buildings, single-family residences, commercial and industrial facilities—in other words, in just about any building that is heated.

Even more problematic than USTs are former USTs, taken out of service because of a change in property usage or a change in the way the property's building was heated. Sometimes, the former fill port or vent pipe still exists. Other times, however, there may be no visual evidence of the existence of former USTs. A large patch in the asphalt or concrete outside of a building may be the only visual clue that a UST had been removed from that spot (although it may just be that the area has been repaved).

6.4.1.2 Above Ground Storage Tanks

An *above ground storage tank* (AST) is a storage vessel with less than 10% of its volume underground. ASTs can be present in the basement of a building, even though the tank is below the ground level, provided that it meets this criterion.

Despite having similar capacities and holding similar substances, ASTs, as shown in Figure 6.6, generally do not represent the same hazard to the environment as USTs, because they do not have a direct pathway into the environment. Instead, they often are suspended above the underlying surface by steel legs or concrete cradles. Releases from an AST not only do not have a direct pathway to the environment, but often the integrity of ASTs is relatively easy to verify. Therefore, ASTs are more likely than USTs to be fixed or replaced before they can impact the environment.

FIGURE 6.6
Above-ground storage tank within secondary containment. Note that the AST is mounted on legs to prevent it from being in contact with the underlying concrete floor. (Courtesy of GZA GeoEnvironmental, Inc.)

6.4.1.3 Drum Storage Areas

Smaller bulk storage containers, such as 55-gallon drums, pose a similar threat to the environment as an AST. Though they store a smaller volume of hazardous substances or petroleum than an AST, they are typically present in groups and therefore pose, as a group, a major concern if in damaged condition or stored improperly. The EP must judge whether a combination of a drum's ability to release a material quantity of its contents and the existence of a material pathway to the environment constitutes a REC.

6.4.1.4 Industrial Establishments

Facilities that store and handle chemicals and/or petroleum products are high on an EP's list of concerns. Often, the EP must make a quick study of what can be a complicated or simply unfamiliar industrial operation. One way to conceptualize an industrial operation is a three-step process regarding hazardous substances and petroleum: materials and chemicals come in to the facility, materials and chemicals get processed at the facility, and products and wastes leave the facility. The places where hazardous substances and petroleum are most often handled are the places where spills are most likely to occur. These places include loading and unloading areas, areas where chemicals are introduced into a process or into machinery, and places where wastes are containerized for eventual disposal (see Figure 6.7). Pathways for wastes to escape the process area—floor drains, floor trenches, wastewater lines—are of particular interest to the inspector.

Typical areas where hazardous substances or petroleum products are stored or used include boiler rooms, mechanical rooms, maintenance and janitorial rooms, and basements. In a facility, the inspector must develop a basic understanding of the processes occurring to understand where the hazardous substances or petroleum products and wastes are located and where the potential migration paths to the environment are located

FIGURE 6.7
Fifty-five-gallon steel drums stored on their sides near a storm drain.

so that AOCs can be evaluated. Soliciting the input of someone knowledgeable of the facility's processes, such as the plant manager or the facility engineer, is crucial to understanding the risks associated with plant operations and is required under the E1527 standard.

A factory may also have smaller containers located at various places within the facility. There also might be machinery that contains reservoirs of petroleum, such as hydraulic equipment (including elevators and lifts) and air compressors. Each area where hazardous substances are stored or used is considered to be an AOCs, and the EP must judge whether the quantity of chemicals stored or used is more than *de minimis* in quantity in determining whether their presence constitutes a REC.

Industrial facilities often keep detailed records of chemical usage and chemical waste disposal. Such records may be in the form of regulatory filings under various statutes, including hazardous waste disposal manifests under RCRA, toxic release inventories, community right-to-know filings, and safety data sheets (SDSs). Sometimes relevant records can be located in the engineering department, the maintenance department, or even the purchasing department. The EP must ask the appropriate questions to the appropriate people at the facility to root out the information needed to complete the Phase I ESA.

6.4.1.5 Dry Cleaners

Dry cleaning establishments are worthy of special mention. Ubiquitous along commercial roads and in shopping centers, the USEPA estimates that there are more than 25,000 facilities in this country where dry cleaning is done on the premises. Dry cleaners, along with gasoline filling stations and auto repair facilities (and printers before the advent of desktop publishing), are the most common users of chemicals that can contaminate the subsurface.

Whereas early dry-cleaning operations used kerosene, a petroleum distillate, and later carbon tetrachloride, a highly toxic petroleum-based solvent, since the mid-1930s the vast majority of dry cleaners have used perchloroethene, known in the industry as perc (although in recent years, dry cleaners have increasingly used non-toxic methods to dry clean clothing). The physical and chemical properties of perc are discussed in Chapter 4.

Releases from dry cleaners can occur due to incidental spillage while loading fresh perc into the equipment, inappropriate handling of waste machine filters, and poor management of the perc-laced condensate that builds up inside the equipment. Leakage from improperly maintained machines or from underground piping also can be a major source of perc contamination. In summary, an improperly operated dry-cleaning facility that utilizes perc as the cleaning agent is neither "clean" nor "dry." Due to its physical properties and relatively high toxicity, even a minor spill of perc could cause a major problem.

6.4.1.6 Septic Systems

Another AOC that warrants close attention is the septic system. Septic systems are present where no municipal sewer system is available to handle sanitary wastes. They are common in rural areas and in some suburban communities. A problem arises if waste water containing petroleum or hazardous wastes is being or has been introduced into the septic system.

A typical septic system, shown in Figure 6.8, consists of a waste pipe leading to an underground holding tank, usually made of concrete. Solids settle out in this tank and may be pumped out periodically by a commercial cleaning service. The liquids are either pumped out by the cleaning service or dispersed into the ground through a series of perforated pipes known as laterals. Laterals serve to spread the waste liquids across a broad

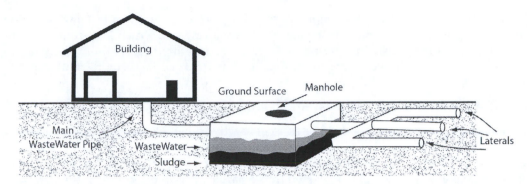

FIGURE 6.8
A typical septic system. The manhole is the access point used to clean out the septic tank.

span of property to avoid liquid pooling in a particular area and making the soils in that area unstable. When petroleum products or hazardous wastes are introduced into a septic system with laterals, the contamination immediately impacts the soils, with a great chance of spreading throughout the area.

6.4.1.7 Electrical Transformers

The high voltage electric current available from power lines must be converted to lower voltage electric current for it to be usable for common purposes. Larger industrial and commercial facilities typically have their own "step-down" transformers to perform the energy conversion. Because of the heat generated during the conversion process, step-down electrical transformers, such as the pad-mounted transformer shown on Figure 6.9, may use dielectric fluid to cool their internal equipment. Dielectric fluids are petroleum-based, and

FIGURE 6.9
A pad-mounted electrical transformer such as this one may use dielectric fluid as a coolant. (Courtesy of GZA GeoEnvironmental, Inc.)

often contained polychlorinated biphenyls (PCBs) prior to 1979, when PCBs were banned in the United States (see Chapter 3). However, transformers in which PCB-containing fluids are present still exist, and those that have been retrofitted with non-PCB-containing fluids may have discharged PCBs to the ground prior to or while being retrofitted.

6.4.1.8 Evidence of a Subsurface Investigations

Subsurface investigations are conducted at properties where contamination is suspected, as described in Chapter 7 of this book. Monitoring wells typically are installed to investigate groundwater. The presence of a monitoring well at a property should be a red flag that subsurface contamination exists, existed in the past, or was expected to exist when someone decided to install the well. The presence of multiple monitoring wells strongly implies that the first monitoring well installed in that area encountered contamination, and that the other monitoring wells were installed to delineate the extent of that contamination, as described in Chapter 8 of this book. Other evidence of a subsurface investigation, such as patched boreholes, also alerts the EP to a potential issue on a property.

6.4.1.9 Controlled Substances

The E1527 standard specifies the need to investigate for "controlled substances" as part of a Phase I ESA, but only if the Phase I ESA is funded with a federal grant under the USEPA Brownfield Assessment and Characterization Program. Controlled substances could be an environmental issue because of the usage of certain polluting chemicals in processes that create methamphetamine and other illegal substances.

6.4.2 Reconnaissance of Adjoining Properties

The. E1527 standard identifies an *adjoining property* as a real property or properties with contiguous or partially contiguous borders with the site. However, the definition also includes a real property or properties that would have contiguous or partially contiguous borders if not for the presence of a public thoroughfare. For instance, on Figure 6.10, the property at 100 Elm Street adjoins 104 Elm Street. It also adjoins 101 Elm Street, because if Elm Street, a public thoroughfare, was not present, then 100 and 101 Elm would share a border. On the other hand, 100 Elm Street does not adjoin 105 Elm Street or 97 Elm Street, since, if Elm Street and First Avenue were not present, these properties would share a corner but not a border.

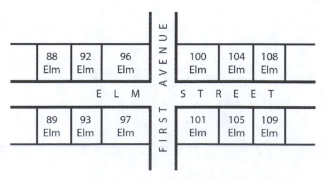

FIGURE 6.10
A schematic diagram of a typical set of properties.

Adjoining properties have special significance because of their ability to impact the subject property. Spills near or on a property can easily migrate onto the next property—property boundaries have legal meaning but no scientific meaning. This is especially true if there is a surface water body, such as a river, available to transport contaminants, or if the contaminants have reached groundwater, as described in Chapter 5 of this book.

If contamination originates on an adjoining property, then the owner of the adjoining property is responsible for cleaning up the contamination, unless it can qualify as a contiguous property owner under the AAI rule. To qualify for this exemption, the property owner must demonstrate through AAI that it did not contributed to the contamination. If the property owner does not qualify for this protection, an extensive investigation involving the installation of boreholes and/or monitoring wells, the collection of groundwater and/or soil samples, and costly analyses of those samples may be warranted. Even then, the results may be ambiguous, especially if the pollutants in question were also used and possibly spilled on the subject property.

6.4.3 Reconnaissance of the Site Vicinity

Nearby facilities also can impact a site via the migration of contaminants in surface waters or groundwater. The evaluation of nearby properties with the potential to impact the site is discussed later in this chapter.

6.4.4 Interviews of Knowledgeable Parties

Of no less importance than the site reconnaissance are the interviews of knowledgeable parties. The following are the goals of the interviews:

- Identify past uses of the site
- Identify specific chemicals present or once present on the site
- Inquire about spills or other chemical releases on or near the site
- Inquire about past environmental cleanups on or near the site

The E1527 standard requires the inspector to interview three categories of people: the key site manager, major occupants, and persons with actual knowledge. The key site manager, as the title suggests, is the person with the most knowledge about the site, about the facility, and about the facility's current and past operations. If the key site manager is relatively new to the site, then the EP is responsible for identifying and interviewing the former key site manager, or someone else at the facility who has comprehensive knowledge of the four issues noted above.

Interviewing major occupants is crucial at facilities where there are multiple occupants, such as shopping centers or multi-tenant office buildings. These occupants have more significant control of the facility and its operations, and tend to have more history at the facility, than smaller tenants. Smaller tenants should be interviewed as well, especially if their operations involve the handling or storage of hazardous substances or petroleum, but the E1527 standard specifies only the major occupants to avoid placing an undue burden on the EP assessing facilities with numerous tenants.

It is critical that the EP identify persons with actual knowledge when trying to construct the history of the property and property vicinity. Actual knowledge is what we might call first-hand knowledge, in that the person does not rely on records or someone else's testimony or memory, but rather on his/her own experiences. While people's memories can often be false or misleading, it is an excellent opportunity for the EP to obtain information not available in the written records. Typical people with actual knowledge of past events on the property are long-time employees or residents and facility operators. The employees need not be upper or even middle management types. In fact, the people "down in the trenches" often have more actual knowledge of "what really went on" than, say, the executive in the 3rd floor corner office.

The E1527 standard requires that at least one contiguous property owner or occupant be interviewed if the subject property is abandoned property or in general disrepair, or if there is evidence of "unauthorized uses or uncontrolled access" to the property. This requirement foresees that the owner of an unused property may not be as aware of events occurring on the property as the next-door neighbor, who may be more likely to witness these events. The standard does not specify whether more than one adjoining property owner needs to be interviewed—this is left to the judgment of the EP. Certainly, if the adjoining property owner interviewed provides little or no knowledge of the property of interest, the inspector should try another adjoining property owner.

6.4.5 User Responsibilities

The 2005 revision to the E1527 standard added a step to the Phase I ESA. Known as "user responsibilities," it recognizes that the person considering the purchase or financing of a piece of real estate has information at his/her disposal that may have bearing on the Phase I ESA.

A typical purchase or financing of a piece of real estate involves many professionals other than the EP. Consulting engineers, surveyors, title search professionals, and others generate reports, maps, and other documents in support of the pending deal. Any of these people may come across information that can assist the EP in the identification of RECs on the site.

The user, as defined under the E1527 standard, is the person who will use the Phase I ESA report in making a business decision. In general, it is the person who retains the EP to perform the Phase I ESA, although sometimes there is an intermediary retaining the EP, such as an attorney. The E1527 standard requires the user to divulge information on at least three topics:

The presence of environmental liens on the property—A lien is a legal claim against an asset that must be paid when the asset is sold. An environmental cleanup lien is typically placed on a property by a federal, state, or local regulatory authority when the property owner has been negligent in performing a mandated cleanup. With the lien in place, the regulatory authority guarantees that the property will not be sold until the lien is satisfied. The presence of an environmental lien on a property is a direct indication that the property has contamination that has not been remediated. The EP is required to search for environmental liens even if the user indicates that there are no environmental liens on the subject property.

Consideration of "specialized knowledge"—In this context, "specialized knowledge" is knowledge either supplied to or generated by the user of the Phase I report that may have environmental implications. For instance, the surveyor or property assessor may encounter a vent pipe, fill port, or some other evidence of the presence of a UST. Perhaps the title search professional comes across a document describing the former site usage as a dry cleaner. Often, confidential information is exchanged in the course of a property transaction or financing that the EP may not know exists. The E1527 standard requires the user to provide any "specialized knowledge" which he or she may have about a property to the EP.

Relationship of purchase price to fair market value of property, if not contaminated—As the old adage goes, if something is too good to be true, it usually is. If a property is being sold at a price well below market value, there may be a good reason—in particular, the property owner may have been compelled to discount the price because of the presence of known or perceived contamination on the property.

6.4.6 Site History Review

Most RECs—indeed, most contamination—occurred in the past, leaving little or no visible evidence in the present. The E1527 standard requires the EP to identify the "history of the previous uses of the property and surrounding area" to identify past uses that may have created present-day RECs. Discovering RECs from historical practices or past events requires the ability to seek out and sift through clues, interpret them, and postulate on the potential occurrence of spills and releases in the past, on the subject property and on surrounding sites. This part of the Phase I ESA, more than any other part, requires the EP to be a detective.

The E1527 standard requires the EP to construct a site history back to 1940 or the site's first development, whichever occurred first, with no less than five-year intervals separating historical information sources.

Evidence of historical spill events or releases can come from numerous sources. Four sources in particular, however, yield the most information to help the EP piece together the history of the site and the surrounding area. They are:

- Historical aerial photographs
- Fire insurance maps
- Local street directories (also known as city directories)
- Local agency records (which also yield information about current operations and are discussed in a separate section below)

These sources of historical data are described below.

6.4.6.1 Historical Aerial Photographs

Aerial photographs were and are commissioned by federal, state, and local governments for planning purposes; by private sector companies to sell to public and private entities for their usage; and by large enterprises for documentation and planning purposes. In some places dating back to the 1920s, historical aerial photographs are readily available through a number of commercial sources. They can often be accessed for free at public agencies such as municipal or county departments of planning, engineering, or transportation.

Aerial photographs provide information on property conditions at the time the photograph was taken. A series of such photos can show the change in property usage over time. They may provide the only evidence of landfilling operations, dumping, or other activities that can contaminate a property that aren't related to buildings or industrial processes, especially in rural areas. For instance, Figure 6.11 is an aerial photograph taken in 1969 along the Allegheny River in Pittsburgh, Pennsylvania. Figure 6.12 is a

FIGURE 6.11
Industrial land usage as seen in a 1969 aerial photograph. (Courtesy of GeoPak Historical Aerials, www.historicalaerials.com.)

FIGURE 6.12
Changes in land usage since 1969, as of 2005. (Courtesy of GeoPak Historical Aerials, www.historicalaerials.com.)

2005 aerial photograph of the same area, which had undergone huge changes in land usage. What had been a highly industrialized area underwent a metamorphosis and had become an area with office buildings and a major league baseball stadium.

Aerial photographs vary widely in quality. The best aerial photographs are clear and have a scale large enough to be useful. However, even the best available aerial photographs typically do not offer a scale better than 1" = 200'. It can be very difficult to discern features on an aerial photograph that are smaller than one acre or so in size (an acre is 43,560 square feet, or slightly larger than 1" × 1" on a 200-foot scale aerial photograph). Depending on the scale of the photograph, buildings may look like dots, so their usage cannot be easily interpreted. However, their presence or absence can be obvious. Integrating aerial photographic interpretations with other data sources can enhance the piecing together of the historic record.

6.4.6.2 Fire Insurance Maps

Fire insurance maps became popular in the United States in the mid-1800s as a way to identify and control risk for the insurance industry. At the time, most structures were constructed of wood, fire-proof construction was in its infancy, and the risk of fire to a structure from within or from its neighbors was very real. A cottage industry arose in which individuals physically visited and inspected properties in urban areas and mapped the properties, identifying their construction and any materials that could cause or contribute to a fire. Fortunately for the EP, the items that were of interest to the purchasers of the fire insurance maps are also of interest to the practitioner preparing the Phase I ESA.

While many companies were active in the fire insurance map business, one company, the Sanborn Map Company, came to dominate the business in the United States by the early 1900s. Therefore, the terms "Sanborn map" and "fire insurance map" are often but incorrectly used interchangeably.

The 1950 Sanborn map shown in Figure 6.13 depicts a gasoline filling station with eight circles, which symbolize underground storage tanks. The 1980 Sanborn map of the same location shown in Figure 6.14 shows the western portion of what had become a college campus in New York City. Other details, such as room usage (swimming pool, amphitheater) are clearly annotated on the map. The map also contains an elaborate system of shorthand notes and symbols. For instance, the number of stories of that part of the building is indicated in Figure 6.13 (2B indicates a two-story building). A key containing an explanation of the shorthand and symbols most often shown on the maps is available from The Sanborn Library, LLC.

6.4.6.3 Local Street Directories

Local street directories were prepared by governmental agencies and private enterprises from the 1800s into the 1960s, when they were supplanted by telephone directories. Rather than listing the residents and businesses in a town alphabetically, they listed them by street address, making them handy references for historical uses of a property and the neighboring area. They are often available in libraries, historical societies, and local governmental offices.

Commercial database providers have geocoded the directories and synthesized the information contained in them to derive city directory abstracts. These abstracts take the data provided in multiple directories and arrange them first by year, then address,

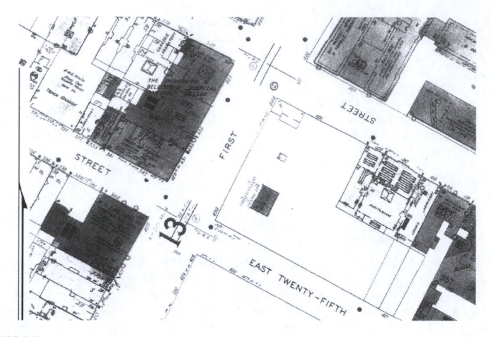

FIGURE 6.13
Sanborn map of First Avenue and East 25th Street in Manhattan in 1950. Compare land usage to that in Figure 6.14. (From The Sanborn Library, LLC. With permission.)

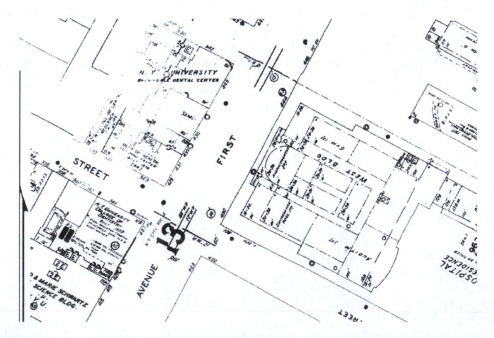

FIGURE 6.14
Changes in land usage as of 1980. No visual evidence remains of the 1950 filling station present in Figure 6.13. (From The Sanborn Library, LLC. With permission.)

74TH ST		
7415	Wendy's Restaurant	854-9008
7416	Paul's Service Ctr	854-3633
7500	Neurologic Headache & Pain Clinic	869-7009
7501	Buy-Rite Liquor	868-7606
7514	Courtesy Cab	868-6111
7514	Manny's Mobil	854-8553

FIGURE 6.15
A portion of a city directory from 1961.

then resident/business. City directories will specify the person who was living on a property or the business that was operating on a property. Figure 6.15 shows a portion of a city directory from 1961.

6.4.7 Local Agency Review

Public agencies are often critical sources of information about a property. Agencies at the local level (city, town, or county) are often more readily accessible than state or federal agencies and are more likely to have documents relating to a given property (unless the property is state- or federally-owned, or otherwise of interest to the state or federal government.)

Tax assessor—The tax assessor offers basic information about a property: its size, its ownership, its legal description, etc. It is a good starting point for the person who is investigating local records, since this information, especially the legal description of the property, is often the way in which records in other departments are organized.

Registrar of deeds—This department contains historical records of property ownership. Information about past owners of the property may provide clues as to industrial usages of the property or other usages that may have been the source of contamination. However, the EP must be cautious in inferring that ownership by an industrial company implies industrial usage. An industrial firm may have had only an office on the property or may not have utilized the property at all. Conversely, a property that had industrial usage may have been owned by a real estate company, a private individual, or an entity related to the industrial company but whose name does not reveal the connection.

Property deeds themselves can contain clues regarding environmental conditions on the property. They may indicate the presence of landfills on the property, or other practices that would suggest the presence of RECs on the property. The inclusion or exclusion of USTs as part of the property transaction could be indicated on a property deed, as could language regarding USTs and other environmental conditions, such as the excerpt below from a deed for a property in Greensboro, North Carolina:

- Install asbestos cap sheet on roof.
- Install gutter and down spouts where necessary.

- Excavate hole of sufficient size outside building...for two 10,000-gallon tanks; tanks to be furnished by Lessee and placed in hole by Lessors in manner to comply with requirements of Building and Fire regulations of the City of Greensboro.

- Install oil-fired heating system...inside building; oil tank (2,000 gal.) to be furnished by Lessee; excavation for burying tank to be done by Lessors; install new door inside furnace room.

Zoning board—Information on the current and former zoning of a property can be obtained at the zoning board. Information on current zoning (e.g., industrial, commercial, residential, or a mixture thereof) usually can be obtained from the tax assessor (see earlier). Although a property that is zoned industrial is not always used for industrial purposes, one can assume that a property that is not zoned industrial is not currently being used for industrial purposes.

Building department—The local building department is often the best source for property-specific information at the municipal or county level. Because jurisdictions generally require the issuance of permits before building construction, renovation, or demolition can be performed, building department records often contain information regarding the history of the current structures and past structures that often cannot be obtained elsewhere. Records of particular interest to the environmental professional are the following:

- Installation, abandonment, and removal of USTs, ASTs, and other equipment associated with heating oil systems

- Blueprints for building construction, which may include the location of hazardous waste and hazardous material storage areas, facility process areas, heating systems (including bulk storage containers and associated piping), and waste water systems (including septic systems)

- Inspection records of facilities, especially records of violations of building code due to issues related to hazardous material or petroleum storage or use

- Removal of lead-based paint or asbestos-containing materials (if these ordinarily out-of-scope issues are included in the Phase I ESA)

Engineering and planning—The engineering or planning department is often a good source for historical aerial photographs (see Section 6.4.6.1). It can also be a source of historical information on public works projects (roadways, water, sewer, electric), or the history of large properties that have since been subdivided.

Fire department—In many jurisdictions, fire code inspectors conduct annual inspections to verify compliance with local fire codes. These people are often valuable sources of information for the EP, not just regarding current practices of storage and usage of organic chemicals and petroleum products, but past practices as well. They are also good sources of information regarding chemical spill incidents, leaking storage tanks, and the like. In many jurisdictions, bulk storage tanks, both USTs and ASTs, must be registered with the local fire department, making the department an important database resource as well.

Health department—Local health departments tend to be focused on public health issues, such as vaccinations, disease prevention, etc. However, they may also maintain records regarding USTs, hazardous waste spills, and the like, in which case they should be researched as part of the Phase I ESA.

Water and sewer department—The local water and sewer departments may have information about when a property was connected to the sanitary sewer system, or when a sanitary sewer line became available to the property. If there is a gap between the time the property was developed and the time when a sanitary sewer line became available, it is possible that a septic system was a component of the original construction. Similarly, if the property was not connected to the municipal water system at the time of development, then a drinking water well may have been present on the property. A drinking water well would be a convenient pathway to the subsurface if contaminants were introduced into the well or found their way into the well via the groundwater pathway.

Environmental/conservation commission—Various jurisdictions may have environmental commissions established to adjudicate local environmental issues that may crop up from time to time. Many of these issues typically are excluded from the E1527 standard such as wetlands, open space, traffic, etc. However, members of these commissions may have knowledge of property conditions involving hazardous waste or petroleum, especially if the property in question is well-known, centrally located, or near a sensitive waterway or other significant environmental receptor.

In many jurisdictions, a formal request must be filed for records access under the federal Freedom of Information Act (FOIA). It can take several weeks to obtain access to these records, so FOIA requests should be submitted as soon as practicable. Even then, this information often is not accessible within the time constraints of the Phase I ESA (see Section 6.5).

6.4.8 Database Search of Regulated Properties

Contaminated neighboring properties can impact the subject property via the migration of chemicals. Contaminants on a nearby property could migrate onto the subject property via surface water or overland runoff. The less visible and more common pathway for chemicals to migrate onto a property is via groundwater (see Chapter 5 for a discussion on the principles of groundwater flow). Also of concern is the intrusion of vapors emanating from subsurface contamination into a building on the subject property, as described in Chapter 10.

When groundwater flow data exist and are readily available, it is relatively straightforward to determine whether a chemical spill on a neighboring property has impacted or has the potential to impact the subject property. In most instances, however, that data does not exist or is not readily available to the EP. Therefore, the EP must make certain assumptions in evaluating whether an off-site spill of a chemical or a petroleum product is a REC.

Table 6.2 lists the federal and state databases that must be researched as well as those that are recommended to be researched for an E1527 Phase I ESA; it also provides the recommended search radius for each database.

The issues represented by these databases may range from something as common as the presence of a UST on the property, to something as complex as a National Priorities List (NPL) site. Generally speaking, ASTM's recommended search radius correlates with an increased likelihood that an off-site source on that database may have impacted the subject property.

TABLE 6.2

Mandatory Public Database Review

Federal Databases	ASTM Recommended Search Radius
National Priorities List (NPL)	1 mile (1.6 km)
Superfund Enterprise Management System (SEMS)	½ mile (0.8 km)
RCRA Corrective Action Database (CORRACTS)	1 mile (1.6 km)
RCRIS-TSD (Resource Conservation and Recovery Information System-Treatment, Storage and Disposal) Database	½ mile (0.8 km)
RCRIS-LQG (Large Quantity Generator) Database	Site and adjoining properties
RCRIS-SQG (Small Quantity Generator) Database	Site and adjoining properties
Emergency Response Notification System (ERNS)	Site only
State Databases	ASTM Recommended Search Radius
SHWS (State Hazardous Waste Sites)	1 mile (1.6 km)
SWL/LF (Solid Waste Landfill or Landfills)	½ mile (0.8 km)
UST (Underground Storage Tanks) Database	Site and adjoining properties
LUST (Leaking Underground Storage Tanks)	½ mile (0.8 km)

Source: ASTM International, Standard Practice for Environmental Site Assessments: Phase I Environmental Site Assessment Process (E1527-13), 2013.

6.4.8.1 Mandatory Database Searches

Unsurprisingly, the NPL database has the largest recommended search radius. Even though contamination from NPL sites can migrate great distances, the recommended search radius is just one mile (1.6 km) to prevent this research portion of the Phase I ESA from becoming too cumbersome. In addition, due to the widespread knowledge of the existence of NPL sites in a given area, it is likely that other means of research, especially interviews with local officials, will enable the EP to identify NPL sites with widespread contamination and assess their potential to impact the subject property. Environmental due diligence practices under CERCLA are described later in this chapter.

The other mandatory federal database with a one-mile recommended search radius is the RCRA Corrective Action, or CORRACTS database. The CORRACTS database lists facilities that are regulated under RCRA and that are undergoing corrective action, ordinarily due to a known release or a material threat of a release. Most RCRA facilities under CORRACTS enforcement are treatment, storage, and disposal facilities (TSDFs), which, by the nature of their operations, store or treat large quantities of hazardous materials or petroleum products. Due diligence practices for CORRACTS facilities are described in Section 6.9.

State hazardous waste sites (SHWSs) go under various acronyms in their states, include a variety of different types of spill cases, and therefore constitute varying levels of environmental concern. Because in some states they are the state equivalents of NPL sites, they are given the same recommended search radius as NPL sites. However, it should be noted that this and the other search radii are recommended rather than mandatory. If the EP determines that SHWSs in a given state should not be given the same consideration as NPL sites, then a smaller search radius can be used while still complying with the E1527 Phase I ESA standard.

With the possible exception of SHWS sites, leaking UST (LUST) sites are the most prevalent sites on most radius searches that have confirmed releases of chemicals or petroleum products. The vast majority of LUST sites involve petroleum products, either a motor vehicle fuel or a heating oil. A study performed by Lawrence Livermore National Laboratory (LLNL) in 1995[1] on leaking underground fuel tanks (LUFTS, as they are known in California) concluded that fuel hydrocarbons (FHCs) rarely migrate more than 250 feet from the source. This limited migration distance is due mainly to passive bioremediation occurring at the downgradient end of the plume by microorganisms that recognize benzene and other petroleum products as food to be consumed (see Chapter 9 for a description of bioremediation). Based on that study, the ASTM E1527 committee considered shortening the search radius for LUST cases.

A development that occurred right about that time put an end to that consideration. The LLNL study did not take into account an additive that had just become widespread in the early 1990s, namely methyl tertiary-butyl ether, or MTBE (see Chapter 4). The MTBE molecule is generally too large for microorganisms to consume, and due to its physiochemical properties has the ability to spread quickly and efficiently throughout an aquifer. Some MTBE plumes were found to extend well over one mile from their source, which permanently changed the thinking regarding LUST search distances. That stated, the ½-mile search radius specified in E1527 does capture the majority of sites that can impact a subject property. Increasing the search distance would be up to the discretion of the EP.

The bane of many Phase I ESA's is the radius search in an urban area, where literally hundreds of sites are present on one or more of the mandatory databases within their recommended search radii. Often it could take several hours to sort through these sites, and even more time to evaluate their ability to impact the subject property. The E1527 standard allows the EP to reduce the search radius for a given database if, in the EP's professional judgment, the deleted information is not significant with respect to the objectives of the Phase I.

Three major databases have a recommended search radius that only entail the subject property and adjoining properties. Two of these databases, RCRA *large quantity generators* (LQGs) and RCRA *small quantity generators* (SQGs), apply to the generation of hazardous wastes on a property (see Chapter 3). The generation of hazardous wastes implies the storage and usage of hazardous materials and/or petroleum products on the property. Therefore, the listing of the site or adjoining properties on this database warns the EP that it is worth researching further to assess the potential impact of these activities. However, since their presence on one of these databases does not imply the occurrence of a spill of hazardous substances or petroleum, they have a relatively low potential to impact the subject property in the absence of a spill report. As with many aspects of the Phase I ESA, this assessment is specific to the time that the Phase I ESA was performed. Future events could change the EP's assessment regarding the threat presented by the off-site property to the subject property.

---

[1] Lawrence Livermore National Laboratory, 1995. California Leaking Underground Fuel Tank (LUFT) Historical Case Analysis. UCRL-AR-122207.

The *registered UST database,* usually managed by a state environmental regulatory agency, is the third major database with the recommended search radius limited to the site and adjoining properties. Most states use the federal requirements for UST registration, which exempt the following USTs:

- Farm and residential tanks with capacities of 1,100 gallons or less holding motor fuel used for non-commercial purposes
- Tanks storing heating oil used on the premises where it is stored
- Tanks on or above the floor of underground areas, such as basements or tunnels
- Septic tanks and systems for collecting storm water and wastewater
- Flow-through process tanks
- Tanks of 110 gallons or less capacity
- Emergency spill and overfill tanks.

Some of these exemptions are significant for Phase I ESAs, since, for instance, the heating oil tanks found in single-family residences and small commercial properties, of which hundreds of thousands exist, can cause significant contamination if they leak. The USTs on adjoining properties are also of concern because they represent documented petroleum (or possibly chemical) storage near to the subject property; depending on their proximity to the subject property, even a relatively small leak or overfill could impact the subject property.

Since "adjoining properties" doesn't constitute a distance, the EP must decide what search radius would incorporate all of the properties that adjoin the subject property. For larger properties or for irregularly-shaped properties, it may be necessary to increase the search radius for these databases to ensure that all adjoining properties are evaluated. For small properties, especially properties in densely populated urban areas, it may be advisable, even necessary, to reduce the search radius for these databases.

6.4.8.2 Additional Database Searches

In addition to the mandatory databases cited in this section, the federal government and state governments, especially if delegated by the USEPA, maintain numerous databases that track a myriad of other environmental activities. While it is not mandatory to review these databases, they should be perused for information that could be useful to the Phase I ESA. For instance, air emissions data may indicate the usage of a certain chemical, or the generation of a certain waste product that may not have been revealed elsewhere in the document trail. Since the E1527 standard does not specify the recommended search distance for non-mandatory databases, the EP must use judgment in selecting an appropriate search radius, based on the nature of the database and its potential importance in identifying RECs on the subject property.

6.4.8.3 Vapor Intrusion Assessment under the E1527 Standard

The 2013 revision to the E1527 standard redefined the concept of contaminant migration to include potential vapor migration from the subsurface into a building on the subject property. Vapor intrusion is the process by which chemicals or petroleum products that have been released to the subsurface enter the gaseous phase and then intrude into a building or other structure. Before this change in the standard, it wasn't clear whether the vapor pathway was a consideration under the E1527 standard; the definition of a REC in previous versions of the E1527 standard seemed to exclude indoor air issues. When evaluating the potential for off-site contamination to impact the subject property, the potential for vapor intrusion from nearby properties must be considered under the E1527 standard.

Please note that vapor intrusion considerations differ from the vapor encroachment survey as described in ASTM Standard E2600. Vapor intrusion and vapor encroachment are described in Chapter 10 of this book.

6.5 Limits of Due Diligence Research

The E1527 standard makes it clear that this document is not meant to be exhaustive. Section 4.5.2 of the standard states:

> There is a point at which the cost of information obtained or the time required to gather it outweighs the usefulness of the information and, in fact, may be a material detriment to the orderly completion of transactions. One of the purposes of this practice is to identify a balance between the competing goals of limiting the costs and time demands inherent in performing an *environmental site assessment* [italics theirs] and the reduction of uncertainty about unknown conditions resulting from additional information.

In general, there are three factors to be considered in deciding whether information, historical or otherwise, is available but would be considered a potential "material detriment to the orderly completion of transactions." The first factor is whether the material is *publicly available*. The E1527 standard does not require the EP to obtain information that is not readily available upon request to public agencies or information services.

A data source is not considered to be *practically reviewable* if an extraordinary review of mostly irrelevant data would be required to extract the necessary information from the data source. A good example of this second criterion would be telephone books. To find out who occupied a certain building at a certain time, the researcher could go through every page of the phone book and get the names of companies and people whose numbers were not unlisted. However, this would be an extremely cumbersome method of research, and one that would not be required under the E1527 standard.

Information that is not *reasonably ascertainable*—the third criterion—is publicly available and practically reviewable, but not obtainable from its source within reasonable time and cost constraints. If files available at a regulatory agency cannot be reviewed until, say, six weeks after the FOIA request has been submitted, the information in that file would not be considered reasonably ascertainable. Similarly, if a bureau or a person intends to charge you several hundred dollars for a review of a file, it would also not be considered "reasonably ascertainable." If it is not publicly available, then the EP must decide whether the potential usefulness of the information outweighs the time and potential cost involved in obtaining the information.

Because the E1527 standard is not exhaustive in its scope, it is inevitable that knowledge gaps will remain after the completion of the due diligence activities. These are known as *data gaps,* which the E1527 standard defines as information gaps resulting from an inability to obtain required information despite good faith efforts to gather such information. If the data gap is due to a gap in the site's historical record that is longer than five years, it is known as *data failure.*

Here are a few examples of data gaps. The latter two examples also show data failure:

- Part of the building/property could not be accessed
- A key person could not be interviewed
- Property owner would not release a report in his/her possession regarding the removal of an underground storage tank
- The historical records reviewed do not identify property usage before 1965
- There are gaps in site history that are greater than five years' time and there is evidence that property usage changed within that time period

Data gaps by themselves do not doom the usefulness of the Phase I ESA, since some data gaps are much more important than others. The E1527 standard requires the EP to decide whether or not the data gap is significant. A data gap would be deemed significant if, in the opinion of the EP, the missing data would have had a reasonable chance of indicating the presence of a REC.

For example, being unable to visit the manufacturing area of a chemical plant would be considered a significant data gap, because the activities that presumably are occurring in that area are likely to have environmental implications. Being denied access to the office area of the same plant would be a lesser concern to the EP, although being denied access to records stored in the office area would be a significant data gap. Being denied records regarding the removal of a UST is not necessarily a significant data gap—those records might be reasonably ascertainable from another data source. An interview with plant personnel who had actual knowledge of the UST removal, may be sufficient to prevent this data gap from being significant.

The significance of historical data gaps has to be considered in the general context of the property history. If the history of the property or environs suggest the occurrence of activities that could impact the environment in the undocumented time period, then the data gap should be considered significant. For instance, say a property has residential usage, then 25 years later is vacant. Neither of these property uses, by itself, would be considered a material threat to the environment. If, however, the properties surrounding that property have or had industrial usage, or if, by its location, the property could have contained a filling station, then the data gap could be considered significant. A significant data gap could be as important as a REC.

Sampling and analysis of media in areas with recognized environmental conditions are not required to satisfy the E1527 standard, but could be conducted to obtain data to address data gaps. Such follow-up activity is strictly at the discretion of the user, who must decide whether to live with the likelihood of subsurface contamination and associated uncertainties in the environmental condition of the subject property. This decision often is made by the user, the user's lender, or some other financial stakeholder in the property transaction.

In some cases, it may be prudent to do a file review, interview additional people, or dig deeper into the documentation before digging into the soil. Although the E1527 standard makes this optional for most off-site properties, it also requires the EP to review files that are relevant to listings of the subject property or adjoining properties in one of the standard regulatory databases. The EP can use professional judgment and decide that a file review is not warranted. Perhaps the EP has sufficient information from other sources to conclude that the condition described in the radius search for a particular regulatory database does not present a material threat to the subject property. However, the E1527 standard requires the EP to justify the decision in the Phase I ESA report.

6.6 Report Preparation

The final product of a Phase I ESA is the Phase I ESA report. This report must include all of the relevant documentation collected in the course of the Phase I activities. The report should include sufficient supporting documentation "to facilitate reconstruction of the assessment by an environmental professional other than the environmental professional who conducted it" (E1527-13, Section 12.2).

The E1527 standard provides a suggested report format; many financial institutions and other frequent users of Phase I ESA reports have other preferred formats. Whatever format is used, the Phase I ESA report must contain, at a minimum, the following elements:

- The identity of the EP and the person who conducted the site reconnaissance, if different than the EP
- The scope of services performed. It should be noted whether the scope of services included items excluded from the E1527 standard, such as an asbestos survey or soil sampling
- A list of findings, which includes RECs, CRECs, HRECs, and *de minimis* conditions
- The EP's opinion as to the potential for the conditions listed in the findings section to impact the subject property
- A list of data gaps, and a discussion as to their significance
- The conclusions of the assessment
- Any deviations from the standard practice
- References used in preparing the Phase I ESA report

The Phase I ESA report must contain a section of a U.S. Geological Survey (USGS) topographic map that shows the location of the subject property. Reports also typically contain a site plan and a photographic log. While these elements are not required under the E1527 standard, they assist the reader in reconstructing the assessment performed and therefore are useful additions to the report. The E1527 standard does not require the report to contain recommendations for further assessment except in the unusual circumstance that "greater certainly is required regarding the identified recognized environmental conditions" (E1527-13, Section 12.6.1).

6.7 Phase I ESAs for Forested Land and Rural Properties

In 2008, ASTM International promulgated a new standard for Phase I ESAs, this one specifically for forested land and rural properties. The standard, issued by the E2247 committee, was designed to address the difficulties in performing a site reconnaissance on such properties due to lack of access, lack of improvements, and, in some cases, their sheer size.

Like the E1527, the E2247 standard practice was designed to comply with the USEPA's AAI requirements, and the USEPA has acknowledged that a user will qualify for the innocent purchaser defense if the E2247 standard is followed. Specifically, it applies to forested land or rural properties that are at least 120 acres in area and that have been forests or rural properties throughout their history. The property can consist of non-contiguous parcels, provided that they share a common history.

The E2247 standard acknowledges that in some cases the property cannot be physically observed, and allows the EP to rely on remote imagery, observations from nearby locations, and even flyovers. The EP is required to verify areas of concern identified by these alternate inspection methods to the extent feasible. Except for the site reconnaissance, the other parts of the E2247 standard, including the report preparation, are similar to the E1527 standard.

6.8 Preliminary Assessments under Comprehensive Environmental Response, Compensation, and Liability Act

The CERCLA, or Superfund, has a six-step process for remediating sites that are on the NPL. The steps are defined as follows:

- Preliminary assessment
- Site inspection
- Remedial investigation
- Feasibility study
- Record of decision (ROD)
- Remedial design/remedial action

This section describes the preliminary assessment (PA), which, although analogous to the ASTM Phase I ESA, differs from it in significant ways. As with the ASTM Phase I ESA, the inspector visits the subject property as part of the PA, although a second site inspection, entailing the sampling and analysis of environmental media, is often conducted as well. This second site inspection is similar to the site investigation and is described in Chapter 7 of this book; the remedial investigation under CERCLA is described in Chapter 8 of this book; and feasibility studies, records of decision, and remedial designs are described in Chapter 9 of this book.

6.8.1 Structure of the Preliminary Assessment under Comprehensive Environmental Response, Compensation, and Liability Act

A PA under CERCLA is a limited-scope investigation whose main purpose is to decide whether a property should be placed on the NPL. The PA is designed to distinguish between properties that pose little or no threat to human health and the environment and properties that require further investigation. The PA also identifies properties requiring emergency response actions under CERCLA.

The PA includes a research component. The researcher will obtain the same general site information that is required in an ASTM E1527 Phase I ESA, such as the facility's name, address, owner, and size; its current and former operations; its current status (active/inactive, etc.); and a description of the environmental setting.

Historical research for a PA under Superfund entails essentially the same data resources as the ASTM Phase I ESA. The first stage site inspection is similar to the site reconnaissance conducted as part of the Phase I ESA. Interviews are conducted with key site personnel and other people with actual knowledge of current and former site operations. However, unlike the Phase I ESA, no "outs" are available if a data source is difficult, expensive, or time-consuming to obtain or difficult or time-consuming to review. The PA under CERCLA is meant to be comprehensive and exhaustive.

Unlike the Phase I ESA, which emphasizes sources and pathways to the environment, the PA under CERCLA considers receptors as well. The PA is organized by pathway: groundwater, surface water, soils, and air. Each category receives a score based on waste characteristics (sources of chemicals or petroleum products), pathways (likelihood of release evaluation), and receptors (known as targets under CERCLA). These three components of an environmental hazard, and the four pathways, are described below.

6.8.2 Evaluation of Waste Characteristics

As part of the site inspection, the inspector will estimate the type and quantity of hazardous wastes associated with all sources at the site. For each source, the quantity of the waste can be evaluated by one or all of four different measures called "tiers": constituent quantity, waste stream quantity, source volume, and source area. The USEPA provides guidance on how to estimate volumes of wastes based on the type of facility—landfills, manufacturing facilities, waste treatment and storage facilities, and so on. All waste sources are evaluated for the PA, regardless of size or the amount of time they were present at the site.

6.8.3 Likelihood of a Release

In a PA, the inspector must evaluate the likelihood of release via one of the four pathways and exposure routes. After evaluating the evidence, the inspector will categorize the likelihood of release as a suspected release or no suspected release. To facilitate this evaluation, the USEPA provides criteria lists for the four pathways. These criteria lists are described in the subsections below.

6.8.4 Evaluating the Four Pathways and Exposure Routes

6.8.4.1 Groundwater Pathway

The principal threat considered in the evaluation of the groundwater pathway is the threat posed to drinking water and to populations relying on groundwater as their source of drinking water. Therefore, the evaluation of the groundwater pathway is primarily concerned with identifying drinking water wells and their users. The USEPA requires the researcher to evaluate drinking water wells within four miles of the property if its occupants/residents use groundwater as their sole or partial water source. The researcher will determine the distance to the nearest drinking water well; whether the potentially impacted wells get their drinking water from multiple aquifers; the size of the population served by potentially impacted groundwater; and the type of population using the wells (whether they are full-time residents, or workers or students, who have less residency and therefore less potential exposure over the long term).

The USEPA requires the researcher to answer the following questions regarding the likelihood of a release to groundwater:

- Are sources of chemicals and petroleum products poorly contained?
- Is the source a type likely to contribute to ground water contamination?
- Is the waste quantity particularly large?
- Is precipitation heavy in the site vicinity?
- Is the water infiltration rate high?
- Is the site located in an area of karst terrain? (see Chapter 5 for an explanation of karst terrain)
- Is the subsurface highly permeable or hydraulically conductive?
- What is the depth to the aquifer(s)?
- Is drinking water drawn from a shallow aquifer?
- Has a nearby drinking water well been closed?
- Are suspected contaminants highly mobile in ground water?
- Does analytical or circumstantial evidence suggest ground water contamination?

There are separate criteria for the "primary target well," which is the nearby well with the greatest likelihood for contamination. The USEPA provides the following questions to guide the researcher in evaluating the primary target well:

- Have any nearby drinking water users reported foul-tasting or foul-smelling water?
- Does the well produce a large quantity of water?
- Is the well located between the subject property and wells that are suspected to be exposed to a hazardous substance?
- Does analytical or circumstantial evidence suggest contamination in the well?
- Does the well warrant sampling?

6.8.4.2 Surface Water Pathway

Surface waters include streams and rivers, lakes, coastal tidal waters, and oceans. This pathway is evaluated if there are surface waters within an overland flow distance of two miles from the site. The researcher must ask the following questions to assess the likelihood of a release to surface water from the subject property:

- Is rainfall typically heavy in the area?
- Is the water infiltration rate into the soils low?
- Are sources poorly contained or prone to runoff or flooding?

- Is a runoff route well defined?
- Is vegetation stressed along the probable runoff route?
- Are sediments or water unnaturally discolored?
- Is wildlife unnaturally absent?
- Has deposition of waste into surface water been observed?
- Is groundwater discharge to surface water likely?
- Does analytical or circumstantial evidence suggest surface water contamination?

The inspector also should consider the following when evaluating the surface water pathway for contamination:

- Distance of the surface water to the probable point of entry for contaminants
- Distance of the surface water to probable human receptors, such as drinking water intakes
- Populations consuming water originating from the surface water body
- Distance of the surface water to ecological receptors
- Threat of contamination in the surface water body to ecological receptors.
- Use of the surface water in the food chain
- Flood frequency
- Flow rate of the surface water body at drinking water intakes
- Has any intake, fishery, or recreational area been closed?
- Do analytical or circumstantial evidence suggest surface water contamination?

6.8.4.3 Soil Exposure Route

Soils are considered an exposure route rather than a pathway because, rather than moving, the receptor comes to them, such as when a human is exposed to soils through direct contact or some other pathway. Areas of suspected contamination are defined by the presence of hazardous substances; however, source areas with more than two feet of clean cover or an impermeable surface are excluded from evaluation of the soil exposure pathway. The researcher must ask the following questions regarding the soil exposure route:

- Is any residence, school, or daycare facility on or within 200 feet of an area of suspected contamination?
- Is any residence, school, or daycare facility located on adjacent land previously owned or leased by the site owner/operator?
- Can hazardous substances present in soils migrate to nearby residences, schools, or daycare facilities?
- Have onsite or adjacent residents or students reported any adverse health effects, exclusive of apparent drinking water or air contamination problems?
- Does any neighboring property warrant sampling?

The inspector also should evaluate the hazard presented by contaminated soils to workers on the facility property and workers on the property of nearby facilities where soil contamination related to the subject property may be present.

6.8.4.4 Air Pathway

The principal threat under the air pathway is the threat of airborne releases of hazardous substances. The targets evaluation is primarily concerned with identifying and evaluating the human population within the four-mile target distance limit (radius) around the site, and sensitive environments within ½ mile. The inspector must answer the following questions regarding the likelihood of a release to air:

- Are odors currently reported?
- Has a release of a hazardous substance to the air been directly observed?
- Are there reports of adverse health effects potentially resulting from migration of hazardous substances through the air?
- Does analytical or circumstantial evidence suggest a release to the air?

The receptors to be evaluated include residential populations, worker and student populations, and sensitive environments, which are defined as terrestrial or aquatic resources, fragile natural settings, or other areas with unique or highly valued environmental or cultural features.

6.8.5 Hazard Ranking System

The final product of the PA is a PA report, and the core of the PA report is the *hazard ranking system* (HRS) score. The HRS is divided into the four hazardous substance pathways to human or ecological receptors: groundwater, surface water, air, and soil exposure. Table 6.3 is the groundwater migration pathway scoresheet developed by the USEPA to calculate the site groundwater pathway score (S_{gw}).

The site score is derived from the square root of the squares of the four pathway scores:

$$S = \sqrt{\frac{S_{gw}^2 + S_{sw}^2 + S_{se}^2 + S_a^2}{4}}$$

where:
  S = Site score
  S_{gw} = Groundwater migration pathway score
  S_{sw} = Surface water migration pathway score
  S_{se} = Soil exposure pathway score
  S_a = Air migration pathway score

A site with an HRS score of 28.50 or greater is eligible for proposal to the NPL.

TABLE 6.3

Groundwater Migration Pathway Scoresheet

Factor Categories and Factors	Maximum Value	Value Assigned
Likelihood of a release to an aquifer:		
1. Observed release	550	
2. Potential to release		
2a. Containment	10	
2b. Net precipitation	10	
2c. Depth to aquifer	5	
2d. Travel time	35	
2e. Potential to release (2a × [2b + 2c + 2d])	500	
3. Likelihood of release (higher of lines 1 and 2e)	550	
Waste characteristics:		
4. Toxicity/Mobility	(a)	
5. Hazardous waste quantity	(a)	
6. Waste characteristics	100	
Targets:		
7. Nearest well	50	
8. Population		
8a. Level I concentrations	(b)	
8b. Level II concentrations	(b)	
8c. Potential contamination	(b)	
8d. Population (lines 8a + 8b + 8c)	(b)	
9. Resources	5	
10. Wellhead protection area	20	
11. Targets (lines 7 + 8d + 9 + 10)	(b)	
Groundwater score		
12. Aquifer score [(lines 3 × 6 × 11)/82,500]	100	
Groundwater migration pathway score		
13. Pathway score (S_{gw}), (highest value from line 12 for all aquifers evaluated)	100	

Note: (a): Maximum value applies to the waste characteristics category and (b): Maximum value not applicable.

6.9 Environmental Due Diligence for RCRA CORRACTS Sites

As noted in Section 6.4.8, RCRA CORRACTS sites are of significant concern because they store or handle large quantities of hazardous materials and petroleum products and have had or are suspected to have had a release. The first step in the environmental due diligence for a RCRA CORRACTS site is called the preliminary review.

The environmental due diligence procedures for a RCRA CORRACTS site differ in two significant ways from the environmental due diligence procedures performed under CERCLA. Firstly, the facility is operating, so hazardous waste and petroleum storage and handling practices can be observed directly. Secondly, the facility most likely is operating

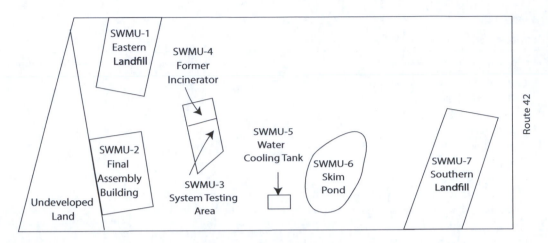

FIGURE 6.16
Solid waste management units at a RCRA corrective action facility.

under a RCRA Part A or Part B permit. Applications for a Part A and Part B permit are comprehensive and contain much of the information needed to complete the environmental due diligence process required by USEPA.

Whereas "areas of concern" are the building blocks in an ASTM-compliant Phase I ESA, the equivalent term for a CORRACTS site is the *solid waste management unit*, or SWMU (pronounced Shmoo). The objective of the preliminary review is to identify the SWMUs at the facility and assess the likelihood that a release has occurred at the SWMU. Figure 6.16 is a map showing the location of the SWMUs at a CORRACTS site. Unlike the Phase I ESA, releases of concern could have been to the air as well as to the surface water, soils, and groundwater at the facility. Also, unlike the Phase I ESA, the preliminary review is concerned with releases of wastes only, not releases of chemicals or petroleum products in use at the facility. The preliminary review differs from the PA under CERCLA in that (1) its primary concern is contamination at the facility whereas the primary concern of the PA under CERCLA is the potential impact on receptors; and (2) there is no ranking system under the CORRACTS process.

The preliminary review includes a site visit, known as a visual site inspection (VSI). The final product of the preliminary review is the RCRA facility assessment report, or RFA report. The RFA report will describe each SWMU, identify the concerns presented by each SWMU, and recommend environmental sampling, as warranted.

6.10 Environmental Consultants and Environmental Due Diligence

Environmental due diligence is very much the turf of the environmental consultant. As defined by the ASTM E1527 and E2247 standards, the environmental professional must have significant work and educational experience. Because of time and budget constraints, however, much of the research and report writing is performed by less experienced staff. For many consultants, the Phase I ESA is the gateway into the environmental consulting profession. Because preliminary assessments under CERCLA are less constrained by time and budget considerations and the subject properties tend to be more complex, mid-level and senior level staff tend to play a larger role in the site reconnaissance and data interpretation, although junior-level staff plays an important role as well due to the magnitude of the effort.

Problems and Exercises

1. A waste storage area contains 6 unlabeled 55-gallon drums that contain liquids. None of the drums have leaked or overflowed. What information would you need to know to decide whether the stored drums represent a recognized environmental condition as per the ASTM E1527 standard?

2. The property owner states that there never were any underground storage tanks (USTs) on the property. However, a fire insurance map from 1998 shows that the property formerly contained a UST. Resolve this data conflict.

3. Petroleum was spilled on a property and the property was remediated. The spill case received a No Further Action letter from the applicable regulatory authority. What information might the case file contain that would warrant claiming that the former spill represents a recognized environmental condition as per the ASTM E1527 standard?

4. Calculate the hazard ranking score for a site that received the following sub-scores: 50 for the groundwater pathway; 0 for the surface water pathway; 25 for the soil exposure pathway; and 20 for the air pathway. Is the site eligible for the National Priorities List?

5. Review the HRS Documentation Record for the Orange County North Basin in California. Indicate which sub-score was primarily responsible for this site's listing on the NPL. Describe why this pathway was determined to be particularly hazardous to nearby receptors.

Bibliography

Abbasi, R.A., 1995. RCRA Corrective Action—A Practical Guide. Environmental Solutions.

ASTM International, 2005. Standard Practice for Environmental Site Assessments: Phase I Environmental Site Assessment Process (E1527-05).

ASTM International, 2013. Standard Practice for Environmental Site Assessments: Phase I Environmental Site Assessment Process (E1527-13).

ASTM International, 2016. Standard Practice for Environmental Site Assessment: Phase I Environmental Site Assessment Process for Forestland or Rural Property (E2247-16).

Hess-Kosa, K., 2008. *Environmental Site Assessment Phase I: Fundamentals, Guidelines, and Regulations.* 3rd ed. CRC Press.

Lawrence Livermore National Laboratory, 1995. California Leaking Underground Fuel Tank (LUFT) Historical Case Analysis. UCRL-AR-122207.

U.S. Environmental Protection Agency, 1991. EPA/540/G-91/013 Publication 9345.0-01A—Guidance for Performing Preliminary Assessments Under CERCLA.

U.S. Environmental Protection Agency, 1986. PB87-107769—RCRA Facility Assessment Guidance.

7

Site Investigations

As with the Phase I environmental site assessment, the next phase of work goes under several different names. It is often called a Phase II environmental site assessment, especially when performed as a follow-up to the Phase I ESA. In this chapter, we refer to the next investigative phase by another commonly used phrase: the "site investigation."

This chapter discusses the investigation of three environmental media: surface water, groundwater, and soils. The investigation of the gaseous medium (soil vapor) is discussed in Chapter 10.

7.1 Initiating the Investigation

There are many reasons to initial a site investigation. It may be triggered by the discovery of environmental concerns, possibly through a Phase I ESA (although a Phase I ESA does not have to precede a site investigation); by the observation of a release of a hazardous chemical or a petroleum product; by the identification of a potential health impact from a release, by the suspected failure of an underground storage tank, or by some other incident or concern involving a hazardous chemical or petroleum product. It may be the next scheduled step in the assessment of a potential NPL site or a site in the Resource Conservation and Recovery Act (RCRA) corrective action (CORRACTS) process. Under CERCLA (Comprehensive Environmental Response, Compensation, and Liability Act), the information obtained during the site investigation is incorporated into the hazard ranking system (HRS) score for the site (see Chapter 6 for explanations of the first stages of environmental investigations for potential NPL sites and CORRACTS sites).

Financial and liability considerations can also initiate a site investigation. Purchasers or lenders may want to understand the potential liabilities associated with a piece of real estate beyond the information provided by a Phase I ESA. This is especially true for industrial properties and other "high risk" properties (see Table 7.1). Insurance companies may also require site investigations as part of their risk management practices. Lastly, the Sarbanes-Oxley Act of 2002 requires corporations to disclose environmental financial liabilities in their accounting statements.

In some jurisdictions, the site investigation can be triggered by regulatory requirements. In Connecticut, for instance, "establishments" that are being purchased or closed must be evaluated in accordance with Connecticut's Property Transfer Program. Areas of concern identified by the preliminary assessment must undergo a site investigation if there is reason to suspect a release of a hazardous substance or petroleum product.

TABLE 7.1

Environmentally Sensitive Industries by North American Industry Classification System (NAICS) Code

211 Oil and gas extraction

212 Mining (except oil and gas)

213 Support activities for mining

237 Heavy and civil engineering construction

311 Food manufacturing (if underground fuel tanks present)

312 Beverage and tobacco product manufacturing

313 Textile mills (not required if sewing, weaving, or hemming only)

314 Textile product mills (not required if sewing, weaving, or hemming only)

316 Leather and allied product manufacturing

321 Wood product manufacturing (if finishing occurs on site)

322 Paper manufacturing

323 Printing and related support activities

324 Petroleum and coal products manufacturing

325 Chemical manufacturing

326 Plastics and rubber products manufacturing

327 Non-metallic mineral products manufacturing

331 Primary metal manufacturing

332 Fabricated metal product manufacturing

333 Machinery manufacturing (not required if assembly only)

334 Computer and electronic product manufacturing (not required if assembly only)

335 Electrical equipment, appliance, and component manufacturing (not required if assembly only)

336 Transportation equipment manufacturing

337 Furniture and related manufacturing (if finishing occurs on site)

339 Miscellaneous manufacturing (only required if hazardous materials are involved)

42311 Automobile and other motor vehicle merchant wholesalers (if service bays present)

42314 Motor vehicle parts (used) merchant wholesalers

4235 Metal and mineral merchant wholesalers

42393 Recyclable material merchant wholesalers

4246 Chemical and allied products merchant wholesalers

4247 Petroleum and petroleum products merchant wholesalers

441 Motor vehicle and parts dealers (if service bays present)

447 Gasoline stations

45431 Fuel dealers (not required for propane or firewood dealers)

481 Air transportation

482 Rail transportation

486 Pipeline transportation

53212 Truck, utility trailer, and recreational vehicle rental and leasing (if repairs, maintenance or vehicle washing are performed onsite)

53241 Construction, transportation, mining and forestry machinery and equipment rental and leasing (if repairs, maintenance or vehicle washing are performed onsite)

53249 Other commercial and industrial machinery and equipment rental and leasing (if repairs, maintenance or vehicle washing are performed onsite)

54138 Testing laboratories

56171 Exterminating and pest control

(Continued)

TABLE 7.1 (*Continued*)

Environmentally Sensitive Industries by North American Industry Classification System (NAICS) Code

---

562 Waste management and remediation services

6221 General medical and surgical hospitals (if fuel tanks are present)

71391 Golf courses and country clubs

71392 Skiing facilities

71393 Marinas

7212 Recreational vehicles parks and recreational camps (if fuel tanks are present or if vehicle repairs or maintenance is performed onsite)

8111 Automotive repair and maintenance

8112 Electronic and precision equipment repair and maintenance (not required if assembly only)

8113 Commercial and industrial machinery and equipment repair and maintenance

8122 Death care services

8123 Laundry and dry cleaning services (if dry cleaning operations have ever existed on site)

812921 Photofinishing laboratories (except one hour)

Source: From Appendix 4 SOP 50 10 5(C) 358. Executive Office of the Presidents, Office of Budget and Management, 2017. Effective Date: October 1, 2010.

7.2 Developing the Scope of Work

Investigating a site can involve one or two areas of concern, all known areas of concern, or no areas of concern in the case of a baseline survey. Regardless of the number of areas of concern and the chemicals or petroleum products involved, many decisions need to be made before chemical data are collected. In general, site investigations rely on sampling and analysis to assess environmental conditions in areas of concern. The discovery of contamination can lead to a remedial investigation (see Chapter 8), which entails many of the methodologies described in this chapter.

7.2.1 Establishing Data Quality Objectives

In a world with unlimited time and money, all investigations would be exhaustive and perfect. Alternatively, in a world where all data are created equal, all investigations could be quick and cheap. In the real world, budget and schedule cannot be ignored, resulting in site investigations ranging in quality between "exhaustive and perfect" to "quick and cheap." We say "between," because even with unlimited time and money, a site investigation will never reach certainty because of the limitation inherent in sampling a medium rather than assessing the entire area.

Establishing *data quality objectives* (DQOs) before designing the site investigation guides the investigator on where the environmental study has to fall on that quality spectrum. Since the core of the site investigation is the collection and analysis of samples, data quality objectives will affect the number of samples collected, the effort expended in collecting those samples, the analyses performed on the samples, and the degree of quality control employed in the laboratory analyses.

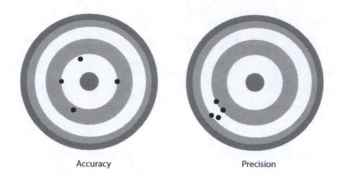

Accuracy   Precision

FIGURE 7.1
The illustration on the left shows good precision but poor accuracy. The illustration on the right shows good precision but poor accuracy.

The criteria most commonly used to specify DQOs and to evaluate available sampling, analytical, and QA/QC options are known collectively as the *precision, accuracy, representativeness, completeness, and comparability* (PARCC) parameters. They are defined as follows:

- Precision—a measure of the reproducibility of analyses under a given set of conditions.
- Accuracy—a measure of the bias in a measurement system.
- Representativeness—the degree to which sampling data accurately and precisely represent selected characteristics.
- Completeness—a measure of the amount of valid data obtained from a measurement system compared to the amount that could be expected to be obtained under ideal conditions.
- Comparability—the degree of confidence with which one data set can be compared to another.

Illustrations showing the difference between precision and accuracy are shown in Figure 7.1.

Because there are tradeoffs between data quality, cost, and schedule, the DQOs should be discussed with the client in the planning process, well in advance of the field activities.

7.2.2 Conceptual Site Model

The sampling and analysis to be performed will depend upon numerous factors, including:

- Area(s) of concern (AOCs) to be addressed
- Hazardous chemicals and/or petroleum products suspected to have been released to the environment
- Type of soils or rock that are expected to be encountered
- Chemical and physical properties of the compounds that may be present
- Area physiography
- Presence of surface waters
- Anticipated groundwater depth and flow direction
- Presence of buildings, subsurface utilities, and other man-made obstructions and preferential pathways in the area of concern

All of these factors can be incorporated into the formulation of what is known as a *conceptual site model*. The conceptual site model utilizes the above information and other site-specific information to predict the fate and transport of a release of the suspected hazardous chemical(s) or petroleum product(s) to the environment. Knowledge of the source of the potential release will determine the sampling locations, while knowledge of the chemicals and petroleum products will determine the analyses to be performed on the samples. A few examples will help elucidate this point.

Example 7.1: Underground Storage Tank

Figure 7.2 is a conceptual diagram of a release from an underground storage tank (UST). The path that the liquids released from the UST will take depends mainly upon the chemistry of the liquid, the hydrogeology of the area of concern, and the presence of man-made pathways, as applicable. If the leak occurs in soils and in the vadose zone, then in most cases the primary movement of the liquid will be downward. However, there will be some lateral spreading, mainly due to the effects of dispersion, but also due to retardation caused by the sorbing capacity of the organic carbon content and other chemical or physical characteristics of the soil. The result of downward movement combined with lateral spreading results in the liquid spreading out in a cone-shaped pattern (although the pattern could be asymmetrical if the soils vary laterally). Fine-grained soils will cause more spreading than coarse-grained soils. Similarly, compounds with low carbon-water partitioning coefficients or low molecular weights will undergo less lateral spreading than compounds with high carbon-water partitioning coefficients or with high molecular weights.

The more volatile petroleum products, such as gasoline, will lose much of their mass to the air in the vadose zone; petroleum products with lower volatility will enter the vadose zone without a significant loss of mass to vaporization. However, the volatile portions that do reach the vadose zone tend to migrate efficiently in the subsurface. Petroleum products generally have low soil-water coefficients, so that much of the downward-moving petroleum will be left behind as molecules adhered to soil. This effect creates an area of residual contamination that can be detected by soil sampling.

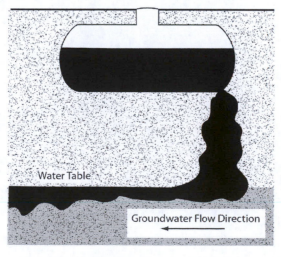

FIGURE 7.2
Conceptual diagram of a petroleum release from an underground storage tank.

When the liquid reaches the water table, it will go into solution, unless it has a high octanol-water coefficient, as is the case for most types of petroleum. If the contents of the UST were a light, non-aqueous phase liquid (LNAPL), such as #2 fuel oil, then the liquid likely would remain separate from the water and float on its surface. It then would move through advection primarily in the direction of groundwater flow, while spreading out due to the effects of dispersion and diffusion.

Example 7.2: Perchloroethene Surface Spill

A surface spill of perchloroethene (PCE) often will volatilize before reaching the soil, although enough of the PCE may reach the subsurface to create a significant problem. Once in the subsurface, PCE, like most chemicals, moves primarily downward through the vadose zone (see Figure 7.3). Due to its very low organic carbon partitioning and soil-water coefficients, PCE will undergo minimal conical spreading in unsaturated soils. PCE remaining in the vadose zone can result in a vapor intrusion issue if there is an occupied building nearby (see Chapter 10).

When the PCE reaches the water table, it will dissolve to the degree allowed by its solubility, and move via advection, dispersion, and diffusion in the saturated zone. Beyond its solubility point, PCE, which has a specific gravity greater than 1, will become a dense non-aqueous phase liquid (DNAPL), moving downward through the water column until it reaches a physical barrier, such as an aquitard. Once in the subsurface, PCE may undergo a series of biotransformations which are discussed in Chapter 8.

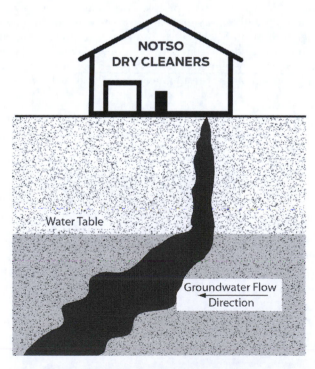

FIGURE 7.3
Conceptual diagram of a petroleum release from a dry cleaning store.

Example 7.3: Surface Spill of Transformer Oil Containing Polychlorinated Biphenyls

The primary concern in a spill of oil from a pre-1979 electrical transformer is polychlorinated biphenyls (PCBs). The PCBs, while soluble in petroleum, are far less soluble in water. This factor, combined with their very low soil-water partitioning coefficient, means that they will not migrate very far from their source in the subsurface, unless they are being transported within a petroleum matrix (see Figure 7.4). With a very high vapor pressure, a small fraction of PCBs will go into vapor phase in the vadose zone, although that small amount can create a vapor hazard.

Only a small portion of the released PCBs will be transported in groundwater due to their very low solubility. With a specific gravity greater than 1, they will act as a DNAPL and move downward through the water column, to the extent that they move at all.

7.2.3 Sampling and Analysis Plan

The quantity and location of the samples to be collected and the analyses to be performed are based on the data quality objectives, the conceptual site model, and the goals of the investigation.

As a rule, the more data available for a given AOC, the more successful the investigation. For sites at which the typical hazards are well-known, such as gasoline filling stations and dry cleaners, much can be assumed about the sources and their potential pathways. For less common contaminants or sites with more variables, more background work often is warranted before an effective sampling program can be developed. Table 7.2 lists some common businesses, the contaminants most often associated with those businesses, and the types of analyses that can be used to investigate for those contaminants.

Proposed sample locations typically are displayed on a scaled map. But sampling points are not just two-dimensional locations—they have a vertical component as well. The selection of sampling locations is a function of the conceptual site model, which will indicate not just the potential geographic location of a given contaminant, but its potential vertical

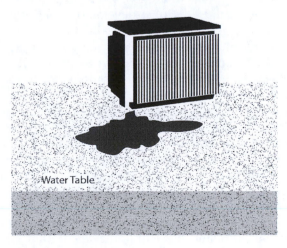

FIGURE 7.4
Conceptual diagram of a petroleum release from an electrical transformer.

TABLE 7.2

Contaminants Associated with Common Businesses

Business	Types of Contaminants	Typical Analyses
Auto service	Petroleum, solvents	PHC, VOC
Auto washing	Petroleum, PAH	PHC, PAH
Auto wrecking/junkyard	Petroleum, metals	PHC, metals
Dry cleaners (on-site cleaning)	perchloroethene	VOC
Foundry	Metals, solvents	Metals, VOC
Gas station	Petroleum, gasoline	PHC, VOC
Gas station (old)	Petroleum, gasoline, lead	PHC, VOC, lead
Machinist	Metals, solvents, petroleum	PHC, VOC, metals
Metal plating	Metals, solvents	Metals, VOC
Painter	Solvents, metals	Metals, VOC
Photo finisher	Solvents, metals	VOC, phenols, silver, zinc
Plastic fabrication	Solvents, metals	VOC, SVOC
Printer	Petroleum, solvents, metals	PHC, VOCs, silver
Railroad	Petroleum, PAH, solvents, paint, fungicides, insecticides	PHC, PAH, VOC, pesticides/PCBs
Sheet metal works	Metals, solvents	PHC, VOC
Welding	Metals, solvents	PHC, VOC

Source: Alaskan Way Viaduct Replacement Project, July 2011. *Final Environmental Impact Statement and Section 4(f) Evaluation*, Final EIS.

PAH—polycyclic (*or* polynuclear) aromatic hydrocarbons; PCBs—polychlorinated biphenyls; PHC—petroleum hydrocarbons; SVOC—semi-volatile organic compounds; VOC—volatile organic compounds.

location as well. In some cases, the sampling point could have a fourth dimension—time—if site conditions change with time, as in a tidally-influenced area or in an area when weather conditions can influence chemical behavior.

Once the investigator has decided where to collect samples and what analyses will be performed, the information can be used to develop a *sampling and analysis plan* (SAP). The SAP will provide information regarding the site and site history, including a summary of environmental information collected to date (if any); a summary of the physical setting, including site topography, geology, and hydrogeology; a description of the project organization (project manager, field supervisor, etc.), including the demarcation of responsibilities; the data quality objectives; the conceptual site model; field methods to be employed; and quality control procedures. It also will provide the objective(s) of the work to be performed, and the regulatory agency that is setting the remediation standards and providing enforcement, if applicable.

The heart of the SAP is a table that provides the sampling locations, the media to be sampled, and the analyses to be performed. An example of such a table for a fictitious manufacturing plant (see Figure 7.5)[1] is shown in Table 7.3. As indicated in the table, three areas of concern are being addressed: a UST, a waste storage area, and an electric transformer. Four soil samples are proposed around the UST, one from each side of the tank. Let's assume that the tank contains #2 fuel oil, that the bottom of the UST is located at 6 feet below the ground surface (also referred to as "below grade"), and that the soils are silty sand. In this situation, collecting soil samples at 10–12 feet below the ground surface should be sufficient to detect a release from the tank. The analysis to be performed,

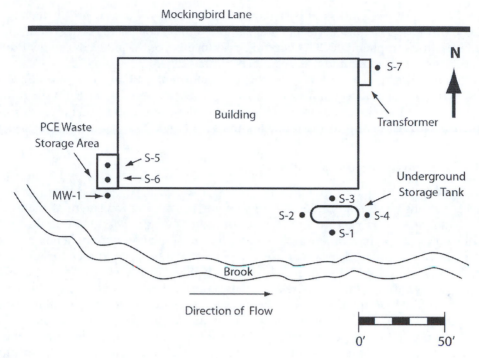

FIGURE 7.5
Schematic diagram of a fictitious factory.

TABLE 7.3

1313 Mockingbird Lane, Anytown, USA, Sampling and Analysis Plan

Sample Name	Media	Location	Depth Below Grade	Analyses
S-1	Soils	South of UST	10'–12'	TPH-DRO
S-2	Soils	West of UST	10'–12'	TPH-DRO
S-3	Soils	North of UST	10'–12'	TPH-DRO
S-4	Soils	East of UST	10'–12'	TPH-DRO
S-5	Soils	PCE waste storage area	0'–2'	VOCs
S-6	Soils		10'–12'	VOCs
S-7	Soils	Electrical transformer	0'–2'	PCBs, TPH-DRO
MW-1	Groundwater	Near UST	Below water table	TPH-DRO

UST—underground storage tank; TPH-DRO—Total petroleum hydrocarbons—diesel-range organics; VOCs—volatile organic compounds; PCBs—polychlorinated biphenyls.

petroleum hydrocarbons, emphasizes the diesel range organics within the petroleum scan, which is appropriate for detecting #2 fuel oil. In addition, a groundwater sample will be collected from a permanent monitoring well, MW-1, that will be installed near the UST (see below for a description of a permanent monitoring well).

Two soil samples are to be collected in the PCE waste storage area. One sample is to be collected near the surface since the storage is at the surface, and the shallowest soils

---

[1] Please note that state and local regulations may dictate specific analyses, or types of analyses, as well as the number and location of samples to be collected. The following discussion is for illustrative purposes only. Local rules and regulations should be reviewed prior to the design and implementation of a sampling and analysis plan.

are the ones most likely affected by a surface spill. However, because PCE travels readily through the soil column, a deeper soil sample will be collected. At this AOC, the full volatile organic compound (VOC) scan can be performed, or the analyses can be limited to PCE and its related compounds, as described in Chapter 4.

The main concern regarding the electrical transformer is PCBs, so a near-surface soil sample will be collected and analyzed for PCBs. However, the dielectric fluids inside the transformer are petroleum-based, so the soil sample also should be analyzed for total petroleum hydrocarbons—diesel-range organics (TPH-DRO).

7.2.4 Quality Assurance Project Plan

Another document required for CERCLA and RCRA CORRACTS investigations, as well as investigations performed under the aegis of various state enforcement agencies, is the *quality assurance project plan*, or QAPP (pronounced 'Kwap'). The QAPP is designed to provide a project-specific "blueprint" for obtaining the type and quality of environmental data needed for a specific decision or use. It includes quality control procedures to be used in the field, as described later in this chapter, as well as quality control procedures to be used in the laboratory and anywhere else where substandard data quality could affect the results of the site investigation. It will define the data quality objectives for the investigation and the conceptual site model. Roles and responsibilities for professional staff will be defined, including the designation of a quality assurance coordinator. The roles of the various subcontractors and vendors will be defined as well, including naming the personnel who will be responsible for quality assurance in the course of the investigation. Calibration procedures for field equipment will be provided as well as sample handling procedures and quality control procedures to be employed at the project laboratory.

7.2.5 Sample Analysis

Knowing the types of analyses to be performed on a sample is only part of the decision-making process. The investigator also has to select the type of laboratory to use (fixed base or mobile), the chemicals to be investigated, and analytical methods to be performed. The data quality objectives, with consideration to time and budgetary constraints, will influence the type of equipment used for the analyses. Some common types of laboratory analyses used in site investigations are described next.

7.2.5.1 Fixed-Base Laboratory Analysis

The U.S. Environmental Protection Agency (USEPA) publication titled "Test Methods for Evaluating Solid Waste (SW-846)" establishes quality control procedures for laboratories to follow. This document is available at http://www.epa.gov/epawaste/hazard/testmethods/sw846/online/index.htm. The Superfund program requires all laboratories to conform to the *Contract Laboratory Program* (CLP), which calls for the employment of an elaborate system of *quality assurance/quality control* (QA/QC) checks. If the data also need to comply with federal or state regulatory requirements, then nothing less than a federal or a state-certified laboratory will suffice. Each regulatory agency dictates the minimum number of quality control samples required for the laboratory analyses.

Quality control procedures employed by laboratories include the following:

- Regular equipment calibrations. Calibrations before and after a series of measurements ensures that the readings provided by the machine are within an acceptable tolerance of accuracy.

- Analysis of various *quality control blanks*. As their name implies, blanks should not contain detectable concentrations of the chemicals of interest. If a chemical of interest is detected in a QA/QC blank, then it is possible that the contamination did not originate in the sample in question, but rather from another sample from the equipment used in the laboratory, or from some other source. Such a circumstance, known as *cross-contamination*, may signify that the data collected is unreliable and must be discarded, or at least designated as "qualified data," which are data of suspect quality. Just as field blanks are important to quality control in the field (as described later in this chapter), so are laboratory blanks important to quality control in the laboratory.

- In a *spike recovery* sample, a known quantity of a chemical is added to a blank. The sample is analyzed, and the analytical result for the added chemical is compared to the quantity that was added to the blank.

- In a *matrix spike* sample, a known quantity of a chemical is added to an environmental sample that has been analyzed for a given chemical. The result of the analysis should equal the amount of the chemical in the environmental sample plus the amount of the spike.

7.2.5.2 On-Site Analysis

Mobile laboratories are preferred by environmental consultants when quick turnaround time is needed. Mobile labs, often located in a portable trailer, often have the appropriate regulatory certifications, so that the data can be used to fulfill regulatory requirements.

Generally dedicated to one project at a time, a mobile laboratory is usually more expensive to employ than a fixed-base laboratory. However, by enabling the field investigator to get rapid feedback on subsurface conditions, a mobile laboratory can more than make up for the additional expense by saving time and unnecessary drilling and sampling costs.

A *laboratory test kit*, such as the test kit shown in Figure 14.7 in Chapter 14, can provide even faster analyses of environmental samples. Within an hour or even minutes, the field investigator can obtain analytical data using a test kit. These types of analyses enable the consultant to make quick decisions in the field and adjust the conceptual site model almost in real-time. Such rapid analyses can shave days, weeks, and even months out of the project schedule, at little additional cost to the project. The degree of reliability in the data depends upon the test kit.

Sometimes the SAP is implemented without modification. Other times, it becomes a "living document," especially when a mobile laboratory or a field test kit is employed. As more knowledge is obtained regarding site conditions, site geology, and the nature of the release, the conceptual site model changes and the SAP changes with it.

7.3 Preparing to Investigate a Site

7.3.1 Health and Safety Considerations

The U.S. Occupational Safety and Health Administration (OSHA) regulates the protection of workers at hazardous waste sites under 29 CFR 1910.120. This section of the OSHA regulations is known as the Hazardous Waste Operations Emergency Response Standard, more commonly called by the acronym *HAZWOPER* (pronounced HAZ'-whopper.)

While at the beginning of a site investigation it may not yet have been established that the property is contaminated, it should be assumed that contamination is present for the purpose of worker safety. Therefore, workers conducting a site investigation should have HAZWOPER training. The HAZWOPER training is a 40-hour course whose primary intent is to protect workers from uncontrolled exposures to hazardous materials. Workers receive training in the recognition of hazardous materials, understanding the health effects from exposure to these materials, and understanding the precautions to be taken to prevent exposure.

HAZWOPER requires the preparation of a site-specific health and safety plan, or HASP, at each suspected hazardous waste site. The HASP contains the following information:

- Potential chemical, physical, and biological hazards at the site
- Lines of communication at the site
- Protective measures to be taken to protect worker health and safety
- What personal protective equipment, or PPE, each worker must use to guard against chemical hazards.

The PPE generally falls into two categories: dermal protection and respiratory protection. To protect their skin from exposure to hazardous wastes, workers will wear protective suits, usually made of Tyvek™, an inert plastic that has numerous other commercial applications. Workers will also wear protective gloves, sometimes more than one type of glove on each hand for extra protection from chemical exposure. When more dermal protection is needed, workers will wear fully-encapsulated suits, similar to the suits worn by astronauts. With no tape to accidentally peel off and expose worker's skin, these suits provide the highest level of dermal protection to the hazardous waste worker. This level of protection is known as "Level A" protection.

Respiratory protection comes in two basic varieties. An air purifying respirator (APR) filters outside air before it is inhaled. In most APRs, the air is filtered by one or two cartridges that are designed to filter out a variety of inhalation hazards, the two most common being organic vapors and particulates. A supplied air respirator (SAR) bypasses the problem of inhalation hazards by supplying certified clean air to the worker. The clean air is stored in compressed gas cylinders, which are either carried on the back of the worker or located elsewhere and connected to the respirator via flexible hosing (see Figure 7.6).

When the chemical hazards on a site are unknown, OSHA requires workers to wear what is known as "Level B" PPE, which includes full dermal protection and an SAR. Level C entails full dermal protection, and an APR if inhalation hazards are expected to be present but not severe. Level D, which involves some dermal protection and no respiratory protection, is typically worn at hazardous waste sites where the chemicals that may be

FIGURE 7.6
A worker in Level A personal protective equipment sampling the contents of a chemical drum. (From U.S. Environmental Protection Agency.)

encountered do not present an immediate hazard to workers if they come into incidental contact with the worker's skin or are present at sufficiently low concentrations to not present an inhalation hazard to the worker. Such sites make up the overwhelming majority of contaminated sites in the United States.

As a precaution, OSHA requires air monitoring of site conditions for workers wearing less than Level B PPE. This precaution is critical since the inhalation pathway typically presents the greatest threat to the safety of the hazardous waste worker.

The two most common parameters for which breathing air is monitored are organic vapors and particulates. Two types of instruments are generally used to test the levels of organic vapors in the breathable air at hazardous waste sites. The most commonly used organic vapor monitor is the *photoionization detector*, or PID, which provides one number for the total concentration of volatile detected in the air. The *flame ionization detector*, or FID, is more cumbersome than the PID, but can provide chemical-specific information to the user.

Dust meters measure the concentration of particulates in the breathable air. Of primary concern with airborne particulates is the presence of hazardous chemicals attached to the dust, making it an inhalation hazard. Other air meters, such as carbon monoxide meters, combustible gas indicators, and the like, are designed to address specific site hazards. Chapter 17 discusses these types of air monitoring equipment.

7.3.2 Utility Mark-outs

The danger posed by buried utilities, especially electric lines and natural gas lines, is a serious hazard in subsurface work. In the United States, contractors that will penetrate 2 feet or more into the subsurface are required to perform a *utility clearance*. In a utility clearance, the various public utilities that have buried equipment in the given locale are notified of the impending subsurface work. Utility personnel then locate and mark-out the buried utility lines, typically using different colored spray paint to signify different buried utilities (see Figure 7.7). There are limitations to the degree that these mark-outs can be relied upon, but woe to the contractor who does not heed the brightly colored markings.

Public utilities are required to perform mark-outs only on public property. In some cases, it may be known that, for example, one or more underground storage tanks are present, but their exact location is ambiguous. To obtain more information about buried objects and lines, when the property owner cannot provide such information, a surface geophysical survey is performed.

7.3.3 Surface Geophysical Surveys

A *surface geophysical survey* involves deploying remote sensing devices on the ground at a property to detect and discern the nature of geological and man-made features without having to actually encounter them directly (another common type of geophysical survey,

FIGURE 7.7
Utility mark-outs in a street. The markings indicate the presence of a buried 1¼″ natural gas line and a buried ¾″ water line.

the downhole geophysical survey, is discussed in Chapter 8 of this book). There are many advantages to surface geophysical surveys:

1. *They are safer than direct contact methods.* Geophysical tools can detect buried utilities on private or public property without risking a chance encounter with electric lines, gas lines, etc.

2. *They can find buried objects and other items of environmental interest.* Geophysical surveys are commonly used to locate USTs and other buried metallic objects, as well as former locations of USTs.

3. *They can provide geological information.* This information can then be used to design your site investigation.

The two most popular surface geophysical tools are the magnetometer and ground penetrating radar. *Magnetometers* are most commonly used to detect buried metal objects. They use *electromagnetic* (EM) methods, in which an electrical current generated by the geophysical equipment induces a current in the buried metallic object. The signal generated by the metallic object is detected by the magnetometer. Magnetometers can be set to different sensitivities, depending upon whether the object of interest is large or small, or buried shallow or deep. Magnetometers are also adept at detecting large changes in geology and the depth to the water table, since water is a good conductor of electricity.

Ground penetrating radar (GPR) uses radio waves to detect density variations in the subsurface (see Figure 7.8). The GPR unit sends radio waves into the subsurface where they reflect off the interfaces between two media that have different densities. The greatest density variations are present at solid-gas interfaces. When the solid structure is man-made,

FIGURE 7.8
Performing a GPR survey on the sidewalk outside of an office building.

for example, a UST, a septic tank, or some other empty or partially empty container, and is filled with liquid and/or gas, the GPR signal will reflect off their interface with the overlying soils. Unlike the magnetometer, GPR can detect USTs and other buried objects that are non-metallic. As with the magnetometer, the instrument can be adjusted to detect large, deep objects or small, shallow objects.

Electrical resistivity works on many of the same principles as EM. An electric current is sent into the earth, and signals are sent back to the machine. The strength of the signal depends upon the resistivity of the formation. This geophysical method is especially useful in locating low resistivity geologic formations, such as those that contain water (water-bearing fractures, karst topography, and other preferred groundwater pathways), and high resistivity formations that don't contain water, such as clay-rich layers. Electrical resistivity can sometimes detect a sufficiently thick layer of non-aqueous liquid, such as petroleum, which has a high electrical resistivity.

Seismic methods make use of energy waves that are sent into the formation. This geophysical method is used to assess the depth, thickness, and composition of geologic strata, the depth to groundwater, and the location and orientation of fractures.

There are two types of seismic methods: refraction and reflection. *Seismic refraction* obtains information from acoustic waves that travel between two geologic layers with differing densities. *Seismic reflection* obtains information from acoustic waves that reflect off these interfaces. These methods are used primarily to obtain information from deeper strata—they are rarely used for investigations of stratigraphy that is less than 50 feet in depth. Acoustic waves are generated by a percussive source, typically a sledge hammer or an explosive charge. An array of listening devices is set up in the area of interest to detect the refractions or reflections from the underlying geologic formations. These listening devices, known as *geophones*, record the acoustic data, which are then uploaded into a computer and interpreted using specially designed software.

7.4 Soil Sampling

There are three basic investigation methods for soils: test pits, boreholes, and soil gas surveys.

7.4.1 Test Pits

The most direct method of soil exploration is the installation of test pits using excavating equipment. Test pits provide a large, clear view of the subsurface in a short period of time. However, it is difficult to determine the depth of the items observed in the test pit, test pits damage the ground surface, and they run a higher risk encountering an underground hazard than less intrusive boreholes.

An excavator (see Figure 7.9) can remove large quantities of soil in a short period of time. Backhoes generally are smaller than excavators and have a smaller bucket than an excavator. Unlike excavators, backhoes have a shovel at the other end to push soil and grade surfaces. Trench diggers have very narrow buckets which allows them to investigate narrow areas.

FIGURE 7.9
Installation of test pits using an excavator, which has an excavating arm and a shovel to move earth and other heavy objects. (Courtesy of GZA GeoEnvironmental, Inc.)

7.4.2 Boreholes

The generally preferred method of subsurface investigation is via the installation of *boreholes* (also known as *borings*). Hand augers (see Figure 7.10) are tools used by contractors or consultants. The user turns the handle, thereby rotating the cutting bit at the bottom of the auger. The cutting bit digs into the soil, and the soil fills the hollow chamber of the cutting bit. Hand augers can dig several feet down in favorable soil conditions.

For boreholes that must penetrate more than a few feet of soil, a drilling rig is used. Drilling rigs come in three varieties: the rotary drilling rig, the direct push drilling rig, and most recently, the sonic drilling rig.

Rotary drilling rigs utilize technology originally developed for the oil industry. One type of rotary drilling rig is the *hollow stem auger rig*. It utilizes a helical screw, like a hand auger (see Figure 7.11). The "flights" of the hollow stem auger revolve in a counterclockwise manner, bringing the penetrated soils, known as soil "cuttings" or "spoils," to the surface. The flights typically are five feet long. When the top of a flight approaches the ground surface, an additional flight is attached to the drill stem and the drilling recommences. Flights will continue to be added to the drill stem until the desired depth is reached.

In the course of drilling with a hollow stem auger rig, soil samples can be retrieved using a *split spoon sampler* (see Figure 7.12). The split spoon sampler consists of a tube that splits into two halves that attach to each other longitudinally and are equipped with a drive shoe and a drive head. Split spoon samplers typically are 2 feet or 4 feet in length. Once at the desired depth, the drill rig operator will attached a split spoon to the drill string and use a drive head known as a hammer to drive the split spoon into the soils.

FIGURE 7.10
Installing a borehole using a hand auger. (Courtesy of EGS, Inc.)

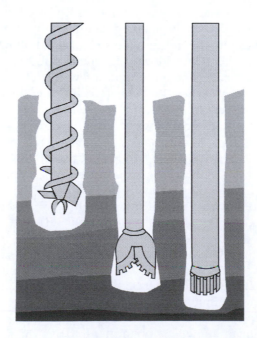

FIGURE 7.11
Different types of drilling equipment. Left: helical flights of a hollow stem auger. Center: Rotary bit for a rotary drilling rig. Right: Solid stem for a cable tool drilling rig.

FIGURE 7.12
Soils retrieved from a split spoon sampler. (Courtesy of GZA GeoEnvironmental, Inc.)

As the split spoon penetrates the soil formation, the hollow barrel fills with soil. Once retrieved, the spoon is opened so that the soil sample can be retrieved.

Drillers will typically count the number of hammer hits it takes to drive the split spoon through a 6-inch interval. The number of hits it takes to drive the length of the spoon is indicative of the competence of the underlying soils. The number of *blow counts*, in what is known as a *standard penetration test*, is used in soil engineering and foundation design to evaluate the load-bearing capacity of a soil formation.

In this chapter's example of a fictitious factory, a hollow stem auger rig would drill down to eight feet below grade at the locations of boreholes S-1, S-2, S-3, and S-4. At that depth, the driller would disconnect the hollow stem augers and attach a four-foot long split spoon sampler to the drive shaft. The rig's hammer would drive the shaft into the ground until it reaches a depth of 12 feet. For boreholes S-5 and S-7, the driller would not use the hollow stem augers at all, but rather attach a 2-foot long split spoon at the surface and hammer the split spoon sampler 2 feet into the ground.

In bedrock or hard soil formations, more horsepower is needed to penetrate the formation than can be provided by a hollow stem auger rig. For such formations, an *air rotary drilling rig* is used (see Figure 7.10). The air rotary rig sends compressed air through the hollow augers, which helps lift the soil cuttings through the flights so that they can be brought more easily to the surface. By doing so, the auger bit can concentrate more of its energy on cutting through rock and hard soil rather than grinding on material that it has already cut. One drawback of the air rotary drilling method is that its pulverizing action makes it difficult to determine the type of rock or soil being penetrated, or whether the water table has been reached.

For even more competent and deeper formations, a *mud rotary rig* often is used. The mud rotary rig sends engineered mud through the hollow augers, which helps to cool the drill bit, so that more power can be supplied to the drill stem without the risk of overheating and failing. This technology, widely used in the oil industry, has limited applications in environmental drilling, since the mud makes it difficult to obtain geologic information or information regarding the location of the water table.

The second type of drilling technology is the percussion method, in which a tool pounds the ground, gutting out a hole. A cable-tool drill rig (shown in Figure 7.10) operates by lifting and dropping a heavy steel drill "string" into the ground. This drilling method can install holes of over 1,000 feet without the use of mud, making it easier to detect groundwater.

A percussion drilling method developed especially for the environmental industry involves using a *direct push drilling rig* equipped with a solid rather than a hollow stem (see Figure 7.13). For this drilling technology, a solid stem of steel is hammered or vibrated down into the round. Generically referred to as a GeoProbe™, after the firm that invented the technology, the direct push drill rig has the advantage of creating a much smaller hole than a hollow stem auger, which is desirable for developed properties. It is also smaller and lighter than a hollow stem auger rig, so it can get into tighter places than a hollow stem auger rig (when this book was written, the smallest rig manufactured by GeoProbe™ was 62 inches [1.32 meters] tall and 23 inches [0.58 meters] wide). The GeoProbe™ rig uses clear acetate liners, which functions in a similar manner to split spoon samplers in retrieving soil samples from a given depth interval.

A third type of drilling rig is the sonic drill rig (see Figure 7.14). The sonic drill rig, like the hollow stem auger rig, utilizes rotary drilling. However, it also employs a high-frequency vibratory action which, like compressed air in an air rotary rig, fluidizes the soil cuttings so that they move out of the way of the drill head. This enables the drill head to push into the deeper formation rather than grind on materials it has already cut through. In bedrock,

FIGURE 7.13
Borehole installations using a GeoProbe™. (Courtesy of GZA GeoEnvironmental, Inc.)

FIGURE 7.14
Sonic drilling equipment. The high-power oscillator and counter rotating rollers set up harmonic standing waves along the drill string. These standing waves push the soil away from the drill string, enabling the drill bit to penetrate the soils. (Courtesy of Summit Drilling Co., Bridgewater, NJ.)

this technique induces fracturing, which weakens the rock, thus facilitating the downward movement of the drill head. Another advantage of sonic drilling is that it generates far less soil and rock cuttings than hollow stem auger or direct push drilling, thereby minimizing handling, containerization, and disposal costs for cutting wastes. Sonic drilling can be combined with the use of compressed air or drilling mud to penetrate particularly competent formations.

Site conditions often dictate the type of investigation equipment that is used. Some equipment cannot work in areas with limited access, have limited depth range, or cannot penetrate hard rock, concrete, or other known obstructions. Natural obstructions include water bodies, steep slopes, wetlands, and soft ground. Man-made obstructions include low ceilings inside buildings, narrow alleys, and the like. Heavy equipment also must avoid getting near overhead electrical lines, which could result in electrocuting the equipment operator.

7.4.3 Soil Observations and Sampling

Once the soil samples are retrieved, the field investigator will look for indicators of contamination. Is there a chemical odor emanating from the soils? Do they appear stained, or otherwise appear unnatural? After the initial inspection, the investigator will break up the soil sample at various intervals and analyze each interval separately for the presence of indicators of contamination using visual and olfactory evidence and interval readings from an organic vapor meter. The investigator notes the soil types encountered and describes them using the classification techniques discussed in Chapter 5.

Soil samples that will be sent to a laboratory for analysis are collected using sampling equipment that does not contain even trace concentrations of the chemicals of concern. Field equipment that is used once and then discarded is known as "dedicated equipment." Field equipment that is being or will be reused must be decontaminated between each sampling event.

Field indicators of contamination often provide new data to the investigator, possibly forcing a reevaluation of the conceptual site model and requiring the collection and analysis of additional soil samples or additional analyses of the same soil samples, thus modifying the SAP. Once samples are collected and the desired information is obtained from the borehole, the borehole is backfilled and abandoned, unless it will be used for other purposes.

Contamination in soils can be very heterogeneous, varying by soil type, depth, and so on. Since the sample that is sent to the laboratory for analysis should be representative of the interval sampled, soils sometimes are mixed in a decontaminated bowl so that a homogenized, presumably representative sample can be sent to the laboratory for analysis. Homogenization is never performed for VOC analyses, since a substantial portion of the VOCs will volatilize into the air during the mixing process and be missing from the sample sent to the laboratory.

Most soil sampling involves the collection of a *discrete sample*, which is a sample from a particular interval in a particular borehole. Another type of soil sample is the *composite soil sample*, which involves combining two or more discrete samples. Composite samples can be collected across depth intervals within a borehole or across boreholes. Sample compositing saves money on analytical costs, but should only be performed if the discrete samples that comprise the composite sample are expected to have similar chemical composition. Compositing soils with differing degrees of contamination can cause the investigator to miss important data regarding site contamination, or suggest that the site is more broadly contaminated than it actually is.

7.4.4 Field Quality Control

Since so much depends upon the quality of the data collected in the field, the USEPA and most states require some level of field quality control procedures.

7.4.4.1 Decontamination

Decontamination of field sampling tools prevents contaminants from one sampling location from adhering to the sampling tool and contaminating the next sample collected with that sampling tool. *Cross-contamination* can result in increased investigation and remediation costs, since it may cause the investigator to misinterpret a clean sampling location as a contaminated sampling location.

When decontamination is performed in the field, it generally follows procedures designated by the USEPA or the applicable state regulatory agency. It usually consists of at least three steps of cleaning to remove gross contamination (dirt and other visible debris), inorganic contaminants, and organic contaminants. The field sampling equipment cleaning and decontamination procedures outlined by ASTM International in its standard entitled, "Practice for Decontamination of Field Equipment Used at Nonradioactive Waste Sites" [D5088-02(2008)]:

Part 1: Remove Dirt and Other Visible Debris
- Laboratory grade glassware detergent plus tap water wash
- Generous tap water rinse
- Distilled and deionized water rinse

Part 2: Remove Inorganic Contaminants
- 10% nitric acid rinse
- Distilled and deionized water rinse

Part 3: Remove Organic Contaminants
- Acetone rinse
- Total air dry or pure nitrogen blow out
- Distilled and deionized water rinse

7.4.4.2 Field Quality Control Samples

To test the quality of the field data, several quality control procedures typically are performed. One type of quality control sample, known as the *trip blank*, consists of laboratory-prepared, chemical-free water stored in an inert glass jar. It is designed to detect whether cross-contamination has occurred during the storage and transport of the samples during their trip from the field to the laboratory. Trip blanks are utilized only for aqueous samples being analyzed for VOCs. The trip blank bottle travels from the laboratory with the empty sample jars and returns unopened to the laboratory with the jars now filled with environmental samples. Like the accompanying samples, trip blanks are analyzed for VOCs.

If the trip blank is contaminated, the contamination could only have occurred by VOCs being released from an environmental sample or from an unrelated source of VOCs, with the water in the trip blank absorbing the fugitive VOCs. If the contaminant present in the trip blank is also present in one or more of the field samples, then its origins are suspect, and the data quality is considered suspect as well. Such contamination is usually marked as "qualified," and this data degradation is taken into account when the laboratory data are interpreted.

Another type of blank sample, known as a *field blank*, is designed to test for contamination emanating from the sampling equipment. Laboratory-prepared, chemical-free water is poured over one of the sampling tools and into an inert glass jar. If there are contaminants on the sampling tool, they will be rinsed by the water into the glass jar, where they will be detected in the laboratory analysis. If field decontamination is performed, then the field blank should be performed over a piece of field-decontaminated equipment. It is not unusual for the field blank to contain nitric acid, acetone, or some other solvent used in the decontamination process. The presence of contaminants in the field blank will also result in a diminution of the data quality, and affect the interpretation of the laboratory data.

Another method of field quality control is the *field duplicate*. In the field, sampled material is placed into two different jars. The duplicate sample is a single "blind" sample in that the field investigator knows which sample is being duplicated but the laboratory does not. Assuming the sample is identical of the original, i.e., the two samples are homogeneous, the analytical results should be identical, within the tolerances of statistical variation. If the analytical results for the blind duplicate vary significantly from the duplicated field sample and sample heterogeneity can be ruled out as the cause, it calls into doubt the accuracy of the laboratory analyses.

7.4.5 Sample Handling Procedures

When environmental samples are sent to an analytical laboratory, they are accompanied by a completed *chain of custody*. The chain of custody provides basic information about the site and the sampling personnel, the types of samples collected, the sample designations, when they were collected, and the analyses to be performed on the samples. It gets its name, however, from the signatures at the bottom of the form. As the samples are handed from the sampler to the laboratory, or to any intermediary, such as the transporter, it is signed by both the person in possession of the samples and the sample recipient. At any time, the last person to sign is deemed to have legal custody of the samples, and the series of signatures forms a chain, which documents that the samples were never left alone where they could be tampered with or otherwise cross-contaminated.

7.5 Soil Gas Investigations

A *soil gas survey* involves the collection and analysis of vapor from the vadose zone. Since the composition of the gas in the vadose zone is influenced by chemicals in the soils and possibly in the underlying saturated zone, soil gas surveys can be used to provide information on the presence, composition, source, and distribution of contaminants in the subsurface.

There are two basic types of soil gas surveys. In an *active soil gas survey*, a hollow probe equipped with a slotted screen at the bottom is inserted several feet into the ground (see Figure 7.15). A pump, attached to the probe by plastic tubing, applies a vacuum to the probe. The change in the partial pressure in the vadose zone results in volatile contaminants going into the gaseous phase. This gas enters the probe through a slotted screen and then is sucked up to the surface and collected in a container. The tip used to drive the

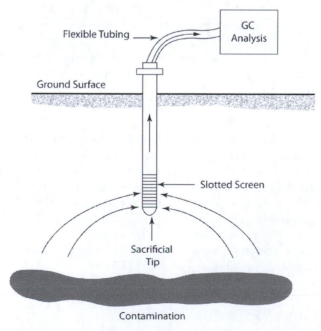

FIGURE 7.15
A schematic diagram of a soil gas probe used in an active soil gas survey.

probe into the ground remains in the ground (is "sacrificed") while the rest of the probe materials are retrieved from the ground and discarded. Because of their reliance on contaminants readily going into the gaseous phase, soil gas surveys are most useful at detecting VOCs, especially VOCs that have high vapor pressures and high Henry's law constants.

Depending on access issues and site geology, a drill rig can install as many as 30 soil gas probes in a day, enabling the rapid collection of subsurface data. Soil gas samples can be analyzed by a fixed-base laboratory, or in the field in a mobile laboratory so that the real-time data can be used to select the next sampling locations. By varying the depth of the soil gas probes, active soil gas surveys can also provide vertical profiling of the concentration of contaminants in the vadose zone.

Active soil gas surveys are only effective in permeable soils, where soil gas can readily migrate through the pore spaces to the gas probes. Clay lenses or clay formations will adversely impact the usefulness of the active soil gas survey. Soil gas surveys also cannot be performed in saturated or near-saturated soils, since soil moisture decreases permeability by blocking vapor flow through the pore space of the soils.

In a *passive soil gas survey*, a sorbent material is placed in the ground and left for an extended period of time (several days to several weeks) to absorb vapors that are migrating to the surface. This investigation method is most useful for investigations involving semi-volatile organic compounds (SVOCs), or when soils prevent sufficient air flow for active sampling. After the sorbent probe is removed from the ground, it is sent to a laboratory where the contaminants are desorbed and analyzed.

Passive soil gas sampling is more effective than active soil gas sampling in getting representative readings in soil with poor permeability or high moisture content. Installing multiple devices at varying depths enables the investigator to get an idea of the vertical profile of the contamination in the area. However, because of the time involved in the field data collection step, passive soil gas surveying cannot be used for rapid site characterization. In addition, the data can be adversely affected by time-related variations unrelated to contaminant distribution.

Both active and passive soil gas surveys can be affected by the presence of preferential pathways in the subsurface. Such pathways, such as utility trenches, highly permeable backfill, or soil cracks, can affect the movement of soil gas and thereby provide misleading information regarding contaminant distribution in the subsurface.

7.6 Groundwater Investigations

As with soil sampling, decisions on the method of collecting groundwater samples balance schedule, budget, and data quality, as described below.

7.6.1 Permanent Monitoring Wells

7.6.1.1 Monitoring Well Installation

The oldest and most common method for collecting a groundwater sample is via the installation of a permanent *monitoring well*. A monitoring well is installed in a borehole that extends below the water table and into the saturated zone. As shown on Figure 7.16,

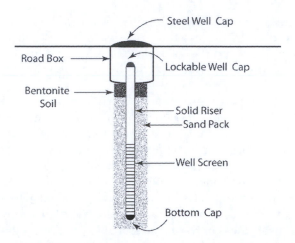

FIGURE 7.16
Diagram of a permanent monitoring well.

the bottom of the well contains a *well screen*, usually composed of steel or polyvinyl chloride (PVC), a hard, durable and, quite importantly, inert plastic (that is, it will not react with the chemicals of concern). For wells designed to test the top of the water table, the well screen bridges the water table, enabling formation water to enter the wellbore. Because, for various reasons, groundwater elevations tend to fluctuate by at least a few feet in many parts of the country, including seasonal variations, diurnal variations due to tidal influences, variations in precipitation, and other factors, the screen should be situated so that the water table will be in contact with the well screen despite the variations in the depth of the water table. Doing so prevents floating LNAPL from rising above the screen, where it can go unnoticed, during periods of high groundwater levels.

The remainder of the monitoring well is constructed so that the groundwater sample collected from that well derives only from the screened interval and nowhere else in the wellbore. A solid plug is attached to the bottom of the screen to prevent infiltration of solids at the bottom of the borehole, and a solid riser is attached to the screen to prevent any unwanted solids or liquids above the screened zone from infiltrating into the screened zone. Surrounding the riser/screen assembly is a sand pack which is designed to capture soil grains and other suspended particles. The sand pack is topped with an impermeable material, usually concrete or *bentonite*, a specially designed clay that swells when it comes in contact with water, thus forming a solid seal to prevent surface infiltration of water and other liquids down the borehole to the screened interval. The top of the well is secured with a cap to prevent unwanted surface infiltration and discourage tampering with the well. When the top of the well is at ground surface, it is known as a flush mount. When the top of the well rises above the ground, it is commonly known as a "stick-up" mount.

7.6.1.2 Monitoring Well Development

Once the well materials have been installed, groundwater is pumped from the well until the discharge is free of sediment. This process, known as *well development*, removes residual suspended solids from the wellbore and the surrounding sand pack so that only clear water remains. Removal of the solids not only provides a purer groundwater sample but avoids false positive readings caused by chemicals entrained on the suspended solids.

7.6.1.3 Monitoring Well Sampling

After the well has been developed and the well materials have settled into place, the groundwater in the well can be sampled. After the well is opened, the field technician uses an organic vapor monitor to detect vapors in the groundwater, the first step in looking for field evidence of contamination in the groundwater. Next, the depth to the water table is measured using a *water level indicator*. This measurement will be used to determine the proper length of tubing to lower into the wellbore. (Water level information also is useful data to the project geologist, as discussed in greater detail in Chapter 8.) The water level indicator is usually a flexible tape with a sensor at the bottom that triggers a sound or lights a bulb when water is encountered. A *water/oil level indicator* can also measure the top of non-aqueous phase petroleum, also known as LNAPL, in a well.

Next, the well is evacuated. A pump that is connected to a battery or AC electrical source via flexible tubing is lowered into the groundwater to pump the standing water to the surface, where it is either discharged to the ground with or without treatment or containerized for future disposal if contaminated. This *well purging* removes the standing water in the well, which has interacted with air from outside the geologic formation and thus may have lost some of its volatile constituents. The purging also allows fresh formation water to flow into the well.

There are two types of purging methods. Conventional well purging methods involves mass removal of three to five well volumes to ensure the complete removal of standing water and the sampling of formation water. Low-flow purging methods involve the removal of standing water from a small portion of the screened interval. This purging method has the advantage of generating less purge water that may need on-site treatment or off-site disposal if it is contaminated.

During well purging, the field technician collects water quality measurements (see Figure 7.17). These measurements are used to confirm that formation water is being collected and to obtain information about the groundwater, which can assist in subsequent investigation and remediation planning. The field technician collects the water quality measurements as the purge water exits the tubing, rather than in the laboratory, since they can change while the samples are being containerized and transported. These field measurements are also used to determine when formation water has entered the borehole, since formation water is expected to have a more consistent and predictable chemistry than standing water in the borehole. The measurements typically include the following:

- **Temperature.** Formation water that hasn't been exposed to the air and that is sufficiently deep to be uninfluenced by the air tends to remain at a constant temperature.
- **pH.** pH is a measure of hydrogen ion concentration. It is also referred to as corrosivity.
- **Dissolved oxygen.** Water that has been exposed to air has absorbed oxygen from the air, whereas formation water that has not been exposed to air is often oxygen deficient.

FIGURE 7.17
Collecting water quality measurements during groundwater sampling. A water level indicator is to the left of the field technician. (Courtesy of GZA GeoEnvironmental, Inc.)

- **Turbidity.** Turbidity measures the quantity of suspended particles in the water. See Chapter 4 for a discussion on turbidity. Fewer suspended particles mean a better opportunity to test the groundwater itself. It is recorded as nephelometric turbidity units (NTUs).

- **Conductivity.** This refers to electrical conductivity rather than hydraulic conductivity. As discussed in Chapter 4, electrical conductivity is a function of the concentration of ions in the groundwater and is related to salinity. It usually is measured in micromhos per centimeter ($u$mhos/cm).

- **Oxidation-reduction potential (ORP).** ORP (also known as E_h) is the measure of the ability of a chemical to either donate electrons (oxidation) or acquire electrons (reduction) when dissolved in water. It is measured in volts (V) or millivolts (mV).

Collection of water quality data is particularly critical for low-flow purging. As its name implies, low-flow purging employs a pump operating at a much lower flow rate than so-called "conventional" well purging. Since one of the objectives of low-flow purging is to generate less waste water, the method lacks the luxury of conventional purging, in which the removal of three to five volumes of water from the borehole essentially guarantees that water from the geologic formation will be sampled. Rather, with low-flow purging, an assumption is made that the formation water is homogeneous. Therefore, once the water quality parameters stabilize, it is assumed that homogeneous formation water rather than heterogeneous standing water has been accessed and can be collected for laboratory analysis. Once the groundwater sample is collected (see Figure 7.18), it is containerized and shipped in the manner described above for soil samples.

FIGURE 7.18
Collecting a groundwater sample using a plastic bailer. (Courtesy of GZA GeoEnvironmental, Inc.)

Another method used to remove suspended particles from a groundwater sample is *field filtering*. Performed in conjunction with either conventional or low-flow purging, field filtering involves passing the groundwater sample through a micro-screen, which captures the suspended particles before they can enter the sampling jar. The disadvantage of field filtering is that chemicals can volatilize due to the extra handling and the turbulence of the water on the micro-screen. Therefore, field filtering should never be performed when sampling for VOCs, and is prohibited for VOC sampling in many states for that reason.

There also is a no-purge option for well sampling. The most common method of no-purge well sampling is with the use of *passive diffusion bags* (PDBs). The PDBs are placed in a wellbore using a string that remains at the ground surface for later retrieval. As groundwater flows through the wellbore, it diffuses through the permeable exterior membrane of the PDB. Eventually the groundwater contaminant concentrations inside the PDB will reach equilibrium with those outside of the PDB. The PDBs are generally utilized for VOC analysis, although some commercially-available PDBs allow for the collection and analysis of other parameters.

The main advantages of PDBs include the lack of generation of waste water that can result in extra costs and the flexibility in selecting one or more depth intervals to sample. In Figure 7.19, two PDBs are deployed in a monitoring well, enabling the field technician to collect two samples from the same borehole and the same time.

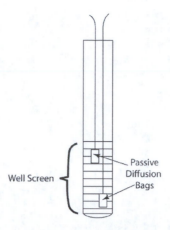

FIGURE 7.19
Positioning of two passive diffusion bags in a monitoring well.

7.6.2 Temporary Well Points

Many site investigations use *temporary well points* rather than permanent monitoring wells to test groundwater conditions at a site. As the name implies, well materials, at a minimum a screen/riser assembly, are placed into the borehole, but removed after groundwater samples are collected rather than set into place. A groundwater sample is collected through the temporary screen, after which the well materials are removed from the borehole.

Temporary well points are less expensive than permanent monitoring wells and can be installed and sampled more quickly. However, temporary well points have several disadvantages. They can only be used within a narrow time frame, typically once, whereas there is no theoretical limit to the number of times permanent monitoring wells can be used. It is more difficult to keep unwanted sediment and other particulates from entering the temporary wellbore than the permanent wellbore that has been constructed with a sand pack and then developed. This factor is especially important for contaminants that tend to sorb to solids, such as SVOCs and metals.

Lastly, as described in Chapter 8, they don't provide accurate depth to water measurements and cannot be used for determining groundwater flow direction in most circumstances. These shortcomings notwithstanding, temporary well points are indispensable for rapid site characterizations.

7.7 Interpreting and Documenting the Results of the Sampling and Analysis

7.7.1 Boring Logs

Field observations of a soil boring are recorded on a *boring log*, also called a *borehole log*. The header of the boring log shown in Figure 7.20 contains information about the date, site name, driller, borehole location, observing geologist, and drilling method. The subsurface strata are described in detail, as are any field indicators of contamination. For each 2 foot sampling interval, the borehole log notes the amount of soil recovery. For instance, for boring S-1, 12 inches of soil were recovered from the 6–8 foot interval. In the next

| ABC CONSULTANTS | | | | 1313 Mockingbird Lane | | BORING NO. | S-1 |
| PROJECT: TOOL & DIE MFG., INC. | | | | Dogpatch, USA | | SHEET | 1 of 1 |

DRILLING CO.	Munster Drilling	BORING LOCATION	South of UST	
FOREMAN	H. Munster	GROUND ELEV.	242 ft.	DATUM None
GEOLOGIST	T. Howell, III	DATE START	4/14/12	END 4/14/12

Drilling method: Hollow-stem auger, split-spoon sampling

Depth (ft)	SAMPLE			Sample Description	Stratum Description	PID Reading	Field Testing
	Split Spoon #	Penetration/ Recovery	Blows /6 "				
	1	24/24	13	Light tan, sandy silt	SILT	0.0	
			11				
		24/24	12				
4			10			0.0	
	2	24/0	8				
			10			0.0	
		24/12	15				
8			6			0.0	
	3	24/24	9	Brown medium- grained Sand	SAND	0.0	S-1
			6				
		24/24	7				
12			5			0.0	

FIGURE 7.20
Example of a boring log.

column, blow counts are recorded in 6-inch intervals. The soils are described in the next column, and a one-word description of the most prominent soil type is provided in the next column. The two columns to the right of the log record PID readings for the sampled intervals. A well-prepared boring log will supply a snapshot of the salient environmental and geological conditions at the location of the boring. Test pit logs are prepared if test pits were installed as part of the site investigation.

7.7.2 Data Reduction and Interpretation

Table 7.4 is an example of an analytical results table for the soil samples collected at our fictitious manufacturing facility. The table is simplified for illustrative purposes. General information provided in the table includes: the sample name, the collection date, the sampling interval, and analytical results for each contaminant class tested. "NS" means that the sample was not analyzed for that parameter. The acronym "ND" means that the laboratory equipment did not detect the chemical. The analytical results are compared to generic remediation standards designated by the regulatory authority. Remediation standards are discussed in detail in Chapter 8.

In this data set, soil sample S-3 contained TPH-DRO at a concentration above the designated remediation standard. PCE and trichloroethylene (TCE), which are related compounds, were the only two VOCs detected. Soil sample S-5 contained PCE at a concentration below its remediation standard. However, the deeper sample S-6 contained PCE at a concentration above its remediation standard, and some TCE as well. Neither S-5 or S-6,

TABLE 7.4

1313 Mockingbird Lane, Anytown, USA, Analytical Results Summary for Soils

Sample Name	Remediation Standard	S-1	S-2	S-3	S-4	S-5	S-6	S-7
Date Sampled		7/1/18	7/1/18	7/1/18	7/1/18	7/1/18	7/1/18	7/1/18
Sample Depth		8'–10'	8'–10'	8'–10'	8'–10'	0'–2'	10'–12'	0'–2'
TPH-DRO	5000	ND	50	6200	ND	NS	NS	150
VOCs								
Perchloroethene	1.0	NS	NS	NS	NS	0.2	4.2	NS
Trichloroethene	1.0	NS	NS	NS	NS	ND	0.7	NS
PCBs								
Aroclor-1242	2.0	NS	NS	NS	NS	NS	NS	0.49
Aroclor-1248	2.0	NS	NS	NS	NS	NS	NS	0.70

Note: All concentrations are listed in milligrams per kilogram (mg/kg).
ND = Not Detected.
NS = Not Sampled.

the soil samples collected around the transformer (refer to Figure 7.4), contained PCBs at a concentration above their remediation standards. If the investigator believes that sampling in this area of concern was adequate, then the investigator will conclude that no further investigation is warranted for this area of concern.

7.7.3 Site Investigation Report

The site investigation report documents the results of the investigation, and provides recommendations, if any, for further investigation. It contains the basic information provided in the sampling and analysis plan, as well as a technical review section.

The technical review section is the heart of the site investigation report. This section describes the field activities performed, the field observations, and the results of the sampling and analyses. It summarizes the field observations that are documented on the boring logs, which are attached as an appendix to the report. The report includes a scaled map that shows the sampling locations and analytical results tables for each medium sampled, such as Table 7.3, above. Any limitations in the investigation (obstructions, field equipment failures, etc.) are identified.

Once the results of the field activities, the field sampling, and the laboratory analyses are described, they are summarized in the findings section. For each area of concern, this section identifies the media affected, the extent of contamination (both horizontal and vertical), and the contaminants encountered. To the extent possible given the scope of the field program, the suspected source of the discharge is identified, and the potential impact of the limitations encountered on the results of the investigation is evaluated.

Ideally, the data collected for the site investigation will support the conceptual site model. If data do not fit the conceptual site model, and there's not a sound technical reason for discounting the anomalous data, then the conceptual site model must be modified and the revised conceptual site model presented in the site investigation report. This process of continual evaluation of the conceptual site model continues throughout the remedial investigation (Chapter 8) and the remedial action (Chapter 9) phases of the project.

7.8 Site Investigations and Environmental Consultants

Environmental consultants perform a variety of functions during site investigations. They supply the brains and a lot of the brawn. In the office, they develop the conceptual site model and sampling and analysis plan and do the research that is used in both documents. In the field, environmental consultants are field technicians, field geologists, and field supervisors. They may use hand augers to collect shallow soil samples or operate test kits or even mobile laboratories. Environmental consultants usually operate the equipment needed to sample temporary well points or permanent monitoring wells. Contractors generally operate the heavy equipment, such as drilling rigs, excavators, and backhoes. If a fixed-base laboratory is utilized, then the technicians at the laboratory will analyze the samples collected in the field.

Problems and Exercises

1. Review the physical properties of a volatile organic compound on the Target Compound List (TCL), a semi-volatile organic compound on the TCL, and a metal on the Target Analyte List (not including the compounds discussed in Section 7.2.2). Describe their expected fate and transport through the vadose zone after being released at ground surface.

2. Can cross-contamination create false positive readings for a chemical? False negative readings? Explain your reasoning.

3. During the installation of a borehole around an underground storage tank (UST) containing #2 fuel oil, black-stained soils were encountered at 4 feet below grade. The bottom of the UST is located at 6.5 feet below grade. Discuss the implications of this finding to the presence of a spill at the facility and the integrity of the UST.

4. During well purging, turbidity is consistently measured above acceptable levels. Discuss possible reasons for this problem and ways that it can be mitigated.

Bibliography

Arkansas Department of Environmental Quality, June 2007. *Arkansas Brownfields Program User's Guide*.

ASTM International, 1993. Standard guide for soil gas monitoring in the vadose zone, D5314-93, Annual Book of ASTM Standards, Philadelphia, PA.

ASTM International, 2004. D6914-04: Standard Practice for Sonic Drilling for Site Characterization and the Installation of Subsurface Monitoring Devices

ASTM International, 2008. D5088-02(2008): Practice for Decontamination of Field Equipment Used at Nonradioactive Waste Sites. ASTM International.

ASTM International, 2011. E1903-11: Standard Guide for Environmental Site Assessments: Phase II Environmental Site Assessment Process.

Benson, R., Glaccum, R.A., and Noel, M.R., 1984. Geophysical techniques for sensing buried wastes and waste migration (NTIS PB84-198449). National Ground Water Association. 236 p.

Driscoll, F.G., 1986. *Groundwater and Wells*, 2nd ed. Johnson Screens.

Interstate Technology & Regulatory Council, Technology, March 2006. Overview of Passive Sampler Technologies.

Kerfoot, H.B. and Barrows, L.J., 1987. Soil gas measurements for detection of subsurface organic contaminants, EPA/600/2-87/027 (NTIS PB87-174884).

Lagrega, M.D., Buckingham, P.L., and Evans, J.E., 2001. *Hazardous Waste Management*, 2nd ed. McGraw-Hill.

New Jersey Department of Environmental Protection, August 2005. *Field Sampling Procedures Manual*. NJDEP.

Occupational Safety and Health Administration web site (www.osha.gov).

U.S. Environmental Protection Agency, August 1987. A Compendium of Superfund Field Operations Methods. EPA 540/P-87/001. Office of Emergency and Remedial Response.

U.S. Environmental Protection Agency, July 21, 2005. Federal Facilities Remedial Site Inspection Summary Guide. Office of Enforcement and Compliance Assurance, Federal Facilities Enforcement Office.

U.S. Environmental Protection Agency, March 1997. Expedited Site Assessment Tools for Underground Storage Tank Sites: A Guide for Regulators. USEPA Office of Underground Storage Tanks, OSWER.

U.S. Environmental Protection Agency, March 2001 (Reissued May 2006). EPA Requirements for Quality Assurance Project Plans, EPA QA/R-5EPA/240/B-01/003.

U.S. Environmental Protection Agency, May 1989. Interim Final RCRA Facility Investigation (RFI) Guidance, Volume I of IV. Development of an RFI Work Plan and General Consideration for RCRA Facility Investigations. EPA 530/SW-89-031. Waste Management Division, Office of Solid Waste.

U.S. Environmental Protection Agency, October 1986. PB87-107769—RCRA Facility Assessment Guidance.

U.S. Small Business Administration, Office of Financial Assistance, October 2010. SOP 50 10 5(C) Lender and Development Company Loan Programs.

8

Remedial Investigations and Remedial Design

Once contamination has been detected on a property, its severity and extent must be determined, with the ultimate goal being its remediation, which is discussed in Chapter 9. The severity and extent of the contamination is evaluated in what typically is known as a *remedial investigation*. In the Resource Conservation and Recovery Act (RCRA) program, it is known as a RCRA facility investigation, or RFI.

Remedial investigation entails the delineation and characterization of contamination in soils, groundwater, sediment, and surface water, as warranted by site conditions. It is performed prior to or concurrently with a *feasibility study* from which a *remedial action work plan* (known as a *corrective measures study* under RCRA) is prepared. This chapter discusses the remedial investigation of contaminated soils and contaminated groundwater and the steps leading to the preparation of the remedial action work plan.

8.1 Remedial Investigation of Soils

8.1.1 Delineation of Soil Contamination

There are several ways to delineate and interpret the horizontal and vertical extent of contaminated soils once contamination has been identified. All of the methods entail soil sampling and analysis, as described in Chapter 7.

8.1.1.1 Single-Point Compliance

The most straightforward way to delineate the extent of soil contamination is to surround, both horizontally and vertically, the known area of contamination with soil samples, which then are analyzed for the contaminants of concern. Having identified known or suspected migration pathways for contaminants in the conceptual site model, the consultant may choose to bias the sampling in a particular direction. The distance between sampling locations will depend upon the nature of the contamination, the distance the contaminants is suspected to have migrated from their point of origin, and the desired degree of certainty in the data.

If the delineating soil sample is contaminated, then additional samples will be collected and analyzed until the area of contamination is surrounded by uncontaminated soil samples, both horizontally and vertically. This method of soil delineation is known as *single-point compliance*.

Continuing the case example from the previous chapter, the map provided in Figure 8.1 shows the locations of boreholes designed to delineate the perchloroethene (PCE) contamination detected in soil sample S-6, which contained PCE at a concentration of 4.2 milligrams per kilogram (mg/kg), with the remediation standard being 1.0 mg/kg. Boreholes

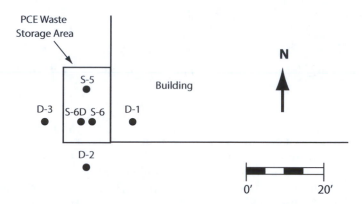

FIGURE 8.1
A map showing the planned delineation of documented soil contamination at a facility.

D-1 through D-3 are designed to delineate horizontally the PCE contamination around soil sample S-6. Borehole S-6D, drilled at roughly the same location as borehole S-6, is designed to provide vertical delineation. The soil samples should be analyzed using an analytical method for volatile organic compounds (VOCs) that includes PCE as well as its related compounds, trichloroethene (TCE), the isomers of dichloroethene (DCE), and vinyl chloride.

8.1.1.2 Identifying a Concentration Gradient

Contamination can be delineated if a *concentration gradient* can be identified. A concentration gradient is a predictable change in concentration over a given horizontal or vertical distance. If successive soil samples exhibit a decreasing concentration trend for a particular contaminant, and this trend is consistent with the conceptual site model, then the horizontal and vertical location at which the soils will comply with remediation standards, known as the *compliance point*, can be estimated. Figure 8.2 shows the delineation of soil

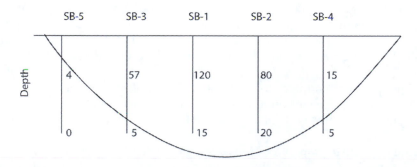

FIGURE 8.2
Using concentration gradients to determine the horizontal and vertical extent of soil contamination. In this diagram, the compliance goal for the selected contaminant is 5 mg/kg.

contamination using a concentration gradient to calculate the compliance points, both horizontally and vertically. The concentration gradient can be used to predict the compliance point, which then can be verified by additional soil sampling.

8.1.1.3 Compliance Averaging

Compliance averaging is another method for delineating the extent of soil contamination. In compliance averaging, the analytical results for a specific chemical in a specific area of concern (and in some cases, in a specific timeframe) are treated as a data set. The data set is then subject to statistical analysis, from which the arithmetic mean and the variance of the data are calculated, and possibly a spatially weighted average of the data as well. The statistical mean or weighted average then is compared with the compliance goal to assess whether the contamination is an acceptable risk to human health and the environment.

ProUCL software, developed by the U.S. Environmental Protection Agency (USEPA), is a software package designed to provide a rigorous statistical analysis of a data set. Among the data generated are the upper confidence limit (UCL), which is the upper limit of the mean or whatever parameter is being measured. The industry standard is the 95% UCL, which means that there is 95% confidence that the parameter, that is, the concentration, does not exceed the benchmark which is the cleanup standard for environmental investigations. ProUCL software also can be used to assess whether the identified contamination represents background conditions, which means that the contamination is due to static site conditions rather than a previous or ongoing spill.

Another, simpler statistical method is the "75%/10x" rule. This rule states that 75% of the data must not exceed the cleanup standard and no one data point can be ten times the cleanup standard for the data set to comply with the cleanup standard. If either of those conditions exists, then the data set does not meet the selected criteria and a cleanup is warranted. The consultant has the option of breaking the data set into compliant and noncompliant parts in an effort to establish compliance for at least part of the data, or collecting additional data that may lead to compliance for the overall data set. Please note that this method assumes that the data collection is unbiased. If additional samples are biased towards uncontaminated or lesser contaminated areas, then arithmetic means are invalid, and a weighted average of the data or some other weighting mechanism would be needed to interpret the biased data set.

Other statistical methods for interpreting data can be employed during the remedial investigation phase of the project.

8.1.2 Field Screening during Soil Delineation

Field screening techniques can provide a quick assessment of the presence of contamination, especially when VOCs are the target of the investigation. For instance, if organic vapor monitor measurements suggest that a soil sample contains a high concentration of organic vapors, the field geologist may choose not to send that sample to a laboratory for analysis. Since the goal is to identify uncontaminated soils, the field geologist can assume that this sample is contaminated, and select new delineation sample locations further from the known zone of contamination, horizontally or vertically.

Field analytical equipment, such as mobile laboratories and laboratory test kits, can provide reliable data in a short period of time, leading to a rapid characterization of the area of contamination. When such methods are not used, it can be useful to collect additional soil samples beyond the first group of samples, to be analyzed by the laboratory only if one of the samples analyzed in the first group of samples is contaminated.

8.1.3 Obtaining Quantitative Soil and Bedrock Data

For many remedial methods, quantitative physical data are needed for an effective remedial design. Some of the methods used to collect physical data are described below.

8.1.3.1 Soil and Bedrock Cores

Collecting direct, high quality physical data for soils requires analysis of a soil sample undisturbed by the drilling process. A *Shelby tube* is a hollow chamber that is driven into the soil formation by a drilling rig. As it is pushed down by the drill rod, the tube fills with soil. Once filled, it is capped at the bottom and sent to a laboratory for analysis. The analysis can provide information about the percentage of each soil type and other soil parameters that can be used to understand contaminant distribution, fate and transport in the subsurface.

Shelby tubes and other types of soil coring also can be used to locate bedrock fractures and estimate their ability to transmit groundwater. While drilling a borehole through bedrock, core samples can be collected. In unfractured bedrock, the core can be retrieved largely intact. If there are bedrock fractures, broken pieces of rock will be retrieved, with the degree of fracturing indicated by the degree to which the core sample is broken. The percentage of competent bedrock to fractured bedrock is described in the *rock quality designation* (RQD). The RQD provides direct information about the presence and quality of fractures in the rock formation. An RQD of 75% of more shows good-quality hard rock; an RQD of less than 50% indicates low-quality weathered rock. The larger the fracture, the greater its ability to transmit water and dissolved contaminants.

8.1.3.2 Borehole Logging

Borehole logging is an indirect, geophysical measuring method that also can provide quantitative soil and bedrock data. It is similar in concept to a surface geophysical surveying (see Chapter 7), except that in borehole logging the measurement devices are lowered into a borehole rather than moved across the ground surface. The measurement device generally is attached by electric wire to a recording device, usually a field computer, located at the surface. In some cases, the tool itself stores the information for eventual uploading onto a computer once it has been retrieved from the borehole.

There are many common tools used in borehole logging. The *electrical log* measures the electrical properties of a geological formation. Most electrical logs measure electrical resistivity (*resistivity logs* perform the same function using different equipment). They can identify saturated vs. unsaturated zones, and the types of soils or bedrock present. Since water conducts electricity, dry formations have higher resistivity than saturated formations, and clean water has a higher resistivity than salt water. Because solids conduct electricity better than gas, sand, with more porosity, tends to have higher resistivity than clay, and sandstone tends to have higher resistivity than shale.

The *natural-gamma ray log* measures naturally occurring radioactive material, including potassium-40, uranium-235, uranium-238, and thorium-232. Since potassium is a major component of most clay minerals, gamma logs (as they're usually called) are useful in identifying clay formations. Together, the gamma log and the electric log form the common *e-log suite*.

Many of the downhole geophysical tools are geared towards identifying bedrock fractures and their properties. The *caliper log* uses mechanical arms to measure the diameter of the borehole. An increase in borehole diameter within a zone is often indicative of the presence of bedrock fractures or bedding planes. *Downhole imaging logs* include optical televiewers, in which a camera collects real-time images of the borehole, and acoustic televiewers, which produce high-resolution images of the borehole walls using high frequency sound waves. Downhole imaging logs can identify bedrock fractures or bedding planes and also provide their three-dimensional orientation, which is critical in the understanding of contaminant transport in complex fractured bedrock formations.

Downhole geophysical tools can be used to obtain information regarding the physical properties of the groundwater within fracture zones as well as the physical properties of the rock itself. A *flowmeter* measures water flow velocity between two zones in a non-pumping well. This information can be used in formulating the groundwater flow pattern within a bedrock aquifer. A *temperature log* measures the temperature of the water and can be used to identify water-bearing fractures, establish the geothermal gradient at a location, and correlate between fractures at different well locations.

8.2 Remedial Investigation of Groundwater

Delineating groundwater contamination is similar to delineating soil contamination, with one important distinction. Contaminants move and transform to a far greater degree when dissolved in groundwater than when in unsaturated soil (see Chapter 7). Some techniques used in the delineation of contaminant plumes in groundwater are described below.

8.2.1 Calculating Groundwater Elevation

Since groundwater movement is the primary transport mechanism for dissolved contaminants, understanding groundwater flow patterns is critical to delineate groundwater contamination. The fundamental measurement in determining groundwater flow direction is *groundwater elevation*, also known as the "static water level," which is the elevation above sea level of the top of the water table.

Groundwater elevation can be measured from permanent or temporary monitoring wells or through the installation and gauging of a *piezometer*, which is a borehole whose primary purpose is to measure groundwater levels. Monitoring wells can be used both as piezometers and as groundwater sampling points. In this chapter, the term "piezometer" applies to all points used for the measurement of groundwater elevation. The groundwater level in a piezometer is gauged using a water level indicator, which is described in Chapter 7.

The depth to water in the well can be converted into elevation above sea level if the elevation of the point where the depth to water was measured is known. The United States Geological Survey (USGS) manages a network of data points that cover the United States and tie into data points in Canada, Mexico, and Central America. Datum points are given in feet above mean sea level (MSL). A surveyor will establish the elevation of the measuring point in the piezometer by comparing that elevation to the nearest datum point.

8.2.2 Calculating Groundwater Flow Direction

The water table derives its name because conceptually its top is a flat surface like the top of a table. It differs from a table in that, for purposes of calculation, it extends an infinite distance in all directions and can have undulations and irregularities. For groundwater to flow, the table must be tilted in one direction. As geometry teaches us, three points define a plane. Therefore, three groundwater elevation points are needed to define the water table.

Figure 8.3 shows the location of monitoring well MW-1, as well as new monitoring wells MW-2 and MW-3. Well MW-2 was installed in the presumed down-gradient direction and well MW-3 was installed in the presumed cross-gradient direction, thereby forming a triangle with which groundwater flow direction could be calculated.

Groundwater elevations were collected from MW-1, MW-2, and MW-3, and Table 8.1 shows the calculations used to derive their groundwater elevations. The groundwater elevations then are plotted on a map and used to calculate groundwater flow direction. Groundwater flow direction is calculated by using triangulation to plot *groundwater contour lines*. Groundwater contour lines, analogous to contour lines on a topographic map, connect locations of equal elevation of the top of the water table (Figure 8.4).

Since the water table was encountered at 75.00′ in MW-1 and at 73.00′ in MW-2, there must be a point between these two piezometers at which the water table elevation is 74.00′ (above MSL is implicit). There are also points between MW-1 and MW-3 at which the water table elevation is 74.00′ and 73.00′. The location of these points can be found by

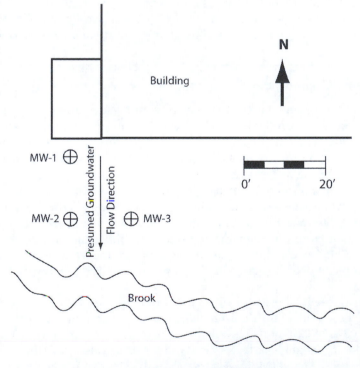

FIGURE 8.3
Groundwater delineation wells MW-2 and MW-3 were installed in the general direction of the presumed groundwater flow, based on surface topography and the presence of a nearby stream.

TABLE 8.1

Calculation of Groundwater Elevations in Monitoring Wells

Well Name	MW-1	MW-2	MW-3
Top of inner casing	89.00'	86,00'	86.00'
Depth to groundwater	14.00'	13.00'	14.00'
Groundwater elevation	75.00'	73.00'	72.00'

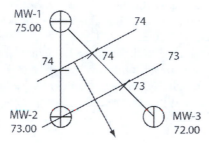

FIGURE 8.4
A contour map of the top of the water table, showing the calculated groundwater flow direction.

interpolation. For instance, the distance of the 74.00' point from MW-1 along a line connecting MW-1 and MW-2 is calculated using the following interpolation formula:

$$((74.00 - E_1)/(E_2 - E_1)) \times D$$

where E_2 is the elevation of the water table at MW-2, E_1 is the elevation of the water table at MW-1, and D is the ground distance between MW-1 and MW-2. A similar interpolation process was used to identify the location of the 73' and 74' contour lines between MW-1 and MW-3. The 74' and 73' contour lines are parallel to each other, as required by planar geometry with three data points.

Just as a ball will roll off a tilted table in the direction of its maximum tilt, so will groundwater flow in the direction of the maximum inclination of the water table. Therefore, identifying groundwater flow direction is found by drawing a line perpendicular to the groundwater contour lines. The slope of the water table is the hydraulic gradient, which is discussed in Chapter 5 of this book.

The implicit assumption in this example is that the three monitoring wells are installed in the same aquifer. If one of the three wells is installed in a different aquifer than the other two, an incorrect groundwater flow direction will be calculated. This problem can be avoided by attempting to identify aquitards that could separate saturated zones, such as what happens if a perched water zone is encountered in one of the wells. Differences in aquifers can sometimes be detected by looking for significant differences in water quality measurements such as pH, dissolved oxygen, etc. Sometimes the groundwater contours suggest that different water has been encountered due to an unusually steep calculated hydraulic gradient, for example, or groundwater flow heading in an unexpected direction.

The depth to groundwater may vary due to seasonal fluctuations, tidal influences, precipitation effects, or for reasons related to micro-geologic factors, such as those discussed in Chapter 5. Such fluctuations can lead to variations in groundwater flow direction. Groundwater flow calculations also can be compromised by the presence of LNAPL floating on top of the water table, which will depress the water table due to its weight.

8.2.3 Contaminant Plume Mapping

Once groundwater monitoring points are installed, groundwater samples can be collected and analyzed to assess whether contamination is present at those locations. Table 8.2 provides the TCE concentrations measured in the three monitoring wells from our example. Not surprisingly, monitoring well MW-1, installed near the source of the TCE release, contains the highest concentration of TCE.

These concentrations can be used to develop an *isopleth map* (also known as an *isoconcentration map*), which is a contour map that shows the extent and severity of a contaminant plume in groundwater. The isopleths connect points of equal contaminant concentrations rather than equal elevation, as with a topographic map or a groundwater contour map. The contour lines are drawn using the process of interpolation, similar to the calculation method described above.

The resulting isopleth map, shown in Figure 8.5, shows a contaminant plume that is elongated in the direction of groundwater flow, as is expected when advection is the primary mechanism of contaminant transport. If advection was the only process affecting the size of a groundwater plume, all groundwater plumes would be shaped like a pencil and oriented in the direction of groundwater flow. However, as discussed in Chapter 5, various forces act to shorten and widen the plume. The degree of eccentricity of the ellipse is controlled by the proportional effect of advection versus these countervailing forces.

In summary, the long axis of the ellipse should more or less point in the direction of groundwater flow, and its length is controlled by the hydraulic gradient and hydraulic conductivity; the amount of time since the initial release to groundwater; and the degree to which the leading edge of the plume is retarded and degraded by biological, chemical, and physical forces.

Complex plumes in a complex geologic environment can require numerous wells for a true grasp of the plume's configuration. Complicated bedrock fracture patterns, such as those shown in Figure 5.10 in Chapter 5, yield complicated groundwater flow and contaminant plume patterns. Complicated fluvial or glacial environments at the time the original soils were deposited, or subsequent alterations of the soils or bedrock, can result in groundwater contour and isopleth maps that show little if any discernable pattern. Adding complexity to the equation are preferential groundwater flow paths caused by natural and man-made phenomena, and man-made obstacles, such as building foundations. Often, computer modeling, which is described in Section 8.2.5, can assist the geologist in analyzing complex groundwater flow patterns.

TABLE 8.2

TCE Concentration in Monitoring Wells, Given in Micrograms per Liter (μg/L)

Well Name	MW-1	MW-2	MW-3
TCE concentration	275	50	5

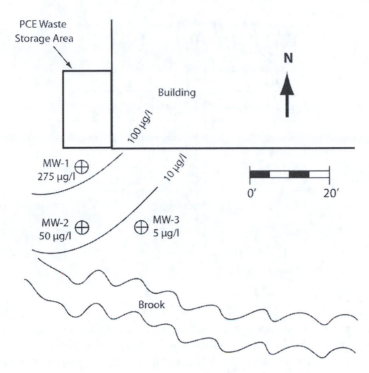

FIGURE 8.5
Isopleth map of perchloroethene (PCE) in groundwater at a hypothetical factory.

Adding complexity to the contaminant plume is the fact that, while it is rendered in two dimensions on a map, it is three-dimensional in reality. In most cases, groundwater contamination must be delineated vertically as well as horizontally, as described next.

8.2.4 Delineation of Groundwater Contamination

Several mechanisms can cause contaminants to migrate downward through a saturated zone. Compounds that have specific gravities greater than 1 are more likely to migrate downward than compounds with specific gravities less than 1. As with soils, groundwater plumes can be delineated horizontally and vertically using single-point compliance, establishing a concentration gradient, by compliance averaging, or other statistical methods. Groundwater data can be collected through permanent monitoring wells or temporary well points. Some methods to delineate groundwater plumes vertically are described below.

8.2.4.1 Vertical Delineation of Groundwater Contamination Using Monitoring Wells

Monitoring wells installed to delineate a contaminant plume vertically may penetrate a shallower contaminated zone along the way. To prevent contamination from one zone to enter another zone, a process known as *cross-contamination*, the shallower aquifer must be sealed off before continuing the drilling activities. This is often accomplished by installing a *double-cased well*, in which a casing is set against the portions of the aquifer above the zone of interest, with a narrower casing set below the upper casing, to which a screen over the zone of interest is attached (see Figure 8.6).

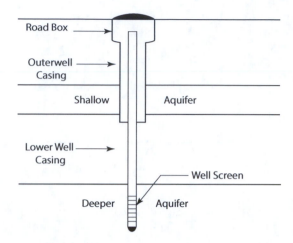

FIGURE 8.6
A schematic diagram of a double-cased well.

   Well screens set at different depths in the same approximate location can enable the geologist to analyze the vertical variations in contaminant concentration. Monitoring wells that are set at different depths within a few feet of each other are known as a *well cluster*. A different method to accomplish the same objective is setting multiple monitoring wells with different screen depths within one large casing, which is known as a *well nest*. Various companies have developed ways to set multiple sampling zones within one well-bore by setting one long well screen in a conventionally-sized borehole. The screens are isolated from each other by water-tight packers that establish the desired sampling intervals. Figure 8.7 shows the various constructions of a well cluster, a well nest, and a multi-level system monitoring well.

8.2.4.2 Membrane Interface Probes

A membrane interface probe (MIP) can provide vertical delineation of VOC contamination in both unsaturated and saturated soils. The MIP is drilled into the subsurface in the same manner as a split-spoon or other sampling device. A heated permeable membrane on the MIP mobilizes VOCs, which diffuse through the membrane into the probe, and then are transported to an analytical device staged at the surface by a carrier gas line. MIPs are particularly effective at providing vertical delineation data since several VOC samples can be collected from a single borehole. MIPs data is used for screening purposes, and should be confirmed with conventionally collected soil or groundwater samples.

8.2.5 Computer Modeling of Groundwater Plumes

Computer modeling software can facilitate the understanding and the visualization of configuration of contaminant plumes and the fate and transport of contaminants in the plumes. Existing geological data can be input into three-dimensional computer models,

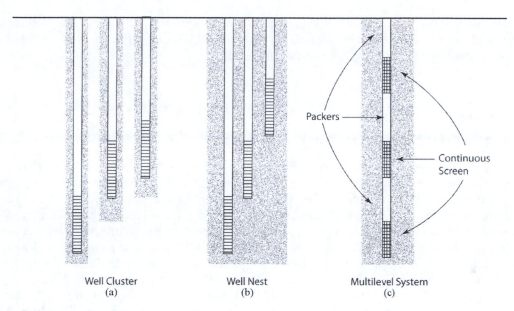

FIGURE 8.7
(a) A well cluster; (b) a well nest; and (c) a multilevel system monitoring well.

which can then be used to simulate the flow of groundwater in an aquifer. This flow data information can then be used to predict the future configuration of the contaminant plumes and its potential impact to receptors.

8.3 Geographic Information Systems

Data can accumulate rapidly for large remedial investigations, creating a need for sound data management. Data regarding borehole locations, sampling depths, and laboratory results need to be well-organized to facilitate retrieval and interpretation. Lost data, or even worse, data plotted incorrectly on maps, can result in project delays, cost overruns, and lawsuits, and possibly unnecessary exposure of a receptor to a toxic pollutant.

Geographic information systems (GIS) are a useful tool to manage project data. At its core, a GIS is a map that is connected to one or more databases. The map is in digital format, usually generated using computer-aided drawing and design (CADD). Databases are often created using relational database software. By activating information contained in a particular database, the map can provide a snapshot of a particular issue of environmental interest about the site.

Different "layers" of data are created in the database. Some of these layers, called *planimetrics*, can be environmentally oriented: soil sampling data, groundwater elevations, locations of ecological receptors, etc. Other layers may be cultural in nature: the location of buildings and roads, for instance, or property boundaries. The user can select which layers will be plotted on the map. The GIS can be linked to software programs, such as automated contouring programs or programs that automatically generate cross-sectional views of geological strata. The flexibility offered by a GIS makes it a desirable method of data management, especially for complicated sites.

8.4 Remedial Investigation Report

The remedial investigation report is similar in function and purpose to the site investigation report (see Chapter 7). It documents the results of the investigation and provides recommendations, if any, for further investigation or remediation. It contains the basic information provided in the sampling and analysis plan, as well as a technical review section (see Chapter 7 for a discussion regarding the contents of the technical review section). It summarizes the data and interpret its implications. If the newly acquired geological and chemical data do not jibe with the existing conceptual site model, then the remedial investigation report should contain an updated conceptual site model.

To aid in the understanding of the data, the remedial investigation report contains groundwater contour maps and isopleth maps showing the concentration distribution of the contaminants. It also contains a *geologic cross-section* of the study area. As shown on Figure 8.8, a geologic cross-section combines the soil and bedrock descriptions obtained from the boring logs and identifies soil and rock layers. When a three-dimensional perspective of the subsurface geology is desired, a *fence diagram* is used (see Figure 8.9). The geologic layers shown on the cross-section and fence diagram should jibe with the observed configuration of the contaminant plumes. For example, it should show an impermeable layer such as a clay layer if that layer has influenced the transport of contaminants in the subsurface.

Chemical data are presented in maps, graphs, and tables. Statistical distribution of data is presented in a form that can be used for a risk assessment, as described in Section 8.5.2.

The RFI report under RCRA contains the elements of the RFI work plan as well as the results of the remedial investigation. The treatment of the data collected is similar to a

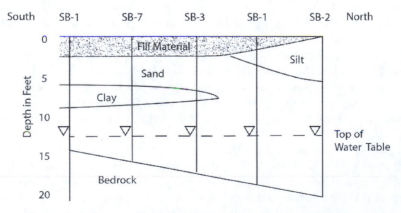

FIGURE 8.8
A typical geologic cross-section.

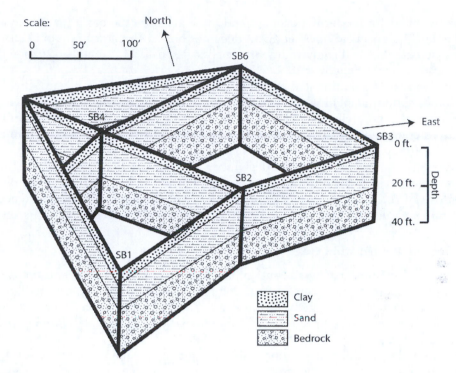

FIGURE 8.9
A typical fence diagram.

remedial investigation report. Remedial actions to be taken in solid waste management units (SWMUs; see Chapter 6), that require remediation are described in a corrective measures study, which is similar to a feasibility study described in Section 8.7.1.

8.5 Establishing Cleanup Goals

In a perfect world, polluted properties would be remediated to "pre-industrial levels," a term used by the New York State Department of Environmental Conservation (NYSDEC). This ideal, however, is often impractical.

Firstly, virtually no previously developed property in the United States is free of chemicals. Chemicals are generated through general human activities, especially through air emissions from combustion of fossil fuels. Once airborne, chemicals can travel thousands of miles from their point of origin. Therefore, in all but perhaps the remotest parts of the United States, chemicals have found their way onto otherwise pristine properties. Even the soils and rocks themselves naturally contain chemicals that are toxic to humans, especially toxic metals such as lead and arsenic. For this reason, an environmental investigation will often detect some quantity, however minute, of hazardous chemicals.

Secondly, the mere presence of contamination should not mean that a property requires remediation. Paracelsus, who was a Swiss physician, alchemist, and astrologer in the first half of the sixteenth century, stated that "Only the dose makes the poison." This concept became the basis for modern toxicology and, by extension, the basis for environmental remediation. If a chemical or other agent does not pose a threat to human health in the quantity detected (or to the environment, which is discussed in Chapter 12), then there is no practical reason to remediate it.

Lastly, removal of contaminated media can by itself create a health hazard, through the migration of contaminants into the air and water, where none previously existed. In the case of naturally occurring elements, such as lead or arsenic, the contamination can't be removed completely. Even in cases where the contamination is caused by human processes, it is often physically impossible to remove all contamination.

Therefore, the first objective in the design of a site remediation is the establishment of remedial objectives for the chemicals of concern. Rather than remediate all properties to "pre-industrial" conditions, *remediation standards* are established to guide the remediation of contaminated media. These standards can be generic or site-specific, as discussed below.

8.5.1 Generic Remediation Goals

Ideally, the remediation goals for a contaminated area will reflect the risk posed by the site contaminants to human health and the environment. The endpoint of the remediation may call for unrestricted future usage, or future usage that does not allow for human occupancy or restricted, non-residential occupancy. The remediation may be site-wide or specific to an area of concern.

Developing site-specific remediation standards for each contaminated property can be time-consuming, expensive, and probably unnecessary in many if not most cases, due to similarities in geology, chemistry, and types of receptors present. For this reason, most regulatory authorities have promulgated *generic remediation standards* for the compounds or analytes required to be tested for the regulation being enforced (Superfund, Clean Water Act, etc.) These standards are designed to protect human health and the environment under conservative scenarios, erring on the side of caution by basing the risk posed by the chemical on ideal fate and transport conditions:

- Soils have high porosity and permeability, enabling free movement of contaminants in the subsurface
- Proximity and direct pathway to potential receptors
- Presence of sensitive receptors, which are receptors (human or ecological) most vulnerable to chemical exposure

As per Paraselsus' theory, generic standards are based a *dose-response relationship*, and are developed using the risk-based principles described in the next section. The dose-response concept assumes that the higher the dose of a certain chemical, the more likely a negative impact of the chemical on human health or the environment.

Often, there are multiple generic cleanup standards to cover multiple generic scenarios. The most stringent generic cleanup standards for soils are reserved for residential properties, school and child care centers, which are usually calculated to protect the health of children. Non-residential standards are less stringent than residential standards because the affected population, assumed to consist primarily of adults, has a larger body mass with which to absorb contaminants, and is present on the site for less time (40 hours per week vs. 24/7). Another commonly used generic standard for soils is designed to protect contaminants in soils from migrating downward to the water table, where they can be ingested by humans directly or indirectly through food grown on the property.

Additionally, generic groundwater standards for drinking water as per the Clean Water Act, are less stringent for non-potable or brackish groundwater and more stringent for protected watersheds and other specially designated water bodies. Surface water and sediment standards are discussed in Chapter 12.

8.5.2 Setting Site-Specific Cleanup Goals Using Risk Assessment

To estimate the threat to human health posed by contaminants on a property, and therefore understand the risk to human health posed by a hazard, a *risk assessment* can be performed. It usually is performed at or near the conclusion of the remedial investigation, when the data set is complete or nearly complete, although a risk assessment can be conducted early in the investigation and undergo adjustments as new data become available.

There are two different approaches to a risk assessment. The most common approach used in a risk assessment is to identify a potentially hazardous condition and then determine the risk it poses to human health. The opposite approach, utilized in risk assessments associated with remediation, is to identify a hazard (e.g., contaminated soil), select a level of *acceptable risk* posed by that hazard, and then back-calculate the level of contamination that is "acceptable" given the risk toleration.

A *baseline risk assessment* evaluates the potential adverse health effects (current or future) caused by contamination on a property. The baseline risk assessment focuses on current site conditions rather than conditions that may exist post-remediation or post-construction. Its baseline is a "no action" scenario in which no actions will be taken to control or mitigate the contaminants. The results of the baseline risk assessment help determine what remedial actions should be implemented at a site by developing site-specific remediation goals, and documenting the magnitude of risk posed by a site.

The baseline risk assessment consists of five steps:

1. Hazard identification
2. Data collection and evaluation
3. Exposure assessment
4. Toxicity assessment
5. Risk characterization

The first two steps are largely accomplished by the Phase I/site investigation/remedial investigation process described in this and the previous two chapters. The data collected for these two steps include not only the concentrations of the contaminants on the site and their spatial distribution, but also their expected fate and transport and potential receptors. Steps 3 and 4, the exposure assessment and toxicity assessment, can be performed concurrently, concluding with the risk characterization.

A *focused risk assessment* analyzes the risk from a particular chemical, or a subset of the chemicals present on a site. This assessment usually starts at Steps 3 and 4, with the particular chemical to be the focus already identified.

8.5.2.1 Exposure Assessment

Step 3, the exposure assessment, evaluates the types of potential receptors, the potential pathways to the receptors, and the estimated dosage to the receptors. Since the baseline risk assessment assumes that no remedial actions will be performed, the exposure assessment must account for current as well as future pathways and receptors.

Typical human receptors to consider in an industrial setting include:

- Facility workers, who may have daily contact with contaminated media
- Emergency utility repair workers who may come into contact with contaminated soils when dealing with underground utility lines
- Construction workers who can contact contaminated media during construction, maintenance, or repair operations on a property
- Trespassers/authorized site visitors
- Office workers
- Children, if the setting is or may become residential, school or a child care center

A *human health risk assessment* focuses on the inhalation, ingestion, and dermal exposure pathways into the human body (environmental risk assessments focus on other biota). The inhalation pathway accounts for the inhalation of contaminants in the gaseous phase as well as contaminants attached to dirt and dust particles. This pathway also considers inhalation of vapors emanating from contaminated water by a human who contacts the water by showering or bathing. The ingestion pathway includes drinking contaminated water or eating food that has absorbed contaminants in the soil directly such as vegetables and fruits, or indirectly such as meats, fish, or dairy products. Children also can ingest contaminants by playing outside in contaminanted soils and placing their dirty hands in their mouths. Table 8.3 provides exposure scenarios used by the NYSDEC in establishing its generic cleanup standards and summarize the prime exposure pathways for a human health risk assessment.

Each exposure route used in a human health risk assessment utilizes a different equation in calculating the *reasonable maximum exposure* (RME). The RME is the highest exposure that could reasonably be expected to occur for a given exposure pathway at a site.

TABLE 8.3

Exposure Scenario Receptors and Pathways

Land Use Category	Unrestricted	Residential	Restricted Residential	Commercial	Industrial
Exposed Population	Adult & Child	Adult & Child	Adult & Child	Adult & Child	Adult & Adolescent
Route of Exposure					
Incidental soil ingestion	√	√	√	√	√
Inhalation of soil	√	√	√	√	√
Dermal contact with soil	√	√	√	√	√
Homegrown vegetable consumption	√	√			
Producing animal products for human consumption	√				
Groundwater protection	√	√	√	√	√
Ecological resource protection	√	√	√	√	√

Source: New York State Department of Environmental Conservation (NYSDEC), 2010.

It should account for uncertainties in both the contaminant concentration and the variability in the exposure parameters (such as exposure frequency and averaging time).

For instance, the equation used to calculate exposure from inhaling airborne dust for an adult in an industrial setting is:

$$\text{Intake (mg/kg/day)} = (EPC_{air} \times IR \times EF \times ED)/BW \times AT$$

where
EPC_{air} = exposure point concentration in air (mg/m^3)
IR = inhalation rate (m^3/day)
EF = exposure frequency (days/year)
ED = exposure duration (years)
BW = body weight (kg)
AT = averaging time (days)

In this equation, the EPC_{air} is determined from the exposure assessment. The inhalation rate used in this equation is provided by the USEPA, which estimates that the average adult inhales 20 m^3 of air per day. This is a conservative assumption, as are the exposure frequency to be considered in this exposure scenario (250 days per year for workers in industrial settings, 365 days per year for residents in residential settings) and the exposure duration (25-year careers for workers in industrial settings, 70-year residencies for people in residential settings). The "standard adult" considered in the risk assessment weighs 70 kg (approximately 154 lbs.). The body weight is in the numerator, because in theory people with greater body weight can tolerate a greater dose of toxins. The averaging time for non-carcinogens is the period of concern for the study. Similar calculations are performed for the other exposure paths.

8.5.2.2 Toxicity Assessment

A *toxicity assessment* entails the acquisition and evaluation of toxicity data for each contaminant present at a site. For this assessment, the consultant will research the latest toxicological studies to obtain current data to be employed in the risk assessment process.

Toxicity assessments separate chemicals into two categories: carcinogens, defined as chemicals with the ability to cause cancer in humans, and non-carcinogens. The USEPA defines five groups regarding human carcinogenicity (www.epa.gov/ttn/atw/toxsource/carcinogens.html):

Group A: Carcinogenic to humans

Group B: "Probably" carcinogenic to humans

Group C: "Possibly" carcinogenic to humans

Group D: Not classifiable as to human carcinogenicity

Group E: Evidence of non-carcinogenicity for humans

In a toxicity assessment, the category of carcinogens includes Groups A and B only. The USEPA's Integrated Risk Information System (IRIS) website (www.epa.gov/iris/search_human.html) lists the chemicals for Groups A, B, and C.

Carcinogenic risk for a chemical is estimated using a *slope factor*, which is an upper bound, approximating a 95% confidence limit, on the increased cancer risk from a lifetime

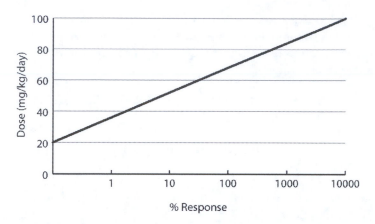

FIGURE 8.10
A graph showing the 95% confidence interval for a known or probable carcinogen.

exposure to a chemical by ingestion or inhalation (see Figure 8.10). Superfund defines an "acceptable risk" that exposure to a chemical will cause an incremental cancer (in addition to the 1-in-4 chance) at a rate ranging from 1×10^{-4} (one in 10,000) to 1×10^{-6} (one in 1,000,000) population, depending on the site.

The risk posed by non-carcinogens at a contaminated site is estimated using *reference doses* (RfDs) for the ingestion and dermal exposure pathways and *reference concentrations* (RfCs) for the inhalation pathway. Reference doses, also known as *no observed adverse effect levels* (NOAELs), are values supplied by the USEPA. They are doses of non-carcinogens below which there is no health effect from human exposure to the chemical. Above this dosage, there is an assumed linear relationship between dosage and health effect (see Figure 8.11). The RfD is equal to the NOAEL multiplied by a safety factor, in many cases a factor of 10. The RfC is derived in a similar manner to the RfD, except that several

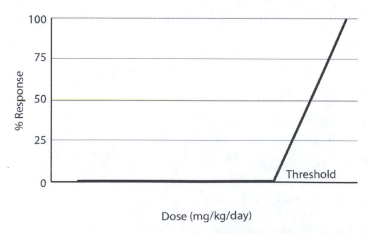

FIGURE 8.11
A graph showing the relationship between dosage and health effect for a non-carcinogen.

inhalation exposure scenarios are considered, with the most toxic scenario for a given chemical used in the toxicity assessment. USEPA has set a non-carcinogenic hazard effect target limit value of 1.

The USEPA recommends the following hierarchy of toxicological sources for Superfund risk assessments (OSWER Directive 9285.7-53, December 5, 2003):

Tier 1—IRIS.

Tier 2—USEPA's provisional peer reviewed toxicity values (PPRTVs)—The Office of Research and Development/National Center for Environmental Assessment/ Superfund Health Risk Technical Support Center develops PPRTVs on a chemical-specific basis. These values are available upon request through the project USEPA risk assessor for Superfund projects.

Tier 3—Includes additional USEPA and non-USEPA sources, such as the California EPA toxicity values (www.oehha.ca.gov/risk/ChemicalDB) and the Agency for Toxic Substances and Disease Registry (ATSDR) Minimal Risk Levels (www.atsdr. cdc.gov/mrls.html).

Since many dozens of chemicals can be present on a contaminated property, it can be impractical to provide a toxicity assessment for each contaminant. Instead, one chemical in a chemical class is chosen to represent the class. Such chemicals, known as *surrogate chemicals*, should be the most toxic, persistent, and mobile of the chemicals within its class, and the most prevalent on the property. If all of these criteria cannot be met, the risk assessor should either select one chemical that fits most of these criteria, or select more than one chemical from the chemical class.

8.5.2.3 Risk Characterization

Once the above steps have been completed, the overall risk from the contamination can be characterized. Where multiple contaminants are present, the cumulative risks posed by those contaminants must be calculated, since estimating the risk posed by one contaminant, even if it is the most toxic and prevalent, will underestimate the risk. However, many safety factors are inherent in risk assessment. Risks are calculated for each pathway and for current and future scenarios, resulting in a *cancer risk* for each pathway and a cumulative *non-cancer hazard index* for the non-carcinogenic contaminants. The calculated risks then would be compared to the established risk limits for risk evaluation conclusions.

For instance, the exposure and toxicity assessments may indicate that the overall cancer risk presented by the contamination is 7×10^{-7} and non-cancer hazard index of 0.3. Since the overall cancer risk is below 1×10^{-6} and the non-cancer hazard index is below the non-carcinogenic hazard effect target limit value of 1, one could conclude that the contaminants pose an acceptable health risk to the exposed populations.

A proper risk characterization will incorporate all major assumptions, scientific judgments, and estimates of the uncertainties present in the data. Once an uncertainty analysis is performed and all site-specific health or exposure studies taken into account, site-specific cleanup standards can be calculated that would be protective of human health within the assumed risk tolerance.

8.6 Remedial Action Design

In a *remedial design*, an action or a set of actions is chosen to remediate the contamination. The design is described in a remedial action work plan. Although it logically follows after the completion of the remedial investigation, the data needed to design the remediation usually is collected during the remedial investigation, and the remedial investigation report and remedial action work plan often are combined into one document. Some of the types of data collected in support of the remedial design are described below.

8.6.1 Waste Characterization

If the remedial action is to include the bulk removal and off-site disposal or treatment of soils or groundwater, then the consultant must determine the means by which they will be treated or disposed of. Waste classification is governed by the Toxic Substances Control Act (TSCA) for PCBs and by RCRA for most other chemicals. Certain wastes are specifically listed under RCRA, meaning that they are hazardous wastes by means of their origin or chemical and physical properties (see Chapter 3). If the waste stream is not listed under RCRA, then it needs to be tested for hazardous waste characteristics. This testing, known as *waste characterization*, entails collecting unbiased samples of each waste stream and analyzing them for many of the various compound classes described in Chapter 4, namely VOCs, semi-volatile organic compounds (SVOCs), pesticides, PCBs, and RCRA metals. Soil samples are composited for the SVOC, pesticide, PCB, and metals analyses. Discrete soil samples are analyzed for VOCs to avoid loss of VOCs in the mixing process.

Because RCRA is concerned about contaminants leaching out of RCRA-permitted landfills, the composite samples are prepared for analysis utilizing the toxicity characteristic leachate procedure, or *TCLP*. The TCLP procedure is designed to extract the portion of the contaminants that would become mobile once in the environment. Much of the contaminant mass will remain bound to the soil. Therefore, a chemical's concentration detected in soil sample analyses is only a fraction of the total concentration of the contaminant in the sample. (The more common methods of extraction are in the USEPA 3500 series. They involve the extraction of all of the chemicals in the sample than the mobile fraction of the chemical.) Because PCBs are regulated under TSCA, the TCLP method is not applied to soils that will be analyzed for PCBs. Rather, waste characterization samples are analyzed for total PCBs.

8.6.2 Groundwater Data Used to Design the Remedial Action

This section describes various data that are utilized in designing remedial actions.

8.6.2.1 Aquifer Analysis

An effective remedial design must demonstrate an understanding the physical, chemical, and biological properties of a contaminated aquifer. Physical properties of interest in an aquifer analysis include, but are not limited to, the following:

- *Hydraulic conductivity*, which is described in Chapter 5 of this book. It is expressed as a velocity, typically in centimeters per second (cm/sec).

- *Transmissivity*, which is the ability of groundwater to move through an aquifer given an ideal hydraulic gradient (i.e., vertical). Transmissivity is equal to the hydraulic conductivity multiplied by the aquifer thickness and can be expressed in cubic meters of water per day (m^3/day).

- *Specific yield* is the volume of water that drains from saturated soil as the water table drops. It is a dimensionless number.

- *Storativity* is the volume of water that drains from the saturated soil as the water table drops for an unconfined aquifer. It is a dimensionless number.

These parameters can be used calculate the potential *capture zone*, which is the area in which contaminants may be recovered, or "captured," by a pumping well in a pump and treat system (see Chapter 9). The most common methods of obtaining hydraulic conductivity are the slug test, the pump test, and the step-drawdown/well recovery test.

8.6.2.2 Slug Test

A *slug test* is most commonly employed on low-yield aquifers. In a slug test, a volume of water or a solid object (a "slug") is introduced into a well, causing the water elevation to rise in the well. This condition is temporary, as the water level will eventually return to its static water level when the water in the borehole water filters back into the geologic formation. The falling water level (a "falling head test") is measured either with a water level indicator or an electronic data logger. The falling head test concludes when the water level in the well returns to the static water level. When the slug is removed, the water level suddenly drops and then rises back to the static water level. Known as a "rising head test," these data can be used to verify the results of the falling head test. Data obtained is then used to calculate the hydraulic conductivity of the geologic formation.

8.6.2.3 Pump Test

For high-yield aquifers, the *pump test* is the most common method of aquifer analysis. The pump test has long been used for the installation of community drinking water wells. In a pump test, water is pumped from the well, thereby creating a *cone of depression* (see Figure 8.12), which is the area that the groundwater vacates during pumping. The change in the height of the water column in the pumping well, known as the *drawdown*, may vary in the course of the pump's operation.

   Prior to the pump test, static water level conditions are established in the pumping well and in nearby piezometers. The well is pumped at a constant rate, and the water levels in the piezometers are gauged. If the geological formation is isotropic, that is, symmetrical in all directions, the amount of drawdown that occurs in a given monitoring well over time will be directly related to its distance from the pumping well. If the geologic formation is anisotropic due to some geologic variation, there will not be a direct correlation between the measured drawdown and the piezometer's distance from the pumping well.

   In a *step-drawdown/well recovery test*, a well is pumped at progressively higher pumping rates and the drawdown at each pumping rate is recorded. Usually four to six pumping steps are used, each lasting approximately one hour. The higher the pumping rate, the larger the cone of depression, and the greater the portion of the aquifer affected.

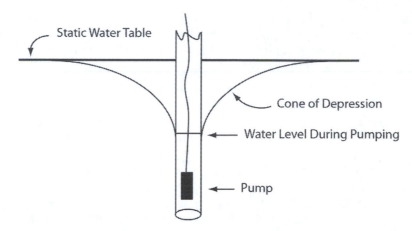

FIGURE 8.12
A schematic diagram of drawdown in a pumping well.

8.6.3 Treatability Studies and Pilot Tests

Other data collected in the remedial design process are used to evaluate remedial actions involving chemical or biological processes. Such tests are known as *treatability studies* (also known as "bench scale tests") and *pilot tests*.

A treatability study involves laboratory simulations of conditions in the contaminated areas. It may study the one contaminant at a time or all of the contaminants. It may use actual site soils to observe the effectiveness of the remedial technology given expected chemical, physical, and biological conditions or use uncontaminated soils with similar physical properties. A treatability study generally precedes a pilot test.

A *pilot test* implements a prospective remedial method over a small portion of the contaminated area. If portions of the contaminated area have different physical properties, more than the one area should be tested or more than one type of pilot test should be performed. As opposed to treatability studies, a pilot test has the advantage of being easily "scaled up" to a full-scale remediation since it already has been constructed in the field and more accurately predicts the effectiveness of the full-scale remediation system than a treatability study. It is more expensive and time-consuming to perform a pilot test, however, and may require permits to conduct, which can delay the testing.

8.7 Remedial Action Selection Criteria

To design a remedial action, one must establish the objective, or *design basis*, for the action. The design basis identifies the chemicals to be remediated and the goals of the remediation, such as whether it must meet residential, non-residential, or site-specific standards, whether it will provide a temporary or permanent solution, and so on. The remedial action also will reflect a balancing of time and cost considerations. In general, there is a trade-off between speed and permanence of the remedial action and its cost. In addition, remedial actions designed for residential settings are more expensive than remedial actions designed for non-residential settings.

8.7.1 Feasibility Study

The process by which a remedial action is selected is known as the *feasibility study*. In a feasibility study, a list of remedial alternatives is developed and screened based on chosen remedial objectives. Since it would take far too long to consider every remedial technology at every site, a representative technology process option for each technology type is usually selected. Estimated costs and timeframes are generated and used in the decision-making process.

8.7.1.1 No Action Alternative

To provide a baseline, the feasibility study must include a no action alternative. In many cases, the no action alternative means no reduction in the extent and severity of the contamination. In other cases, however, no action is needed because remediation is occurring naturally. *Natural attenuation* is the transformation of contaminants into benign or more benign chemicals due to natural processes such as chemical oxidation or biodegradation.

Most contaminants will naturally transform in groundwater, especially within the aerobic portion of an aquifer. Contaminants most prone to natural attenuation include non-chlorinated VOCs and petroleum-related compounds. The primary line of evidence that natural attenuation is occurring is the decrease in the extent and/or concentration of contaminants within a plume. Downward trends can be documented using ProUCL software or other statistical methods described in Section 8.1.1.3. Because it may take years for a contaminant to naturally degrade to acceptable levels, establishing a downward trend in contaminant concentration can take years as well. Less direct but more timely evidence that natural attenuation is occurring can be obtained through an analysis of geochemical conditions in the aquifer. For instance, if the contaminant will transform into a benign chemical when it reacts with oxygen and there is there is sufficient dissolved oxygen content in the groundwater, then it can be assumed that natural attenuation will occur, provided there are no barriers to the oxidation process.

8.7.1.2 Determining Applicable or Relevant and Appropriate Requirements

At Superfund sites, consultants must determine the *applicable or relevant and appropriate requirements*, commonly known as *ARARs*, for a Superfund cleanup. ARARs are environmental regulations not promulgated under Superfund that need to be considered in the remediation of a Superfund site. These requirements are mandated by another federal bureau or by the state or local government in which the Superfund site is located, and can run the gamut regarding their technical and administrative complexity and their costliness.

For example, at the federal level, ARARs relating to the Safe Drinking Water Act include the maximum contaminant levels (MCLs) for the chemical of concern. The ARARs also could include permitting for the injection of chemical agents designed to remediate groundwater contamination into an aquifer if the groundwater in the area is used for potable purposes. The ARARs relative to the National Environmental Policy Act (NEPA) may entail the protection of threatened and endangered species and their habitats. The ARARs relating to the Clean Water Act may regard potential encroachment on waterways and regulated wetlands. The ARARS relating to the Clean Air Act could include emissions of dust or gases during remediation. Waste disposal requirements established by the RCRA will dictate hazardous waste generation, storage, and disposal during the cleanup.

State-specific cleanup standards, documentation requirements, and permit require-ments directly bear on the cleanup to be performed. There may be state requirements regarding landfill closure and post-closure monitoring. There also may be state or local requirements regarding such construction issues as dust suppression, waste water storage and disposal, and noise restrictions that must be considered in the remedial design.

Although ARARs go under a specific name in the Superfund program, the concept applies to all remediations. All the various studies, restrictions, permits, and fees that will apply to a potential remedial action must be considered in the feasibility study.

8.8 Record of Decision under Superfund

After the feasibility study has been completed, the USEPA will issue a record of decision (ROD) for a site that is on the National Priorities List (NPL). The ROD summarizes all of the information gathered to date about the site, including the extent and severity of the contamination and its potential impact to receptors. It summarizes the community's involvement (a required part of the Superfund process) in the development of the docu-ments leading to the issuance of the ROD. The ROD then presents the results of the human health risk assessment and the ecological risk assessment, discussed in Chapter 12. It uses these risk assessments as a springboard for the feasibility study.

The remedial alternatives evaluated as part of the feasibility study are present and eval-uated. The evaluations consider the alternative's:

- Protection of human health and the environment
- Compliance with ARARs
- Long-term effectiveness and permanence
- Ability to reduce the toxicity, mobility, or volume of the contaminants
- Short-term effectiveness
- Implementability
- Cost
- State acceptance
- Community acceptance

Estimating the anticipated cost of each remedial alternative involves first developing an estimate of the capital costs to be expended. Capital costs include all required expendi-tures to implement the remedial action. The costs include labor, equipment, and mate-rial costs as well as contractor markups and agreed-upon profit margins based on either project multiplier or percentage of overall project costs (see Chapter 2 for a discussion on project budgeting). Permitting fees and labor costs for office support work are included in the capital cost estimate. Lastly, the capital cost estimate includes contingency costs for the various tasks comprising the remedial action. Contingency costs are unanticipated costs that may occur in the course of the project. They usually are calculated as a percentage of the total cost of a given task.

The second cost category entails operations and maintenance (O&M) costs. The O&M costs include all expenditures needed to ensure or verify the continued effectiveness of

TABLE 8.4

Example of a Cost Comparison of Alternatives under Superfund

Alternative	Capital Cost	Annual O&M Costs	Timeframe	Net Present Worth Cost (Rounded, @5% Discount)
1. No Action	$0	$39,987	30 years	$40,000
2a. Soil Vapor Extraction with Limited Air Sparging	$2,001,999	$976,135	5 years	$2,978,100
2b. Thermal Enhanced Soil Vapor Extraction with Limited Air Sparging	$5,285,201	$976,135	5 years	$6,261,300
3. Limited Soil Excavation with Soil Vapor Extraction	$2,722,209	$738,963	5 years	$3,461,200
4. *In Situ* Chemical Oxidation with Soil Vapor Extraction	$2,627,391	$738,963	5 years	$3,366,400

Source: Record of Decision Walker Machine Products, Inc. Superfund Site, Colliersville, Shelby County, TN.

the remedial alternative. As with the capital costs, annual O&M costs include labor, equipment, and material costs as well as profit for the consultant and the contractors. A timeframe in which these costs are expected to be expended is calculated based on the technical information obtained during the feasibility study. The total anticipated O&M cost is then calculated by using a discount factor, which is similar to an interest rate and recognizes the time value of money, that is, today's dollar, if left uninvested, will be worth less than $1.00 in future years. The default timeframe for O&M activities post-Superfund cleanup is 30 years. The discount rate is applied to the annual O&M expenditures and that number is then added to the capital cost estimate to generate an overall cost estimate for the remedial alternative. As an example, Table 8.4 provides an example of a cost comparison of remedial alternatives at a Superfund site in Tennessee.

8.9 Remedial Investigations, Remedial Design, and Environmental Consultants

Environmental consultants performing a remedial investigation essentially perform the same functions in the field as they do in a site investigation (see Chapter 7). In the office, there are far more tasks for the consultant. As the data because more voluminous and more complex, environmental consultants will employ data management and data modeling software to understand the distribution, fate, and transport of the contaminants. Visualization of the extent of the contaminant plume can be enhanced with maps, diagrams, and cross-sections, which can be generated by hand or using commercial software packages. Environmental consultants may employ sophisticated statistical methods to interpret the data, and complicated risk assessment procedures to assess the hazards posed by the contamination to human beings to determine remediation goals. For this reason, a consulting "team" often is needed to design complex remediations.

In the remedial investigation phase, an environmental consulting firm may use some or all of the tools in its tool box, that is, geologists, chemists, toxicologists, ecologists, and engineers, to move the project from the investigation phase to the remediation phase, which is described in Chapter 9.

Problems and Exercises

1. The default cleanup standard for benzene is 1 milligram per kilogram. The following benzene concentrations were detected in an area of concern:

   84 mg/kg

   1 mg/kg

   92 mg/kg

   21 mg/kg

   1 mg/kg

   27 mg/kg

   54 mg/kg

   5 mg/kg

   Does the data set comply with the "75%/10x" rule?

2. Monitoring well MW-2 is 30' north of MW-1 and 30' east of MW-3. The elevations of the inner casings of the wells are as follows: MW-1 = 564.15'; MW-2 = 563.18'; MW-3 = 565.51'. The water table was measured at 7.81' in MW-1, at 7.09 in MW-2, and at 8.31' in MW-3. Calculate the direction of groundwater flow.

3. Calculate the hydraulic gradient for the water table described in Question 2.

4. Find a Record of Decision on the internet for a Superfund site. List at least one ARAR associated with the Clean Water Act; the Safe Water Drinking Act; the Clean Air Act; and two state statutes.

Bibliography

Freeze, R. A., and Cherry, J.A., 1979. *Groundwater*, Prentice-Hall.

Guidotti, T.L., 1988. Exposure to hazard and individual risk: When occupational medicine gets personal. *Journal of Occupational Medicine* 30, 570–577.

Hemond, H.F., and Fechner-Levy, E.J., 2000. *Chemical Fate and Transport in the Environment*, 2nd ed. Academic Press.

Lagrega, M.D., Buckingham, P.L., and Evans, J.E., 2001. *Hazardous Waste Management*, 2nd ed. McGraw-Hill.

National Oceanic and Atmospheric Administration, 2017. National Geodetic Survey: Frequently Asked Questions. www.ngs.noaa.gov/datums/newdatums/FAQNewDatums.shtml

New Jersey Department of Environmental Protection, March 2012. Monitored Natural Attenuation Technical Guidance. Version 1.0.

New York State Department of Environmental Conservation, May 2010. DER-10: Technical Guidance for Site Investigation and Remediation.

Nyer, E.K., Regan, T., and Nautiyal, D., Spring 1996. Developing a Healthy Disrespect for Numbers. *Ground Water Monitoring Review*, pp. 59–64.

Nyer, E.K., and Gearhart, M.J., Winter 1997. Treatment Technology: Plumes Don't Move. *Groundwater Water Monitoring & Remediation*. National Ground Water Association, pp. 52–55.

Patton, D.E., 1993. "The ABCs of Risk Assessment." *EPA Journal* 19, 10–15.

Sterrett, Robert J., 2007. *Groundwater and Wells*, 3rd ed. Johnson Screens.

U.S. Environmental Protection Agency, October 1986. RCRA Facility Assessment Guidance. Office of Solid Waste. PB87-107769.

U.S. Environmental Protection Agency, August 1987. A Compendium of Superfund Field Operations Methods. EPA 540/P-87/001. Office of Emergency and Remedial Response.

U.S. Environmental Protection Agency, May 1989. Interim Final RCRA Facility Investigation (RFI) Guidance. Office of Solid Waste. EPA 530/SW-89-031.

U.S. Environmental Protection Agency, November 1989. The Remedial Investigation Site Characterization and Treatability Studies. Directive 9355.3-01FS2.

U.S. Environmental Protection Agency, December 1989. Risk Assessment Guidance for Superfund, Volume I. Human Health Evaluation Manual. EPA 540/1--89/002. Office of Emergency and Remedial Response.

U.S. Environmental Protection Agency, July 2000. A Guide to Developing and Documenting Cost Estimates During the Feasibility Study. EPA 540-R-00-002, OSWER 9355.0-75.

U.S. Environmental Protection Agency, December 2002. Supplemental Guidance for Developing Soil Screening Levels For Superfund Sites. OSWER 9355.4-24. Office of Emergency and Remedial Response.

U.S. Environmental Protection Agency, December 2003. Hierarchy of Toxicological Sources for Superfund Risk Assessments. OSWER Directive 9285.7-53.

U.S. Environmental Protection Agency, September 2004. Site Characterization Technologies for DNAPL Investigation. EPA 542-R-04-017. Office of Solid Waste and Emergency Response.

U.S. Environmental Protection Agency, September 2005. Introduction to Hazardous Waste Identification (40 CFR Parts 261). Solid Waste and Emergency Response (5305W). EPA530-K-05-012.

U.S. Environmental Protection Agency, 2006. National Recommended Water Quality Criteria. Office of Science and Technology (4304T).

U.S. Environmental Protection Agency, September 2011. Exposure Factors Handbook – 2011 Edition. EPA/600/8-89/043. Office of Emergency and Remedial Response.

U.S. Environmental Protection Agency, October 2015. ProUCL Version 5.1.002 Technical Guide. EPA/600/R-07/041

U.S. Environmental Protection Agency, October 2017. Best Practice Process for Identifying and Determining State Applicable or Relevant and Appropriate Requirements Status. Office of Land and Emergency Management. OLEM Directive 9200.2-187

U.S. Environmental Protection Agency, September 2018. Record of Decision Walker Machine Products, Inc. Superfund Site, Colliersville, Shelby County, TN.

Weiner, S.A., 1993. Developing Cleanup Standards for Contaminated Soil, Sediment, & Groundwater: How Clean is Clean? Specialty Conference Series, 1993. Federal Water Environment.

9

Remedial Actions

Remediation is the culmination of an often laborious and time-consuming process of identifying and evaluating areas of concern, developing and refining the conceptual site model, obtaining hydrogeological, geochemical, physical, and biological data, and, not to be overlooked, documenting, documenting, and documenting the observations and findings of the investigations.

Remedial actions broadly fall into two categories: passive remediation and active remediation (although some remedial technologies blur the distinction between the two categories). *Passive remediation* relies on natural forces to remediate the contamination. There may be human intervention to remove or limit the pathway of the contaminants to humans or the environment while passive remediation occurs. *Active remediation* involves applying a physical, chemical, and/or biological process to the contaminants to decrease their concentration to acceptable levels. It can occur *ex situ* (on the ground surface) or *in situ* (in place, where it currently is situated). The commonly-used passive and active remedial methods are described in this chapter.

9.1 Remediation by Pathway Removal

This section describes remedial actions that remove the pathway of contaminants to the human body and the environment. Since the contaminants themselves are not remediated, the pathway removal action must remain in place for as long as the contamination exists.

9.1.1 Engineering and Institutional Controls

Engineering controls are physical barriers that separate contaminants from the human body or the environment. The most common type of engineering control is an impermeable surface cap over contaminated soils that removes the direct contact pathway to humans. A cap may consist of asphalt, concrete, impermeable soils such as clay, or a *geosynthetic liner* comprised of high-density polyethylene (HDPE) or a similar material. These liners are commonly used in the permanent closure of solid waste landfills (see Section 9.5). A surface cap is preferable to a passive engineering control such as fencing around the contaminated area, since fences and other such controls are relatively easy to circumvent, and may be more expensive.

In some circumstances, surface engineering controls may not need to be constructed. If sufficiently thick, uncontaminated soil above the contaminated zone will act as an engineering control. The USEPA considers two to three feet of clean fill or uncontaminated soil to be adequately protective at brownfields sites that will have non-residential usage, and ten feet of clean fill or uncontaminated to be adequately protective at brownfields sites that will have residential usage. This, however, varies from state to state.

Another type of engineering control is the *vertical containment barrier*. This barrier, which is usually a concrete slurry wall or steel sheet piling, prevents the migration of contaminants present in the soils or groundwater towards a receptor, such as a drinking water well or a surface water body.

To ensure the long-term effectiveness of the engineering controls, an operations and maintenance (O&M) program is devised and implemented. A typical O&M program calls for regular inspections of the engineering controls and repair if they are compromised or in danger of failure. Since vertical containment barriers cannot be inspected visually, their effectiveness can only be assessed by monitoring downgradient locations for evidence of a breach.

In the long run, current and future owners of the property need to be aware of the presence of engineering controls on their property and the importance of not damaging them. This is the function of an *institutional control*, which is a legal notice of the existence of the engineering control. The principal type of legal notice is the *deed restriction*, which is attached to the property deed and indicates the location and nature of the contaminated area.

Another type of institutional control involves restricting the use of groundwater on a property. If groundwater is contaminated, then it should not be used for potable or irrigation purposes. This can be accomplished by reclassifying the groundwater in the contaminated area, or restricting its usage. Restricting groundwater usage does not usually entail engineering controls since it would be easy to circumvent any such controls by drilling a new groundwater withdrawal point.

Institutional and/or engineering controls together are known as *activity and use limitations* (AULs).

9.1.2 Monitored Natural Attenuation

Often confused with AULs is *monitored natural attenuation* (MNA), which involves the reduction in the toxicity of dissolved contaminants by natural degradation. MNA is applied to contaminants that will degrade naturally by chemical and/or biological reactions (see Chapter 8).

For MNA to be the chosen remedial action, it first must be established that natural attenuation is occurring. See Chapter 8 for ways to assess the presence of natural attenuation in an aquifer. Once its presence has been confirmed, a long-term monitoring program is put in place. A long-term monitoring program typically will include periodic collection of groundwater samples from monitoring wells. It should include at least one well within the contaminant plume, ideally the well with the highest concentrations of contaminants. It also should include a well, known as a *sentinel well*, which is located between the contaminant plume and a receptor. The receptor may be a drinking water well or the location of a sensitive population such as a day care center. The receptor may simply be the area not under an institutional control and therefore the potential future location of a well to be used for drinking or for irrigation. The sentinel well should be located sufficiently far from the receptor so that the receptor would have sufficient time to take precautionary measures if the sentinel well becomes contaminated. As a rule-of-thumb, the sentinel well should be at least two years upgradient of the receptor, with the velocity of the plume having been calculated during the remedial design. If there is no receptor within this distance from the source, it may be prudent to install a sentinel well at this location to verify that the contaminants are not migrating downgradient and are under control.

The sentinel well should be sampled at least every two years, or more frequently if it is less than two years upgradient of the receptor. Wells within the contaminant plume should be monitored periodically to verify that natural attenuation is still occurring.

The groundwater samples within the contaminant plume should be monitored for the contaminants of concern as well as their degradation products. They also should be monitored for dissolved oxygen, if oxidation is the process driving natural attenuation, or for indigenous microbes if bioremediation is the driver.

It should be noted that the U.S. Environmental Protection Agency (USEPA) and most other regulatory authorities do not accept dispersion and dilution as MNA processes.

9.1.3 *In Situ* Vitrification

In situ vitrification is the process by which the contaminated media are heated to extremely high temperatures, in some cases above 1,000°C. The result is the conversion of the porous, permeable soils or sediments into a solid mass of glass. The contaminants are either destroyed by the process, or sealed within the vitreous mass, rendering them unable to migrate and therefore harmless. This process can be implemented uniformly across a wide range of contaminants, although there must be some accommodation made for off-gases from the vitrification process (especially for polychlorinated biphenyls (PCBs), which can be converted into extremely-hazardous dioxins if the thermal destruction process is not executed properly).

Temperature is the most useful parameter for verifying the success of the vitrification process, the assumption being that the vitrification process will be successful if the temperature inside the area being remediated stays at or above the target temperature. For an added level of security, monitoring wells or soil vapor extraction points can be installed around the area of remediation and sampled periodically to verify whether contaminants are escaping from the vitrified soils.

9.1.4 *In Situ* Solidification

In situ solidification is similar in concept to *in situ* vitrification, since the process creates a solid mass in the subsurface, thus removing their pathway to receptors. Rather than utilizing heat, this process utilizes agents such as concrete mix to convert the soils into a solid mass. Once solidified, the success of the solidification process can be verified by collected core samples of the solid mass and testing the core for compressibility, shear strength, porosity, and permeability. As with *in situ* vitrification, monitoring points can be set up outside of the remediated area for periodic long-term testing to verify the effectiveness of the remedial action.

9.2 Remediation by *Ex Situ* Source Removal

9.2.1 Removal of Light, Non-aqueous Phase Liquid

Non-aqueous phase liquid (NAPL), which commonly is referred to as *free product*, is an ongoing source of contamination, leaching a constant stream of contaminants into the soils and groundwater. Until all NAPL is removed, the dissolved contaminants cannot be addressed effectively.

Most cases of NAPL involve LNAPL, and most LNAPL is a form of petroleum. Because LNAPL floats on the top of water, most technologies for the remediation of LNAPL involve physical removal (DNAPL [dense non-aqueous phase liquid] removal, as a rule, is far more difficult than LNAPL removal). The simplest method of LNAPL remediation is known as "skimming," in which a recovery well, which is similar to a monitoring well but with a wider aperture for ease of access is installed within the LNAPL-impacted area. LNAPL can be manually removed from this withdrawal point using ordinary pumps or a bailer, or with pumps that are activated when they encounter a liquid with high resistivity (as opposed to water, which has a much lower resistivity than LNAPL). In some cases, interceptor trenches are constructed through or downgradient of the impacted area below the depth of the LNAPL. The LNAPL naturally flows into the trench, where it is then recovered using a pump or a bailer.

Bioslurping is a vacuum-enhanced method of LNAPL recovery. In bioslurping, a seal is created on an extraction well and a tube is installed within the LNAPL. A vacuum is applied to the tube, which acts like a straw in sucking the LNAPL to the surface. It also introduces air into the subsurface, encouraging the growth of microorganisms in the same manner as bioventing, which is discussed in Section 9.3.7.

A brute force method of LNAPL removal entails using a pump to remove all liquids, LNAPL as well as water, from the contaminated aquifer. This method generates large quantities of wastewater which either can be sent off-site for disposal or can be sent through an oil/water separator and reintroduced into the geologic formation once free of petroleum. Figure 9.1 is a schematic diagram of a simple gravity oil/water separator. As a side benefit of pumping water, the water table is depressed, and LNAPL will be induced to flow towards the "cone of depression," thus enhancing LNAPL recovery.

The above methods can be combined with various *in situ* (in-place) technologies, such as the injection of surfactants (see Section 9.3.3) or a co-solvent (see Chapter 5) into the subsurface to liberate the LNAPL from soils and other solids and enable it to flow towards the recovery location.

DNAPL remediation is generally performed by removal methods, such as groundwater pump-and-treat (see below). There also are emerging technologies designed to remediate DNAPL. More information is available on USEPA's Clu-In web site (www.clu-in.org).

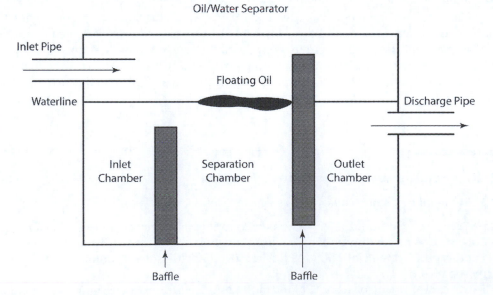

FIGURE 9.1
Oil and water entering an oil/water separator separate out naturally, since oil is less dense than water. The oil exits the oil/water separator through a pipe near its top and the water exits the oil/water separator through a pipe near its bottom. The oil floats on the top to be recovered and sent off-site disposal or treatment.

9.2.2 Soil Remediation by Excavation

The technologically simplest way to remediate contaminated soils is to dig them up and send them off-site for treatment or disposal. Excavating equipment can be larger than the machine shown in Figure 9.2, or as small as a golf cart or even a hand shovel. If there are uncontaminated soils above the zone of contamination, these soils, known as the *overburden*, can be stockpiled separately and used as backfill after the contaminated soils have been excavated. Backfill from an off-site source can be emplaced into the excavation unless the excavation area will be used for other purposes, such as construction of a new building.

9.2.2.1 Community Air Monitoring

Contaminants leaving the site while soils are being excavated can harm neighbors or passers-by. Environmental consultants may monitor the ambient air for the presence of contaminants in a process known as *community air monitoring*. Contaminants of concern in the gaseous phase invariably are volatile organic compounds (VOCs). Contaminants of concern that are bound to dust and other airborne solids can run the gamut; they may be PCBs, semi-volatile organic compounds (SVOCs), or metals. Therefore, air monitoring may test for VOCs in gaseous phase or for particulates.

FIGURE 9.2
An excavator loading contaminated soils onto a dump truck. (Courtesy of GZA GeoEnvironmental, Inc.)

Air monitoring is performed along the property boundary in the downwind position and usually other locations as well. Readings are obtained using either hand-held monitoring devices or automatic data loggers. Automatic data loggers have the advantage of storing data that can be downloaded and interpreted but may be more expensive and cannot be moved around the site easily.

If air readings exceed established maximum thresholds, then the work practices need to be evaluated. If the concentration of airborne particulates is too high, then the soils may need to be wetted or the soil piles need to be covered. If the airborne concentration of VOCs is too high, then vapor suppression measures may need to be implemented, or workers may need to upgrade their respiratory protection.

Disposal of contaminated soils must comply with Resource Conservation and Recovery Act (RCRA), Toxic Substances Control Act (TSCA), and all applicable state and local laws (see Chapter 8 for a discussion on waste classification sampling and analysis). Even if the waste classification results indicate that the soils are not hazardous, they still need to be removed from the site or treated since they are contaminated. These soils generally are designated as contaminated, non-hazardous soils, and can be disposed of at Subtitle D landfills that are licensed to accept contaminated soils.

9.2.2.2 Verification Sampling

Although the extent of contaminated soils should have been defined in the remedial investigation (see Chapter 8), vertification soil sampling is warranted where feasible. Verification soil samples, also known as *post-excavation soil samples*, should be collected from the sidewalls and bottom of the excavation in a sufficient quantity to confirm the uncontaminated nature of the remaining soils. The samples should be analyzed for the contaminants that had been present in the soils that were remediated. The excavation can be backfilled with clean fill or soils, or the treated soils, once it has been verified that the post-excavation soil samples have acceptable concentrations of the target contaminants.

9.2.3 Groundwater Remediation by Pump-and-Treat

The mass removal of contaminated groundwater is known colloquially as *pump-and-treat*. As its name implies, pump-and-treat involves pumping contaminated groundwater to the surface and removing the contaminants from the groundwater. In Figure 9.3, contaminated groundwater is pumped to the surface through an extraction well and piped into a waste water treatment system. Once separated from the contaminants, the treated groundwater can then be injected into the subsurface through an injection well or into a nearby surface water body or sewer system.

There are several options available for the disposal of the separated contaminants. If the contaminant is inorganic, then it can be removed from the groundwater by means of *floccuation*. A common practice in the waste water treatment industry, floccuation involves the addition of a chemical additive to the water stream that will form insoluble solids, such as hydroxides, sulfides, or carbonates, that precipitate from the water. The solids are then recovered and disposed of or treated off-site. Organic compounds can be captured by running the groundwater through *granular activated carbon* (GAC), which is a form of carbon with a large surface area that allows it to adsorb a wide range of compounds.

Air stripping involves removing VOCs from the groundwater by aerating the groundwater, which will result in volatilizing the VOCs. Aeration can occur in a "packed tower," which is a device in which the contaminated groundwater is sprayed through

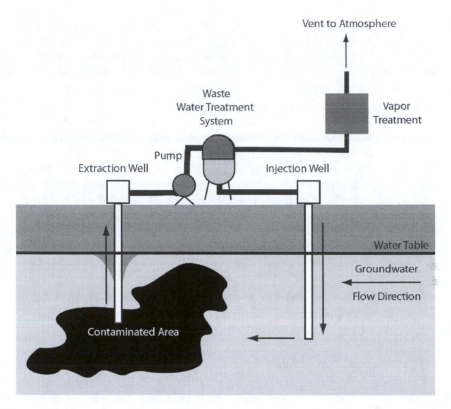

FIGURE 9.3
A schematic diagram of a typical pump-and-treat groundwater system.

a nozzle over packing inside the tower, and a fan blowing air upwards into the sprayed groundwater. The decontaminated groundwater is collected from a sump at the bottom of the tower.

The VOCs either can be released in gaseous form to the atmosphere without treatment, if allowable under the Clean Air Act and state and local regulatory agencies; or they can be captured using GAC; destroyed by biological or chemical processes. VOCs also can be destroyed by ultraviolet oxidation, in which the contaminated groundwater is treated with oxygen, typically in the form of ozone or hydrogen peroxide, and ultraviolet rays. The energy from the ultraviolet rays spurs oxidation of the contaminants, resulting in their destruction. This process is effective with simple aromatic compounds such as BTEX (benzene, toluene, ethylbenzene, and xylenes), and ethenes such as perchloroethene (PCE) and trichloroethene (TCE).

Many tests typically are performed to monitor the operation of a pump-and-treat system. To confirm that the pumping wells have the proper drawdown and therefore the expected capture zone, groundwater in the monitoring wells within the zone of influence is gauged. The extraction wells must be periodically monitored for the buildup of calcium, iron, and other fouling agents that can reduce groundwater inflow into the extraction well and therefore reduce pumping rates and system effectiveness. Collecting samples of the treated effluent periodically evaluates the effectiveness of the treatment portion of the pump-and-treat system. Wells within the groundwater plume also should be sampled to verify the pump-and-treat system's effectiveness in reducing contaminant levels within the plume.

In addition to contaminant removal, a groundwater pump-and-treat system also can be used to contain the contaminant plume. Pumping groundwater from within or upgradient of the contaminant plume can slow or prevent the downgradient migration of the plume. Obtaining physical characteristics of the soils and the aquifer will assist the geologist in determining the minimum pumping rate needed to retard or stop the plume's migration.

One of the main drawbacks of groundwater pump-and-treat is the enormous amount of effort and cost it takes to handle the water. Consider an aquifer that contains a VOC, say benzene, at a concentration of 1 milligram per liter (mg/L), which has a federal maximum contaminant level (MCL) of 0.005 mg/L. Almost all of the effort expended involves removing water; the contaminants comprise just 0.01% of the removed liquid. Pumping groundwater to the surface, moving it through a treatment system, and discharging it requires huge amounts of energy, with associated huge costs. A tiny fraction of this cost involves the contaminants themselves.

Pump-and-treat is not only a very expensive method to treat contaminated groundwater, but one that in most scenarios will fail to achieve the regulatory objectives. As a rule, the more complex the site's geology and chemistry, the more likely the pump-and-treat system will fail to achieve its cleanup goals (National Academy Press, 1994). Among the many reasons for the failure of pump-and-treat systems in achieving groundwater quality standards in contaminated water are the following:

- Heterogeneities in the geologic formation can create preferential groundwater flow patterns, bypassing contaminants located in less permeable formations.

- Migration of contaminants into inaccessible portions of the aquifer will result in the bypassing of contaminants in the groundwater withdrawal process.

- Contaminants will sorb onto soil and rock particles and not be moved by pumping the aquifer because pumping does not break the ionic bonds between contaminant and particle.

9.2.4 Steam-Enhanced Extraction

Several remedial technologies are designed to address the sorption of contaminants onto soil and rock particles in a pump-and-treat system. *Steam-enhanced extraction* involves the introduction of steam through injection wells. Heating the contaminants will cause them to volatilize and then be extracted from the subsurface through hot water recovery wells. The steam also helps mobilize the sorbed contaminants, enabling them to be removed from the subsurface. Sorption of contaminants can also be addressed by injecting surfactants into the geologic formation rather than steam (see Section 9.3.3).

9.2.5 Remediation in Biopiles

Remediation using a biopile is a combination of *ex situ* source removal in *in situ* bioremediation, which is discussed in Section 9.3.6. Contaminated soils are excavated and staged on-site. Microorganisms with the capability of consuming the contaminants are mixed into the soil pile, usually as part of an aqueous slurry. The soils are aerated concurrently, increasing the amount of oxygen available to the microorganisms. This process is less effective in cold weather, where the cold slows the metabolism of the microorganisms.

9.3 Remediation by *In Situ* Source Treatment

Due to the high costs associated with soil excavation and pump-and-treat systems and the frequent ineffectiveness of pump-and-treat systems in accomplishing their objectives, and the complications involved in soil remediation when there are physical obstructions such as buildings or underground utilities, numerous *in situ* technologies have been developed.

VOCs are particularly amenable to *in situ* remediation because their low molecular weight and low partition coefficients tend to make them more mobile than other contaminants and their plumes more widespread, both horizontally and vertically, than other contaminant plumes. Several *in situ* technologies that are frequently employed at sites are discussed in the following sections. Summaries of these methods and other in situ methods are available at USEPA's cleanup information web site: http://clu-in.org/remediation.

9.3.1 Soil Vapor Extraction

Soil vapor extraction (SVE) remediates VOCs in the unsaturated zone by taking advantage of their volatility. An SVE system induces air flow through the contaminated portion of the subsurface and makes use of the ideal gas law to affect remediation (see Figure 9.4).

As with many *in situ* remediation techniques, SVE requires favorable geologic and chemical conditions to work. The soils must have adequate permeability to allow for proper air movement (also, there shouldn't be isolated zones of low permeability—the air flow regime in the SVE could selectively bypass such zones, leaving them unremediated). They also should have low organic content, since VOCs will sorb onto surfaces with high

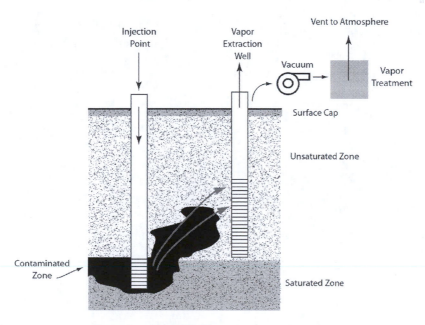

FIGURE 9.4
A schematic diagram of a soil vapor extraction (SVE) system.

organic content. The contaminant of concern should easily go into gaseous phase, making the method less amenable to the heavier VOCs or VOCs with a boiler point higher than 302°F (150°C) (Hutzler et al., 1989).

The movement of VOCs through the soil is partly controlled by diffusion, which is described by Fick's First Law (see Chapter 5). In designing an SVE system, the amount of vapor flow, or flux, induced by the SVE system can be estimated by understanding the contaminant distribution in the subsurface and the diffusion coefficient, which can be estimated by performing grain size analysis and permeability testing on the soils. The vapor flow also can be calculated by running a pilot test in the area to be remediated. A pilot test may consist of one *injection well* through which air is injected, and one or more *vapor extraction wells* from which soil vapor is removed.

A schematic map of an SVE system that might be employed to remediate the TCE in the soils at a facility is shown in Figure 9.5. Boreholes are installed upgradient and downgradient of the contaminated zone. The upgradient boreholes are converted into injection wells by installing a screen across the zone of interest. A compressor pushes air through the injection wells and into the geologic formation. The downgradient boreholes are converted into vapor extraction wells, which also are equipped with screens across the zone of interest. Vacuum pumps attached to the vapor extraction wells remove air from the geologic formation. The combination of push and pull of the air creates air flow across the formation, and the partial vacuum created induces the vaporization of the VOCs. Once in the vapor phase, the VOCs are removed from the contaminated zone through the vapor extraction wells.

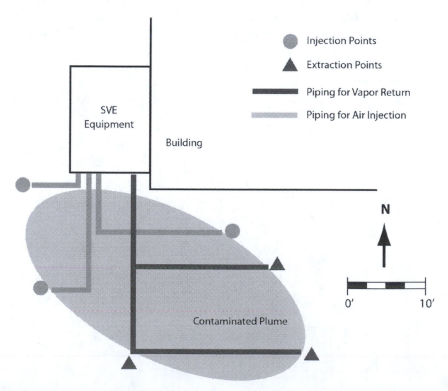

FIGURE 9.5
A schematic map plan of an SVE system at a fictitious manufacturing facility.

Construction of the SVE system involves the following steps:

1. Installation of the injection wells and the vapor extraction wells.
2. Construction of trenches and the placement of underground piping which connect the injection wells and the vapor extraction wells to the SVE equipment.
3. Installation of the SVE equipment (compressors, vacuum pumps, etc.) The SVE equipment is often located in one unit, which is sometimes brought to the site preassembled. Some type of capture mechanism is usually included in the SVE system, such as GAC described above, unless the VOCs can be vented to the atmosphere untreated.

In addition, an impermeable cap or low permeability cap or formation should be present on or near the surface for the SVE system to prevent "short-circuiting." Short-circuiting occurs when air flow bypasses the zone of interest by finding a preferential pathway in the subsurface, or due to the inadvertent creation of a preferential pathway to the surface (if the surface cap is imperfect or damaged). Cap construction is discussed in Section 9.1.1.

To monitor the SVE system, pressure differentials at the injection wells and the vapor extraction wells are measured. Samples of the removed air may be collected to confirm the removal of contaminants from the geologic formation and the absence of short-circuiting in the subsurface. Over time, the contaminant concentrations in the air effluent should decrease. The system can be shut down once no contaminants are being recovered from the soil vapor or the contaminant concentrations are sufficiently low to indicate that the remedial goals have been reached. Soil samples may be collected to confirm the calculations.

9.3.2 Air Sparging

An *air sparging* (AE) system is similar in concept to an SVE system, except that it is designed to remediate contaminants in the saturated zone (see Figure 9.6). Where VOC contamination is present in both the unsaturated and saturated zones, dual vacuum extraction systems, consisting of both SVE and AS components, are often employed, at a cost savings to the project.

In an AE system, injection wells, installed upgradient of the contaminated area, are screened across the contaminated portion of the saturated zone rather than the vadose zone as with an SVE system. The injection of air through the injection wells creates bubbling in the groundwater, sending VOCs into the vapor phase. Once in the vapor phase, the VOCs are removed from the formation through the vapor extraction wells that are located downgradient of the contaminated area. Off-gas treatment, if performed, is similar to that described for SVE systems.

Construction and monitoring of an AE system is similar to the construction and monitoring of an SVE system. The same concerns regarding short-circuiting of the air flow regime applies to AE systems as well as SVE systems.

9.3.3 Soil Flushing

In *soil flushing*, surfactants are injected into the saturated zone through injection wells located upgradient of the contaminant plume. Injecting a surfactant into the aquifer lowers the interfacial tension between the contaminants and water. By doing so, the surfactant liberates sorbed contaminants, which are then recovered in recovery wells, along with the

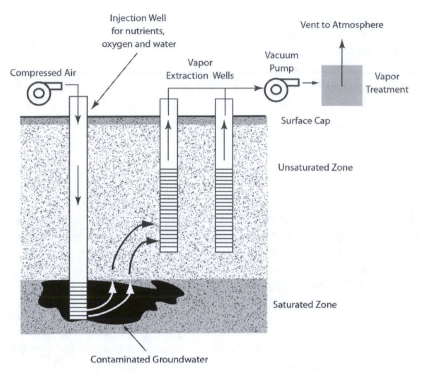

FIGURE 9.6
A schematic diagram of an air sparging system.

surfactant, which can be distilled and reused. Sometimes a co-solvent is injected into the aquifer to enhance soil flushing. Co-solvents may be organic compounds such as alcohols, or ketones, such as acetone. However, some of these co-solvents are contaminants themselves, so they must be used with caution.

9.3.4 Thermal Remediation

Thermal remediation utilizes heat to either mobilize or destroy the contaminants. The process by which contaminants are heated just enough to mobilize them and facilitate their removal from the soils is known as *thermal desorption*. The volatilized contaminants are then either collected or thermally destroyed. A thermal desorption system therefore has two major components; the desorber itself and the off-gas treatment system. The amount of heat needed to mobilize a contaminant depends on its chemical properties, with VOCs being the easiest compounds to treat.

9.3.5 *In Situ* Chemical Treatment

In situ chemical oxidation (ISCO) and *in situ chemical reduction* (ISCR) involve the injection of reagents into the contaminated area to induce oxidation or reduction of the contaminants, thus transforming them into harmless or less harmful compounds. These processes often are coupled with *in situ* bioremediation, which also can create an oxidizing or reducing environment (see Section 9.3.6). Table 9.1 summarizes the process by which contaminants are transformed to benign chemicals *in situ*.

TABLE 9.1

Four Options for *In Situ* Biochemical Remediation

	Bioremediation	Chemical Remediation
Oxidation	Bioremediation by oxidation	Chemical oxidation
Reduction	Bioremediation by reduction	Chemical reduction

9.3.5.1 In Situ *Chemical Oxidation*

In situ chemical oxidation (ISCO) involves the addition of an oxidizing agent to chemically transform a contaminant. The typical final products of ISCO are carbon dioxide and water, as with the chemical breakdown of methane, shown in the equation below.

$$CH_4 + 2O_2 \rightarrow CO_2 + 2H_2O$$

In this equation, the two oxygen atoms in the oxygen molecule transfer, or "donate" electrons to the carbon atom. The carbon atom bonds in the methane molecule releases the four hydrogen atoms in the methane molecule and bonds instead to the two oxygen molecules, thus transforming into carbon dioxide. The freed hydrogen atoms "accept" the freed electrons and also bond with oxygen atoms to become two water molecules. The processes of electron donation and acceptance are complementary, since electrons cannot remain free in ordinary conditions. Table 9.2 provides formulas showing the chemical oxidation of some common gasoline-related contaminants and the amount of oxygen needed to break down one gram of each contaminant.

Figure 9.7 shows the ISCO process. In this diagram, compounds containing electron acceptors are injected into either the saturated or unsaturated zones.

The various oxidizing compounds commonly used have advantages and disadvantages. The strength of the oxidizer is measured by the number of volts generated by one gram of the oxidizer in the oxidizing process. Table 9.3 shows the oxidation potential of five common oxidizing agents compared to oxygen. These injectants have abundant oxygen in their molecular structure, enabling them to generate large quantities of free oxygen.

One might think that the strongest oxidizer, which on Table 9.3 is the hydroxyl radical, would be the preferred oxidizer. However, strong oxidizers tend to work quickly and then dissipate quickly. Because ISCO relies on direct contact between the contaminant and the

TABLE 9.2

Oxidation Stoichiometry for Common Gasoline-related VOCs

Contaminant	Oxidation Reaction	Oxygen Requirement (Gram of O_2 per Gram of Contaminant)
Methyl tertiary-butyl ether (MTBE)	$C_5H_{12}O + 7.5O_2 \rightarrow 5CO_2 + 6H_2O$	2.7
Benzene	$C_6H_6 + 7.5O_2 \rightarrow CO_2 + 3H_2O$	3.1
Toluene	$C_6H_5CH_3 + 9O_2 \rightarrow 7CO_2 + 4H_2O$	3.1
Ethylbenzene	$C_2H_5C_6H_5 + 10.5O_2 \rightarrow 8CO_2 + 5H_2O$	3.2
Xylenes	$C_6H_4(CH_3)_2 + 10.5O_2 \rightarrow 8CO_2 + 5H_2O$	3.2
Naphthalene	$C_{10}H_8 + 12O_2 \rightarrow 10CO_2 + 4H_2O$	3.0

Source: U.S. Environmental Protection Agency, How To Evaluate Alternative Cleanup Technologies For Underground Storage Tank Sites A Guide For Corrective Action Plan Reviewers, EPA 510-B-17-003, October 2017.

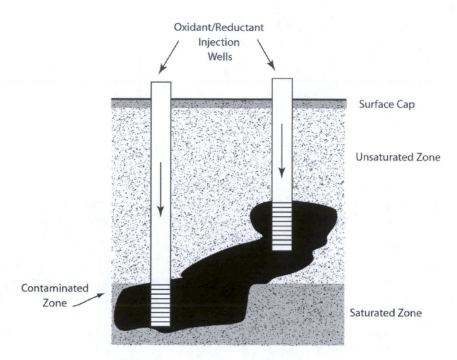

FIGURE 9.7
A schematic diagram of *in situ* chemical oxidation (ISCO).

TABLE 9.3

Oxidizing Potential for Five Common Oxidizing Agents Compared to Oxygen

Oxidizing Agent	Chemical Formula	Oxidation Potential (Volts)
Hydroxyl radical	OH^-	2.8
Sulfate radical	SO_4^-	2.6
Ozone	O_3	2.1
Hydrogen peroxide	H_2O_2	1.8
Permanganate	MnO_4^-	1.7
Oxygen	O_2	1.2

Source: U.S. Environmental Protection Agency, How To Evaluate Alternative Cleanup Technologies For Underground Storage Tank Sites A Guide For Corrective Action Plan Reviewers, EPA 510-B-17-003, October 2017.U.S. Environmental Protection Agency, How To Evaluate Alternative Cleanup Technologies For Underground Storage Tank Sites A Guide For Corrective Action Plan Reviewers, EPA 510-B-17-003, October 2017.

oxidizing agent, there is an incentive to slow down the oxidation process. Allowing the oxidizing agent to last longer in the subsurface allows it to reach a larger swath of the contaminant plume and therefore more contaminant molecules. Therefore, oxidizing agents with lower oxidation potential often are preferred for VOCs, although strong oxidizers are needed to break down polycyclic aromatic hydrocarbons (PAHs) and various recalcitrant compounds.

Carbon dioxide, one of the byproducts of the oxidation process, becomes carbonic acid when it reacts with water, which lowers the pH. Low pH can be hazardous and can harm the microorganisms that are bioremediating the contaminants concurrently. To avoid the acidification of the aquifer, buffers often are introduced into the aquifer along with the oxidizing agent. Most commercially available ISCO products contain built-in catalysts, or catalysts that are applied simultaneously to encourage the desired molecule-on-molecule interaction. Other potential negative effects of introducing common oxidizing agents into the aquifier include:

- Ozone is a caustic substance that can be dangerous to human health if inhaled. In addition, ozone, being a gas at normal temperatures and pressures, is hard to control once it has been introduced into the subsurface.
- Hydrogen peroxide can volatilize if excessive heat is generated.
- Permanganate can cause metals to precipitate out of solution, including manganese dioxide, which can clog pore spaces and decrease the permeability of the soil formation.
- Oxygen, like ozone, is a gas at normal temperatures and pressures, and is hard to control once it has been introduced into the subsurface.

In addition, subsurface conditions rarely are ideal due to the presence of geologic complexities and chemicals that can interfere with the process through their usage of the oxygen intended to remediate the contaminant. As a result, the quantities of the active ISCO ingredient injected into the formation will far exceed the theoretical quantity needed (and the quantity of the contaminant to be remediated). Bench testing can be conducted to confirm the amount of oxidant required for field implementation.

The ISCO process is monitored by periodic sampling of monitoring wells or well points in the remediation area. In addition to evaluating whether the concentrations of the contaminants are decreasing, the consultant also will test for groundwater for dissolved oxygen (DO) concentrations, oxidation/reduction (redox) potential, and pH. High DO concentrations and high redox potential means that the oxidizing agents are still at work. Similarly, high pH may indicate that carbon dioxide is being generated, more evidence that the desired chemical reactions are taking place.

9.3.5.2 In Situ *Chemical Reduction*

In situ chemical reduction (ISCR) remediates contamination in the subsurface by creating reducing conditions that are designed to remove halogens and other negatively charged ions from the molecular structure of a contaminant. In ISCR, the contaminant "accepts" electrons provided by a reducing agent, thereby transforming the contaminant into a harmless or less harmful molecule.

The most common negatively charged ion dealt with by ISCR is chlorine, a halogen. Removal of chlorine from the molecular structure of a contaminant is a process known as *reductive dechlorination*. It transforms common, toxic chlorinated solvents such as PCE and TCE, into harmless compounds, such as ethene, as shown in Figure 9.8. In essence, the dechlorination process reverses the method in which the PCE was manufactured, which was by adding chlorine atoms to ethene, a two-carbon alkene structure. Reversing the process requires substitution of a chlorine atom with a hydrogen atom, thus converting the PCE into TCE. This process is repeated until all four chlorine atoms have been removed from the molecular structure.

Reductive dechlorination is aided by the presence of an electron donor substrate, which is a food source for microorganisms (an electron acceptor is oxygen or another substance that microbes use in digesting the electron donor substrate). This food source could be lactate, acetate, glucose, or some other simple organic compound that microorganisms can metabolize. The halogens, in this case chlorine, are freed to form HCl and other such byproducts.

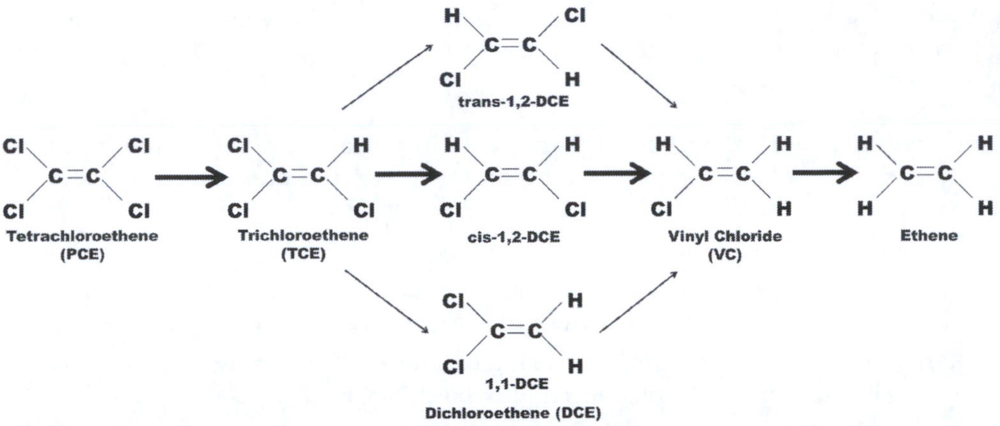

FIGURE 9.8
The reductive dechlorination of perchloroethene. (Courtesy of GZA GeoEnvironmental, Inc.)

Chemical reduction can be enhanced by the introduction of certain non-toxic metals, such as iron, into the contaminated zone. Iron is an electron donor that transfers electrons, thus enhancing the reductive process. *Zero-valent iron* (ZVI) is a specially engineered form of iron that provides a surface that enables hydrogen atoms to replace chlorine atoms in the molecular structure of a halogenated contaminant. The ZVI particles are miniscule, typically ranging from 25 to 75 microns, thus creating significant surface area in the aggregate when injected into the aquifer in slurry form.

9.3.5.3 Chemical Treatment of Metal Contaminants

The simplest example of chemical treatment of contaminants involve metals and metal compounds. With the establishment of anaerobic conditions in the subsurface and the introduction of reducing agents, metals will reduce into lower valence states. For instance, under anaerobic conditions, the highly toxic hexavalent chromium (Cr^{+6}) will transform into the much less toxic Cr^{+2} or to the metallic form of Cr^{0}.

For metals that are toxic in their lowest valence states, the goal of *in situ* metals remediation is to take the metal out of solution through precipitation (see Section 9.2.3). Precipitation of metals can be stimulated by manipulating pH and creating either oxidation or reduction conditions to precipitate the metal contaminants. Typical oxidizing agents include potassium permanganate, hydrogen peroxide, hypochlorite and chlorine gas. Typical reduction reagents include alkali metals, such as sodium and potassium, sulfur dioxide, sulfite salts, and ferrous sulfate.

9.3.6 *In Situ* Bioremediation

In situ bioremediation involves stimulating the biological activity portion of the remediation by introducing microorganisms into the contaminated area, a process known as *bioaugmentation*. These microorganisms may be indigenous to the area, thus enhancing the established population, or foreign to the area, in some cases genetically engineered. In either case, the microorganisms have properties that are desirable to the remediation of the contaminant of concern. As with chemical injections, *in situ* bioremediation involves the injection of water containing nutrients and alternate electron acceptors (food for the microorganisms) to stimulate the organism's growth and metabolic activity (see Figure 9.9).

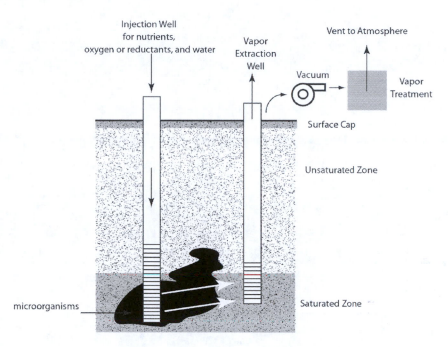

FIGURE 9.9
A schematic diagram of *in situ* bioremediation.

Many organic contaminants, particularly chlorinated compounds, may be toxic to natural bacteria. Bacteria have been bioengineered that withstand exposure to chlorine and bioremediate chlorinated compounds. Also, classes of naturally occurring bacteria, known as *methanotrophs* and *dehaloccoides*, are naturally resistant to chlorine, and are cultured in laboratories for use as agent for *in situ* bioremediation.

Pilot or bench scale testing is important for bioremediation, especially as a complementary process for ISCR. Vinyl chloride, the last step in the dechlorination process before attaining non-harmful ethene, is highly toxic to humans. It also is more stable than those compounds, so its creation increases the difficulty in completing the dechlorination process. Small scale testing prior to rolling out the full remediation would assess whether the proposed bioremediation technique will achieve full dechlorination, or stall at dichloroethene (DCE) or vinyl chloride.

Aerobic microorganisms need oxygen to metabolize organic matter. Adding oxygen will stimulate the microorganism's ability to metabolize the contaminant by oxidation. Alternatively, anaerobic microorganisms need hydrogen to metabolize the contaminant by reduction. Oxidation and reduction, however, cannot take place simultaneously in the same area, due to the differing exigent chemistries involved.

9.3.7 Bioventing

Bioventing encourages the metabolic activities of indigenous or introduced microorganisms. Air is pumped directly into the contaminanted portion of the unsaturated zone through injection wells (see Figure 9.10). Periodically, nutrients are added to the injected air, which enhance the natural biodegradation process. Vapor recovery wells screened in the unsaturated zone are used to recover vapors that may contain contaminants, although

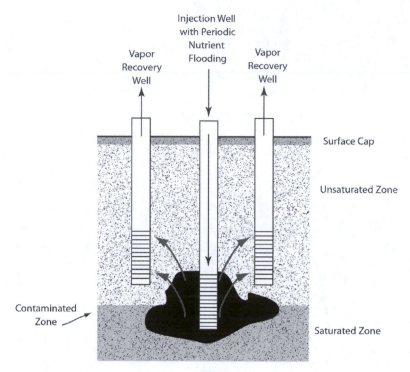

FIGURE 9.10
A schematic diagram of a bioventing system.

this aspect of the remediation is of secondary concern. As with many other *in situ* remediation methods, an impermeable cap should be installed above the zone of remediation to avoid short circuiting.

9.3.8 Permeable Reactive Barrier

A *permeable reactive barrier* is a "wall" constructed in the subsurface through which groundwater can flow (see Figure 9.11). This process, sometimes known as "funnel and gate," requires an in-depth understanding of the geologic and hydrogeologic conditions in the area of the contaminant plume.

The reactive medium contains chemicals designed to react with the contaminants dissolved in the groundwater, creating harmless substances such as carbon dioxide and water. Any of the chemical, biological, or physical agents discussed elsewhere in this chapter can be installed into a permeable reactive barrier. For instance, the barrier may contain GAC to capture organic contaminants, certain metals that will react with the contaminants, or act as catalysts by causing the dehalogenation of contaminants, zero-valent iron, microorganisms, or electron acceptors or donors.

A reactive barrier is constructed by digging a trench deep enough and wide enough to intercept the entire contaminant plume. Once constructed, the trench is backfilled with the reactive material. Impermeable slurry walls are sometimes constructed along the flanks of the reactive barrier, creating a channel or a funnel that directs the contaminant plume towards the reactive wall.

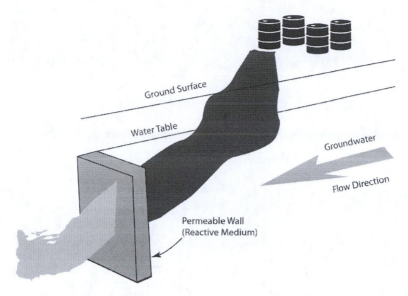

FIGURE 9.11
A schematic diagram of a permeable reactive barrier.

9.4 Performance Monitoring

Determining when the remediation has been completed is usually decided in the same manner that the extent of contamination was determined—by the collection and analysis of media samples. Similar to the investigation techniques described in Chapter 8, soil or groundwater samples are collected and analyzed for the compounds of concern.

As with the remedial investigation phase of the project, the compounds of concern should be well-known in the remediation phase of the project. At this stage, the extent of the contamination should be well-defined as well. Therefore, for soils, samples could be targeted in three dimensions (plan location as well as depth). The analytical results would then be compared with the pre-established cleanup standards for the project. As described in Chapter 8, the analytical results can be interpreted using (1) single-point compliance, which means that the analytical results for each compound in each sample must be less than the cleanup standard; or (2) compliance averaging, in which the analytical results of a specific chemical from the samples collected are treated as a single data set. The one compliance method described in Chapter 8 that does not apply to remediation is the concentration gradient, since there cannot be an area with contamination above the cleanup standard for the remediation to be deemed complete.

Verification of the completion of groundwater remediation also can rely on the single-point compliance. However, because contaminant concentrations in groundwater can fluctuate due to varying groundwater levels and seasonal conditions, compliant groundwater results should be verified in at least one round of confirmation sampling.

Deciding when groundwater remediation has been completed can lack a straightforward answer in certain cases. As often occurs in pump-and-treat systems, contaminant concentrations can become stable, or asymptotic, after a periodic of decreasing concentration. This occurs when contaminants stuck in the less accessible pore spaces of the

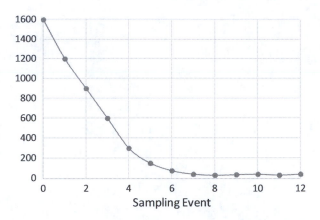

FIGURE 9.12
An example of chemical concentrations becoming asymptotic over time.

geologic formation balance out the ongoing mass removal of contaminants, resulting in no net change in concentration. This effect results in the concentration graph becoming asymptotic over time, as shown in Figure 9.12. In such cases, the pump-and-treat system may have reached a point of diminishing returns, and must either be turned off temporarily to allow the contaminants in the subsurface to equilibrate, reconfigured, or shut down completely. Alternatively, the established remediation goals have become technically impracticable, in which case they need to be revisited and reassessed.

9.5 Landfill Closures

The closure of uncontrolled landfills presents special circumstances to the remediation engineer due to the sheer quantity and variety of wastes involved. Landfill closures typically involve three main elements to control contaminant migration from the landfill: capping, venting, and leachate collection.

9.5.1 Landfill Capping

Capping involves the installation of an impermeable cap across the landfill. The cap should be designed to account for settling of the underlying materials, and needs to be monitored periodically to ensure its integrity. Caps are constructed in the same manner as described in Section 9.1.1. Figure 9.13 shows the installation of an HDPE liner during a landfill closure.

9.5.2 Methane Venting System

Organic materials deposited in the landfill continue to degrade even once waste deposition ceases. Once the landfill is capped, oxygen can no longer penetrate into the landfill, resulting in the creation of an anoxic environment. This results in anaerobic degradation or organic matter, which creates methane. Methane is explosive, and therefore must be controlled to avoid building up to dangerous concentrations within the landfill.

FIGURE 9.13
An HPDE liner is being installed over prepared subgrade as part of a cement kiln dust (CKD) landfill closure project in upstate New York. (Courtesy of Abscope Environmental, Canastota, NY.)

To control the buildup of methane within a landfill, a *methane venting system* is constructed. This system consists of a permeable gas collection layer, built into the surface cap, which is designed to collect methane that is migrating to the surface. It is constructed beneath an impermeable barrier to enhance gas collection. This layer is analogous to the permeable layer constructed beneath a building structure as part of a radon mitigation system (see Chapter 18). Vents are installed to channel the methane to the surface. They may be situated in the gas collection layer, beneath this layer within the landfill, or both. These vents may be passive or active, in which case vacuum pumps attached to the vent pump the methane to a surface collection system. In some cases, the recovered methane can be used as fuel for a cogeneration energy system.

9.5.3 Leachate Collection System

As the wastes in the landfill settle, the wastes tend to dry out, creating free liquids that, due to the impermeable cap, cannot vaporize to the atmosphere. Instead, then condense and travel downward in the landfill, and, if left uncontrolled, will carry contaminants outside of the landfilled area. To control free liquids generated within a landfill, a controlled landfill, hazardous or otherwise, is designed with a *leachate collection system*. This system consists of underground piping that is designed to capture leachate and transport it to a wastewater treatment system.

9.6 Remediation and Environmental Consultants

Consultants play many roles in the remediation of soils and groundwater. Remediations that involve heavy machinery used for trenching, excavation, etc. are usually performed by contractors, with consulting engineers observing the field activities to ensure that they follow the work plan. Consultants will be in the field to verify that the field work is conforming to their remedial design and provide perimeter air monitoring services. Follow up work, involving post-excavation sampling, groundwater monitoring, etc., is usually performed by consultants.

Consultants design remedial actions and then oversee their implementation. Contractors typically construct the remediation systems, such as soil vapor extraction and air sparging systems. Contractors typically will perform chemical injections, although consultants sometimes will do it themselves. Consultants will verify the effectiveness of the remedial action through the collection of post-excavation soil samples, groundwater samples, effluent air samples, and so on.

As in most aspects of environmental investigations and remediations, the paperwork, remedial action work plans, permit applications, and remedial action reports (prepared at the conclusion of the remedial activities) is the province of environmental consultants.

Problems and Exercises

1. Figure 9.5 is an isopleth map of perchloroethene (PCE) in groundwater. If maximum contaminant level (MCL) in the state where the facility is located is 1 microgram per liter (μg/L), describe or show the location of a monitoring well in which the PCE concentration is expected to comply with regulations.

2. Discuss the pros and cons of utilizing air sparging vs. *in situ* chemical oxidation for (1) a shallow gasoline spill with limited areal extent and for (2) a deep gasoline spill with a wide areal extent.

3. Soils at a former industrial facility are contaminated with naphthalene. What information do you need to know to evaluate the potential technologies to be used to remediate the naphthalene contamination?

4. Groundwater at a former industrial facility is contaminated with cadmium. What information do you need to know to evaluate the potential technologies to be used to remediate the naphthalene contamination?

Bibliography

Evanko, C.R., and Dzombak, D.A., October, 1997. Remediation of Metals-Contaminated Soils and Groundwater. Technology Evaluation Report TE-97-01. Ground-Water Remediation Technologies Analysis Center.

Hutzler, N., Murphy, B., and Gierke, J., 1989. *State of Technology Review: Soil Vapor Extraction Systems.* EPA/600/2-89/024.EPA Risk Reduction Engineering Laboratory, Cincinnati, ON.

Interstate Technology Regulatory Council (ITRC), December 2009. Evaluating LNAPL Remedial Technologies for Achieving Project Goals. ITRC.

Koenigsberg, S.S. (editor), 2004. *Cost-Effective Groundwater Remediation: Selected Battelle Conference Papers 2003–2004*. Regenesis.

National Research Council, 1994. *Alternatives for Ground Water Cleanup*. National Academy Press.

New Jersey Department of Environmental Protection, March 2012. Monitored Natural Attenuation Technical Guidance. Version 1.0.

New Jersey Department of Environmental Protection, October 2017. In Situ Remediation: Design Considerations and Performance Monitoring Technical Guidance Document. Version 1.0

New York State Department of Environmental Conservation, May 2010. DER-10: Technical Guidance for Site Investigation and Remediation.

Nyer, E.K. et al., 2001. *In Situ Treatment Technology*, 2nd ed. CRC Press.

Testa, S.M., and Winegardner, D.L., 2000. *Restoration of Contaminated Aquifers*, 2nd ed. CRC Press.

U.S. Environmental Protection Agency, September 1998. Technical Protocol for Evaluating Natural Attenuation of Chlorinated Solvents in Ground Water. EPA/600/R-98/128. Office of Research and Development.

U.S. Environmental Protection Agency, November 2010. Engineering Controls on Brownfields Information Guide. EPA-560-F-10-005.

U.S. Environmental Protection Agency, 2011. An approach for evaluating the progress of natural attenuation in groundwater. National Risk Management Research Laboratory. EPA/600/R-11/204.

U.S. Environmental Protection Agency, 2011. Close out procedures for the national priorities list sites. OSWER 9320.2-22.

U.S. Environmental Protection Agency, 2013. Guidance for evaluating completion of groundwater restoration remedial actions. Office of Solid Waste and Emergency Response. OSWER 9355.0-129.

U.S. Environmental Protection Agency, 2014. Groundwater remedy completion strategy. Office of Solid Waste and Emergency Response. OSWER 9200.2-144.

U.S. Environmental Protection Agency, 2014. In Situ Thermal Treatment Technologies: Lessons Learned. EPA 542-R-14-012.

U.S. Environmental Protection Agency, October 2017. How To Evaluate Alternative Cleanup Technologies For Underground Storage Tank Sites A Guide For Corrective Action Plan Reviewers. EPA 510-B-17-003.

U.S. Environmental Protection Agency. http://clu-in.org/remediation.

Watts, R.J., 1997. *Hazardous Wastes: Sources, Pathways, Receptors*. John Wiley & Sons.

Wiedemeier, T. et al., 1999. Technical Protocol for Implementing Intrinsic Remediation with Long-Term Monitoring for Natural Attenuation of Fuel Contamination Dissolved in Groundwater. Air Force Center for Environmental Excellence.

10

Vapor Intrusion Investigation and Mitigation

The previous three chapters, which describe site investigations, remedial investigations, and remediations, cover three media—surface water, soils, and groundwater. The vapor intrusion pathway merits its own chapter because vapors behave differently in the gaseous phase than do contaminants in the solid or liquids phases, the techniques used to investigate vapor intrusion are unique to this pathway, and the techniques used to mitigate a vapor hazard are different than the techniques used to remediate surface water, soil, or groundwater contamination. Vapor intrusion essentially is an indoor air hazard, but because the origins of the vapors are different, the basis of the vapor intrusion investigation and fundamental portions of the mitigation of the vapor intrusion hazard differ from indoor air, which is discussed in Chapter 17 of this book.

Vapor intrusion is a relative latecomer to the club of environmental concerns, gaining national attention in the 1990s when a study by the Massachusetts Department of Environmental Protection brought the issue to the attention of the regulatory community. It did not become part of the hazard ranking system (HRS) under Superfund until 2017, more than 30 years after the HRS was developed. As of 2018, over 35 states had developed technical guidance regarding the investigation of the vapor intrusion pathway.

10.1 What Is Vapor Intrusion?

Vapor intrusion, is the presence of toxic vapors emanating from a contamination source in the vadose zone (see Chapter 4 for a description of a vadose zone) to the breathing space inside a building. Indoor air contamination, as covered in Chapter 17, is caused by a contamination source that is inside of or part of the subject building. Vapor intrusion (VI), as the term is commonly used, refers to contamination caused by volatile organic compounds (VOCs), which are defined in Chapter 3. Methane, another common VOC that can create indoor air problems, typically is created by the anaerobic degradation of organic matter deposited in a landfill. Radon, another common indoor air contaminant, is naturally occurring and is discussed in Chapter 18. Because they are not directly related to contamination caused by the release of hazardous substances or petroleum, methane and radon typically are not chemicals of concern in a standard vapor intrusion study. The inhalation pathway for semi-volatile organic compounds (SVOCs), including polychlorinated biphenyls (PCBs), can be a concern in certain circumstances. However, VOCs are the common concern in most vapor intrusion studies.

10.2 Conceptual Site Model of Vapor Intrusion

Vapors emanating from the soils or groundwater underneath or near a building enter the building by volatilizing into the gaseous phase and migrating into the building, as shown in Figure 10.1. The pathway for chemical vapors to reach a receptor can be very complicated, more complicated than the other contaminant pathways.

First of all, chemical vapors emanating from the subsurface must have a viable pathway to the building. Impermeable soils, such as clays and fine silts, can be effective barriers to vapor migration. Bedrock fractures, coarse soils, or karst formations, on the other hand, provide excellent conduits for vapors to follow. The temperature of the soils or groundwater will affect Henry's constant (described in Chapter 5), which in turn will affect the ability of a chemical to go into the gaseous phase. The distance between the contaminated soils or groundwater and the building must be considered. This distance is vertical as well as horizontal—the further the vapors need to rise through the geological formations, the less likely they are to encounter the building in question and the more the vapors dilute with uncontaminated soil gas.

Once at or near the building, vapors must find a way inside the building. (Vapors leaving the subsurface and entering ambient air will become too diluted to present a hazard to humans.) External weather conditions such as temperature and barometric pressure will affect the ability of a chemical to go into the gaseous phase. The air flow regime inside a building needs to be understood as well. Vapors will have difficulty entering a building if it is under positive pressure, whereas building depressurization will encourage vapor intrusion. Lack of air circulation could allow vapors to build up, especially if the chemical is denser than the surrounding air. Alternatively, a robust air circulatory system inside a building can disperse vapors throughout the building, which could result in either diluting the vapor concentration to relatively harmless levels or creating a building-wide health hazard. Temperature or pressure gradients within the building will cause vapor migration inside the building.

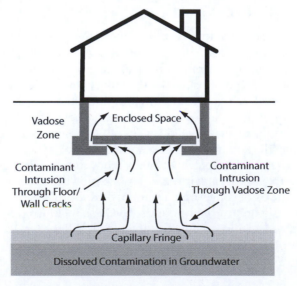

FIGURE 10.1
Conceptual diagram of vapor intrusion. (Courtesy of U.S. Environmental Protection Agency.)

The conceptual site model must consider the chemicals of concern, their physical properties and their sources. Unlike other environmental hazards, the vapor intrusion hazard affects only humans. There are different exposure scenarios to consider depending on the current and future building use. As with most receptor scenarios, residential usage, and usage by young children or immuno-compromised populations pose the greatest concerns.

10.3 Identifying Vapor Intrusion during Due Diligence

10.3.1 Vapor Intrusion under the ASTM 1527 Standard

As discussed in Chapter 6, the ASTM E1527 standard recognizes that the presence or material threat of vapor intrusion into a building qualifies as a recognized environmental condition. As vapor intrusion became an increasing concern to the regulated community for business reasons as well as for CERCLA (Comprehensive Environmental Response, Compensation, and Liability Act) liability, ASTM realized there was a need to develop a standard outside the realm of the Phase I ESA for the vapor intrusion pathway.

10.3.2 Vapor Encroachment under the ASTM 2600 Standard

In 2008, ASTM International issued its first standard practice regarding vapor intrusion. The standard is named "Standard Guide for Vapor Encroachment Screening on Property Involved in Real Estate Transactions," or E2600. The standard is separate from the E1527 standard for Phase I environmental site assessments (ESAs) and is not required for a prospective purchaser to qualify for an innocent purchaser defense under the U.S. Environmental Protection Agency's (USEPA) "All Appropriate Inquiries" rule (see Chapter 6 for an explanation of this requirement). The version of the E2600 standard discussed in this chapter was issued in 2015 (E2600-15).

Whereas the primary goal of a Phase I ESA is the identification of a recognized environmental condition, or REC, the goal of a vapor encroachment screening, or VES, is the identification of a *vapor encroachment concern*, or VEC. The E2600-15 standard defines a VEC as "the presence or likely presence of chemical of concern vapors in the subsurface of the target property caused by the release of vapors from contaminated soil and/or groundwater either on or near the target property." The reader should note the parallelism between the definition of a REC and the definition of a VEC.

The E2600-15 standard establishes a two-tier approach in assessing whether a property is being impacted by vapor intrusion, as described below.

10.3.2.1 Tier I of a Vapor Encroachment Screen

A Tier I screen is a due diligence approach similar in scope to the Phase I ESA. The objective of the Tier I screen is to identify VECs on the target property. The environmental consultant must assess whether nearby properties or the target property have chemicals that can create a VEC. To accomplish this goal, the E2600 standard requires a two-phase search of neighboring properties. To assess the significance of releases or likely releases of petroleum-related VOCs, the standard requires a search of facilities that may use, store, or may have used or stored petroleum-related VOCs that covers a 0.1 mile (0.16 kilometer) radius from the targeted property. Because certain non-petroleum-related VOCs, especially chlorinated VOCs (CVOCs) such as perchloroethene (PCE) and trichloroethene (TCE), typically travel much further distances than petroleum-related VOCs, the standard requires a search of facilities that may use, store, or may have used or stored non-petroleum-related VOCs that covers a 0.33 mile (0.54 kilometer) radius from the targeted property. These two radii are called *critical distances* under the E2600 standard.

The standard allows these critical distances to be decreased if the investigator has special knowledge about the geologic conditions in the area or if the direction of groundwater flow is known. On the other hand, if there are unusual conditions that can exacerbate vapor migration into a building, such as an earthen floor or an interior dry well, then the critical distances may need to be increased for a specific property.

Information sources for a Tier I screen are similar to the Phase I ESA information sources. The reader will note that the E2600 standard uses many of the same terms as the E1527 standard.

- Interviews of knowledgeable parties;
- A reconnaissance of the neighborhood to identify "high risk" properties, such as dry-cleaning establishments, gasoline stations, and industrial facilities;
- Reviews of "reasonably ascertainable" and "readily available" environmental records available on the target property and at local and state agencies;
- Historical record sources, especially records that indicate property usage, such as fire insurance maps and city directories;
- General geological information regarding soil type, groundwater depth, and groundwater flow direction. Because the vapor intrusion pathway is affected by subsurface conditions perhaps more than the other pathways, an in-depth understanding of area geology school, and hydrogeology is essential to an effective vapor encroachment screen.
- A standard radius search. A radius search, as described in Chapter 6, would help identify neighboring properties that would fall into the "high risk" category for vapor concerns. The databases searched are the same databases searched for a Phase I ESA. The search distances, however, are different, as shown in Table 10.1.

A Tier I screen looks for potential pathways for vapor migration and considers the future use of the property. If the future use is unknown, then the investigator is required, under the standard, to make a conservative assumption, i.e., that a sensitive population (residential, school, day care, etc.) will occupy the building.

The conclusions of the Tier I screen must be conservative. If a VEC exists or cannot be ruled out, then a Tier II screen must be performed (called a Tier III assessment in the USEPA guidance document).

10.3.2.2 Tier II of a Vapor Encroachment Screen—File Review

A Tier II screen involves non-invasive data collection, at a minimum. It includes review of site investigations and remedial investigations conducted on neighboring properties available at the site or in regulatory files to obtain a more thorough understanding of the threat posed by the neighboring property. If, after the file review, vapor encroachment remains a possibility, invasive sampling is warranted. Such sampling is part of the Tier II vapor screen under the ASTM E2600 standard, but in general is called a "vapor intrusion survey."

TABLE 10.1

Minimum Search Distances for a Vapor Encroachment Survey

	ASTM Recommended Search Radius	
Federal Databases	**Chemicals of Concern**	**Petroleum Hydrocarbons**
National Priorities List (NPL)	0.33 miles (0.54 km)	0.1 miles (0.16 km)
Superfund Enterprise Management System (SEMS)	0.33 miles (0.54 km)	0.1 miles (0.16 km)
RCRA Corrective Action Database (CORRACTS)	0.33 miles (0.54 km)	0.1 miles (0.16 km)
RCRIS-TSD (Resource Conservation and Recovery Information System-Treatment, Storage and Disposal) Database		
Federal Institutional Control/Engineering Control Registry	Site only	Site only
RCRIS-LQG (Large Quantity Generator) Database	Site only	Site only
RCRIS-SQG (Small Quantity Generator) Database	Site only	Site only
Emergency Response Notification System (ERNS)	Site only	Site only
State or Tribal Databases		
SHWS (State/Tribal Hazardous Waste Sites)	0.33 miles (0.54 km)	0.1 miles (0.16 km)
SWL/LF (Solid Waste Landfill or Landfills)	0.33 miles (0.54 km)	0.1 miles (0.16 km)
UST (Underground Storage Tanks) Database	Site only	Site only
LUST (Leaking Underground Storage Tanks)	0.33 miles (0.54 km)	0.1 miles (0.16 km)
State/Tribal Institutional Control/Engineering Control Registry	Site only	Site only
	Site only	Site only
State/Tribal Voluntary Cleanup Sites	Site only	Site only
State/Tribal Brownfields Sites		

Source: ASTM International, Standard Guide for Vapor Encroachment Screening on Property Involved in Real Estate Transactions, E2600-15, 2015.

10.4 Vapor Intrusion Survey Triggers

As indicated above, a vapor intrusion survey could be triggered by the results of a Phase I ESA or a VES. It also can be triggered by the identification of VOCs in the groundwater on or near the subject property. The mere identification of VOCs in the groundwater, however, is not sufficient to trigger a vapor intrusion survey. The VOCs need to be sufficiently close to the subject building and at a sufficient concentration in the groundwater to present a health hazard to building occupants should the VOCs infiltrate into the subject building. Determining the likelihood of this scenario has two parts—the pathway from the contaminated groundwater to the building and the pathway through the building to its occupants.

10.4.1 Generic Horizontal Trigger Distances

As per USEPA guidance, a VOC source less than 100 feet (30.5 m) vertically or laterally from a building should trigger a vapor intrusion survey. Some states have shortened the triggering horizontal distance for petroleum-related VOCs to 30 feet (10.1 m), with no consideration for the vertical distance to be traveled by the chemical. Technical guidance developed by other states have two separate criteria for VOCs related and unrelated to petroleum, with vertical distances sometimes included in the formula. Soil type is also a consideration in a few states when assigning critical distances.

10.4.2 Calculating Site-Specific Trigger Distances

Instead of using generic search distances, site-specific search distances can be calculated using attenuation factors. An *attenuation factor* is a unitless number that calculates the expected concentration of a given contaminant based on its concentration in the subsurface and its generic trigger concentration, as follows:

$$C_{indoor} = (C_{water} \times H) \times \alpha_{gw}$$

where:

C_{indoor} is the contaminant concentration in the indoor air
C_{water} is the contaminant concentration in groundwater
H is Henry's constant, and
α_{gw} is the attenuation factor between groundwater and indoor air.

Generic attenuation factors recommended by Kansas Department of Health and Environment are shown in Table 10.2. The shorter the pathway to the indoor air, the closer the attenuation factor is to 1. Since a crawl space provides a direct pathway into the breathable air, it is assigned an attenuation factor of 1.

10.4.3 Triggering Concentrations of Volatile Organic Compounds in Groundwater

The second part of the equation is determining the threshold concentrations of VOCs in the groundwater that will trigger a vapor concern. The USEPA and many states have developed lookup tables with generic values based on conservative exposure scenarios. In May 2018, the USEPA released an updated version of its vapor intrusion screening level (VISL) calculator. The VISL lists generic triggering concentrations for 388 chemicals, including all chemical categories (SVOCs, pesticides, metals, etc.) as well as VOCs. It assumes an ideal pathway for a chemical to enter the building, that is, an unobstructed pathway with minimal attenuation in the subsurface and a direct pathway into the building. The VISL can be downloaded from https://www.epa.gov/vaporintrusion/vapor-intrusion-screening-level-calculator.

TABLE 10.2

Generic Attenuation Factors

Media	Attenuation Factor
Groundwater	0.001
Groundwater (fine-grained soils)	0.0005
Sub-slab soil gas	0.03
Shallow external soil gas (near source)	0.03
Crawl space air	1

Source: Kansas Department of Health and Environment, Topeka, KS, August 2016.

10.4.4 Establishing Indoor Air Target Values

To calculate site-specific exposure scenarios, the USEPA has developed a spreadsheet based on the Johnson and Ettinger model, which is widely accepted as the method to calculate site-specific chemical concentration triggers. The model can be accessed at the following web site: https://www.epa.gov/vaporintrusion/epa-spreadsheet-modeling-subsurface-vapor-intrusion#model. The inputs into this model include:

- Maximum chemical concentration in the groundwater;
- Depth to groundwater;
- Groundwater temperature;
- Building information such as depth and thickness of the foundation, and the rate at which air is circulated and exchanged with outside air;
- A target risk for carcinogens, typically 1×10^{-6} or 1×10^{-5}, and a target hazard quotient of 1 for non-carcinogens (see Chapter 8 for a discussion on risk factors associated with carcinogens and non-carcinogens).

Some states utilize formulas similar to the risk assessment formulas provided in Chapter 8. As a rule, the more that is known about the site-specific conditions, the more accurately the environmental professional can calculate site-specific and chemical-specific information to decide whether to conduct a vapor intrusion survey.

10.5 Vapor Intrusion Investigation

A vapor intrusion investigation consists of three steps: (1) the pre-sampling survey; (2) sample collection; and (3) sample analysis.

10.5.1 Pre-sampling Survey

The objective of the pre-sampling survey is to identify the potential for background contamination caused by chemicals inside the building creating false positive readings during sampling and analysis. The pre-sampling survey involves taking an inventory of chemicals that are present inside the building; looking for potential chemical sources in the building, such as dry cleaning, new carpeting, or other materials that can give off gases, etc. Chemicals should be removed prior to collecting the air samples to the extent feasible. If they cannot be removed, such as the case of new carpeting, the investigator should note the chemicals normally associated with the object and take that into account when interpreting the analytical results from the samples collected.

The pre-sampling survey also should attempt to identify *preferential pathways* into and within the building. Preferential pathways can include utility openings in the foundations for services such as natural gas lines, water lines, sewer lines, and electric lines. Preferential pathways also could include sumps and other openings designed to manage water intrusion. This pathway could be particularly important for buildings with high water tables that can bring the contaminants right up to the building's foundation floor.

Cracks in the building foundation will provide easy access for vapors to enter a building. There are natural preferential pathways as well, such as bedrock fractures, as noted above.

Once potential sources of VOCs have been removed and preferential pathways into the building have been identified, it is time to collect air samples. Three types of samples are collected during the sampling phase of the vapor intrusion investigation: indoor air samples, sub-slab soil vapor samples, and ambient samples. These sample types are described below.

10.5.2 Indoor Air Sampling

Indoor air samples are designed to test the quality of the air that building occupants inhale. An indoor air sample is collected by placing a *summa canister* on the lowest floor of the building at strategic locations, such as nearest the plume or near floor openings, such as sumps. In non-residential settings, indoor air samples should be collected while the building's air circulation system is not operating to simulate the worst-case scenario, which is vapors concentrating at a particular location, with maximum exposure to the workers at that location. Part of the worst-case scenario is a closed building. Therefore, outlets to the outside should be sealed during the VI survey. Many northern states recommend performing VI surveys during the winter months, when there is minimal air flow through windows and doors and the heating system is in operation. Samples should be collected near building openings, where sensitive receptors are located, or where a large quantity of receptors are located, again to simulate worst-case conditions. The number of samples to be collected is a factor of the building size. Figure 10.2 shows the locations of indoor air samples in a typical vapor intrusion survey.

10.5.3 Sub-slab Sampling

The type of sample that is associated only with vapor intrusion studies is the *sub-slab soil vapor sample*. The sub-slab soil vapor sample provides the intermediate step between the subsurface contamination and the indoor air. For a vapor intrusion pathway to be complete, the contamination must migrate upward from the subsurface, penetrate the building, and enter the occupied portion of the building. The sub-slab soil vapor sample provides the nexus between the source and the receptor.

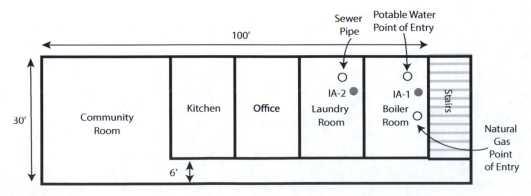

FIGURE 10.2
Typical indoor air sampling locations.

FIGURE 10.3
A summa canister being used for sub-slab vapor sampling. (Courtesy of Dermody Consulting, New York.)

Sub-slab sampling (see Figure 10.3) involves drilling a narrow hole through the foundation floor of the building and installing a probe rod into the hole. The rod is connected to a summa canister using flexible tubing and the soil vapors will flow into the summa canister once the connection is made since the interior of the summa canister starts out as a vacuum. Ideally, the hole is drilled in the part of the building closest to the potential VOC source or preferential pathway, although variations in building construction should also be taken into account when locating sub-slab sampling points.

An air-tight seal must be formed around the tubing leading from the probe rod to the summa canister. Otherwise, ambient air might enter the tubing and contaminate the sample. To assess whether ambient air has entered the tubing, a tracer compound introduced to the base of the soil probe so that it flows into the summa canister. If it is detected in the indoor air sample, then the canister will contain soil vapor. It is important that the tracer gas is not a contaminant of concern at the facility.

10.5.4 Ambient Air Sampling

An *ambient air sample* performs the same quality control function as a field blank in a subsurface investigation (see Chapter 7). It is collected outside the building away from obvious sources of volatile chemicals such as filling stations, dry cleaners, or gasoline-powered equipment. It is collected over the same time interval as the indoor air sample. If a compound is detected in both the ambient air sample and an indoor air sample, it may not be related to vapor intrusion conditions but rather to external conditions unrelated to the subsurface.

10.5.5 Near Slab Sampling

A *near slab soil gas sample* is collected as a substitute for a sub-slab sample. As its name implies, near slab soil samples are collected outside but close to the subject building, ideally no more than 10 feet (3.1 m) from the building's foundation. A near slab soil gas sample is less desirable than a sub-slab sample because can only offer an approximation of vapor buildup beneath the building, since it is not within the footprint of the building. Also, vapors are less likely to build up at the location of a near slab sample since there is no foundation to act as a barrier to migration. Despite its shortcomings, near slab samples sometimes are warranted, especially when the owner of the subject building does not grant access to the building or there are unsafe conditions inside the building.

10.5.6 Vapor Testing on Vacant Land

In certain situations, the threat of vapor intrusion needs to be assessed although no building is on the property. A prospective purchaser or developer of a property may want this information for due diligence purposes or to assess whether vapor mitigation measures, as described later in this chapter, would be warranted in a proposed new construction. In such scenarios, the presence of volatile vapors in the subsurface would be assessed by conducting a soil gas survey, as described in Chapter 7. The closer the sample can be collected to the VOCs in the groundwater, both horizontally and vertically, the more likely vapors will be detected that can intrude into the planned building.

10.5.7 Sample Analysis

A fixed-based laboratory typically analyzes the air samples, although the analysis can be performed by a mobile laboratory. The chemicals of concern are VOCs, but the list of VOCs targeted in a vapor intrusion survey differs from the Target Compound List (TCL) of VOCs discussed in Chapter 4 because the TO-15 list of VOCs is a subset of hazardous air pollutants (HAPs) established under the Clean Air Act whereas the TCL VOC list was developed for the Superfund program. The VOCs on the TO-15 list are, in general, the more toxic and/ or more volatile of the VOCs on either the TCL or the full HAPs list (see Table 10.3 for a listing of targeted VOCs in a TO-15 analysis).

10.5.8 Interpreting the Results

Once the results of the laboratory analysis are obtained, they are compared to generic guidance values, known as *indoor air screening levels* (IASLs), or site-specific values calculated using the methodologies described above. Just because a chemical's concentration in an indoor air sample is greater than IASL does not mean that a vapor intrusion condition exists. The pathway must be complete, which means that three things must hold: the chemical must be present in the subsurface, in the sub-slab soil gas sample, and in the indoor air sample, all at concentrations greater than their applicable screening values. Table 10.4 provides a reference by which to interpret vapor sampling results.

TABLE 10.3

TO-15 Analytes

Compound	CAS No.	Included in TCL VOCs?
Acetone	67-64-1	√
Acetonitrile	75-05-8	
Acrylonitrile	107-13-1	
3-Chloropropene (Allyl chloride)	107-05-1	
Benzene	71-43-2	√
Benzyl chloride	100-44-7	
Bromodichloromethane	75-27-4	√
Bromoethane (Ethyl bromide)	74-96-4	
Bromoform	75-25-2	√
Bromomethane	74-83-9	√
1,3-Butadiene	106-99-0	
n-Butane	106-97-8	
Chlorobenzene	108-90-7	√
Chloroethane	75-00-3	√
Chloroform	67-66-3	√
Chloromethane	74-87-3	√
Carbon disulfide	75-15-0	√
Carbon tetrachloride	56-23-5	√
2-Chlorotoluene	95-49-8	
Cyclohexane	110-82-7	√
Dibromochloromethane	124-48-1	√
1,2-Dibromoethane	106-93-4	√
1,2-Dichlorobenzene	95-50-1	√
1,3-Dichlorobenzene	541-73-1	√
1,4-Dichlorobenzene	106-46-7	√
Freon 12 (Dichlorodifluoromethane)	75-71-8	√
1,1-Dichloroethane	75-34-3	√
1,2-Dichloroethane	107-06-2	√
1,1-Dichloroethene	75-35-4	√
1,2-Dichloroethene (cis)	156-59-2	√
1,2-Dichloroethene (trans)	156-60-65	√
1,2-Dichloropropane	78-87-5	√
1,3-Dichloropropene (cis)	542-75-6	√
1,3-Dichloropropene (trans)	10061-02-6	√
Freon 114 (1,2-Dichlorotetrafluoroethane)	76-14-2	
1,4-Dioxane	123-91-1	√
Ethyl acetate	141-78-6	
Ethanol	64-17-5	
Ethylbenzene	100-41-4	√
4-Ethyltoluene	622-96-8	
n-Heptane	142-82-5	
Hexachloro-1,3-butadiene	87-68-3	
n-Hexane	110-54-3	
Isopropyl alcohol (2-Propanol)	67-30-0	

(Continued)

TABLE 10.3 (*Continued*)

TO-15 Analytes

Compound	CAS No.	Included in TCL VOCs?
Isopropylbenzene (Cumene)	98-82-8	√
Methylene chloride	75-09-2	√
2-Hexanone (MBK)	591-78-6	√
2-Butanone (MEK)	78-93-3	√
4-Methyl-2-pentanone (MIBK)	108-10-1	√
Methyl methacrylate	80-62-6	
Methyl-tert-butyl ether (MTBE)	1634-04-4	√
Naphthalene	91-20-3	
Propylene	115-07-1	
Styrene	100-42-5	√
Tertiary butyl alcohol (TBA)	75-65-0	
1,1,2,2-Tetrachloroethane	79-34-5	√
Tetrachloroethene	127-18-4	√
Tetrahydrofuran	109-99-9	
Toluene	108-88-3	√
1,2,4-Trichlorobenzene	120-82-1	√
1,1,1-Trichloroethane	71-55-6	√
1,1,2-Trichloroethane	79-00-5	√
Trichloroethene	79-01-6	√
Freon 11 (Trichlorofluoromethane)	75-69-4	√
Freon 113 (1,1,2-Trichloro-1,1,2-trifluoroethane)	76-13-1	√
1,2,4-Trimethylbenzene	95-63-6	
1,3,5-Trimethylbenzene	108-67-8	
2,2,4-Trimethylpentane (Isooctane)	540-81-1	
Vinyl acetate	108-05-4	
Bromoethene (Vinyl bromide)	593-60-2	
Vinyl chloride	75-01-4	√
Xylene (para & meta)	1330-20-7	√
Xylene (ortho)	95-47-6	√

If the contaminant detected did not originate in the subsurface, then the investigator needs to determine its source. The investigator should cross-check the detected chemicals with the findings of the pre-sampling survey. Presence of the chemical in the sub-slab sample but not the vadose zone might indicate "short circuiting" in the chemical pathway, which could occur when a chemical inside the building migrates through a preferential subsurface pathway and is detected in the sub-slab sample. The investigator also should cross-check the results of the ambient air sample with the indoor air sample(s). If the chemical is present in the ambient air sample then outside cross-contamination could be the cause of the exceedance in the indoor air sample(s).

TABLE 10.4

Meaning of Vapor Intrusion Sampling Results

Is Chemical Present In:			
Indoor Air Sample	**Sub-slab Sample**	**Subsurface**	**Interpretation**
No	Yes	Yes	No vapor intrusion issue inside the building.
Yes	No	Yes	No vapor intrusion issue inside the building. However, an indoor air quality issue may exist. Check for other sources of indoor air pollution, such as chemical storage or usage inside the building.
Yes	Yes	No	An indoor air quality issue may exist, but it is not related to the known subsurface contamination. No vapor intrusion issue inside the building.
Yes	Yes	Yes	Contaminant is present in the subsurface, in soil gas just beneath the foundation, and in the indoor air. Pathway is complete. A vapor intrusion issue exists inside the building.

Because indoor air is affected by a myriad of factors, chemical concentrations in indoor air samples can vary significantly. The investigator must keep in mind that false negatives can be as likely as false positives. Sometimes it is advisable to redo the indoor air tests—often, the results do not repeat. This may not be practical, however, if the regulatory agency with jurisdiction requires conclusions to be based on worst-case scenarios.

Some states have established screening levels for indoor air sample results that would indicate the possibility of acute exposure for building residents. What constitutes an acute exposure scenario varies from state to state. This condition may call for immediate mitigation measures, as described in Section 10.6.

If a vapor intrusion condition is discovered in a building, there is an increased likelihood that vapor intrusion conditions exist in neighboring buildings, since they would have most if not all of the same factors (groundwater contamination, favorable geology, pathways into structures, etc.) as the impacted building.

10.6 Vapor Intrusion Mitigation

If acute exposure conditions exist, then immediate actions need to be taken. Pathways into the building should be sealed and air purifying units can be deployed. An *air purifying unit* typically is equipped with granular activated carbon (GAC) filters that capture the volatile vapors. These filters need to be changed regularly, so this remedial solution is usually temporary.

Permanent solutions for the mitigation of vapor intrusion conditions entails two procedures—sealing the foundation, and channeling vapors that have accumulated beneath the building slab.

10.6.1 Vapor Barrier Design and Construction

A *vapor barrier* is a gas-impermeable layer that is intended to prevent contaminants from entering the building. In most cases, vapor barriers are installed for new construction, although existing buildings can be retrofitted with vapor barriers. The most common vapor barrier consists of *high-density polyethylene* (HDPE) that is engineered not only to prevent vapor intrusion but to remain inert and not react with the contaminant. In new construction, the HDPE layer is installed before the foundation is poured. It is fitted around the utility conduits into the building, and a sealant is applied so that there are no gaps between the conduits and the HDPE layer. The effectiveness of the seals can be assessed by smoke testing, which entails injecting smoke beneath the vapor barrier and observing whether smoke emanates from beneath the barrier. Figure 10.4 shows the installation of a vapor barrier over the concrete slab of a building under construction.

Liquid vapor barriers can accommodate new construction as well as some existing structures. Most liquid vapor barriers are emulations that are applied to a foundation. Once hardened, they provide a continuous barrier that is engineered not to crack or otherwise degrade over time.

10.6.2 Sub-slab Depressurization System Design and Construction

A *sub-slab depressurization system* (SSDS) is an engineered system designed to recover vapors from beneath a building and outside its subsurface walls and channel the vapors safely away from the building so that they cannot be inhaled by humans. This system is basically

FIGURE 10.4
Installation of a vapor barrier.

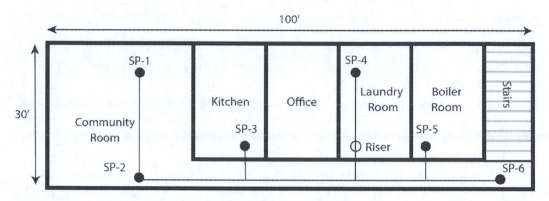

FIGURE 10.5
Schematic diagram of an active sub-slab depressurization system.

a radon mitigation system, which is described in Chapter 18. An SSDS can be designed for new construction as well as existing buildings. It consists of a collection system of piping that is installed beneath the building foundation. The pipes are manifolded together and attached to a riser, which channels the vapors out of the building. The riser portion of the construction is necessary because channeling the vapor out of the building on the first floor of the building puts humans at risk who are walking outside the building near the vent pipe.

The SSDSs can be either passive or active. In passive systems, the vapors find their way through the piping system by means of diffusion. This process is slow and inefficient, but may be all that is warranted for less severe vapor intrusion conditions. Active systems use a fan or fans to create negative pressure regimes that pull vapors through the pipes and out the exhaust. Active systems are more expensive to install and maintain than passive systems, but may be necessary depending on the threat posed by the vapor intrusion condition. Figure 10.5 shows a schematic diagram of a sub-slab depressurization system.

10.7 Vapor Intrusion and Environmental Consultants

Environmental consultants play several roles in the investigation and mitigation of vapor intrusion. Environmental consultants conduct vapor encroachment screens and do the initial sampling for vapor intrusion. They perform the calculations to determine maximum acceptable concentrations of contaminants. Sub-slab testing can be conducted by either an environmental consultant or an environmental contractor. Environmental engineers design vapor mitigation systems and perform engineering inspections at critical junctures in the installation of vapor barriers and SSDSs. Professional engineers typically will certify the construction of the SSDS and monitor its operation and maintenance.

Problems and Exercises

1. Why does vapor intrusion only consider the inhalation pathway into the human body and not other fauna?

2. Trichloroethene is present in groundwater beneath a building at a concentration of 100 micrograms per liter (μg/L). Using the generic attenuation factors shown in Table 10.2, what is the expected TCE concentration inside the building if the soils consist of sand? If they consist of clay?

3. Discuss the importance of preferential pathways in assessing whether vapor intrusion conditions exist inside a building.

4. Discuss the importance of preferential pathways in the design of a sub-slab depressurization system.

Bibliography

ASTM International, 2013. Standard Practice for Environmental Site Assessments: Phase I Environmental Site Assessment Process. E1527-13.

ASTM International, 2015. Standard Guide for Vapor Encroachment Screening on Property Involved in Real Estate Transactions. E2600-15.

Environmental Quality Management, Inc., 1997. User's Guide for the Johnson and Ettinger (1991) Model for Subsurface Vapor Intrusion into Buildings.

Environmental Quality Management, Inc., 2004. User's Guide for Evaluating Subsurface Vapor Intrusion into Buildings.

GeoSyntec Consultants, Inc. web site. https://www.geosyntec.com/vapor-intrusion-guidance.

Interstate Technology & Regulatory Council, 2007. Vapor Intrusion Pathway: A Practical Guideline.

Kansas Department of Health and Environment, Topeka, KS, August 2016.

Massachusetts Department of Environmental Protection, October 2016. Vapor Intrusion Guidance: Site Assessment, Mitigation, and Closure. Policy #WSC-16-435.

New Jersey Department of Environmental Protection, 2018. Vapor Intrusion Technical Guidance. Version 4.1.

U.S. Environmental Protection Agency, 2005. Johnson and Ettinger (1991) Model for Subsurface Vapor Intrusion into Buildings.

U.S. Environmental Protection Agency, 2010. Review of the Draft 2002 Subsurface Vapor Intrusion Guidance. Office of Solid Waste and Emergency Response.

U.S. Environmental Protection Agency, 2013. Guidance for Addressing Petroleum Vapor Intrusion At Leaking Underground Storage Tank Sites. Office of Underground Storage Tanks.

U.S. Environmental Protection Agency, 2015. OSWER Technical Guide for Assessing and Mitigating the Vapor Intrusion Pathway from Subsurface Vapor Sources to Indoor Air. Office of Solid Waste and Emergency Response. OSWER Publication 9200.2-154.

Washington Department of Ecology, 2018. Guidance for Evaluating Soil Vapor Intrusion in Washington State: Investigation and Remedial Action.

Section III

Land Development and Redevelopment

11

Brownfields

11.1 Urban Decay and Urban Renewal

The Comprehensive Environmental Response, Compensation, and Liability Act (CERCLA), commonly known as Superfund (see Chapters 3 and 6 for background information about Superfund), affected the economy of the United States in ways not envisioned by its creators. The threat of liability, especially the onerous threats of strict, joint, and several liability costs, put a chill on real estate owners, developers, financers, and insurers who had financial interests in urban areas. Such areas were having difficulty attracting investment capital due to blighted, deteriorating infrastructure, the additional costs associated with demolishing old buildings compared to the availability of undeveloped, unencumbered properties outside of the urban core, and many other reasons unrelated to environmental degradation. Fear of inadvertently incurring Superfund liability added another layer of concern to entities interested in investing in old, urban areas. By creating a stigma regarding properties that have actual or perceived contamination, Superfund quite possibly accelerated the process of urban decay and complicated governmental efforts at urban renewal.

11.2 The Brownfields Act

The *Small Business Liability Relief and Brownfields Revitalization Act of 2002*, commonly known as the "Brownfields Act," addressed some of the shortcomings of the original Superfund law. The term "brownfields" was coined to identify properties in the hazy middle ground between unpolluted properties colloquially called "green fields" and heavily polluted properties. Depending on how it is defined, there could be hundreds of thousands of brownfields in the U.S.

11.2.1 Formal Definition of a Brownfield

According to the Brownfields Act, a brownfield is "a property, the expansion, redevelopment, or reuse of which may be complicated by the presence or potential presence of a hazardous substance, pollutant, or contaminant." The Act excludes the following types of known contaminated properties, to avoid conflicting with environmental priorities under other federal statutes:

- Properties that are listed or are proposed for listing on the National Priorities List (NPL)
- Properties that are subject to a unilateral administrative order, consent decree, or a similar type of legal proceeding under a federal environmental statute
- RCRA corrective action facilities (as described in Chapter 3)
- Title C solid waste disposal facilities as defined by RCRA
- Properties that are managed by the United States for a Native American tribe
- Portions of a property with releases of polychlorinated biphenyls (PCBs)
- Portions of a property with leaking underground storage tanks that are being remediated using funds from the federal Leaking Underground Storage Tank Trust Fund.

As with the earlier chapters on investigation and remediation, brownfields typically are real estate, that is, properties that can be bought or sold. The appearance in the definition of brownfields of terms such as "expansion," "redevelopment," and "reuse" identifies them by their economic rather than their technical or scientific condition. The definition implicitly states that brownfields are properties that currently are unused, underutilized, or not being used to their best economic advantage. As such, they are contributors to urban blight and its associated sociological problems.

11.2.2 Practical Definition of a Brownfield

For practical purposes, brownfields can be separated into two categories. The first category includes properties for which property expansion redevelopment (which will be called "redevelopment" for the rest of this chapter) likely is economically viable. Typical economic evaluations of such a property are shown in Tables 11.1 and 11.2. The complications in redeveloping the property consist of uncertainties in timing, budget, liability relating to the presence or potential presence of contamination.

   The second category includes properties whose redevelopment is complicated because of economic considerations. Such properties, common in areas of urban blight, might be worthless or even have a negative worth (colloquially known as an "underwater" or "upside-down" property) due to environmental contamination. Consider the same property as evaluated in Table 11.1, but, as shown in Table 11.3, with a much higher likelihood of contamination.

TABLE 11.1

Economic Evaluation of a Viable Brownfield Property Redevelopment with Diminished Value Due to Threat of Contamination

Property value, no environmental impacts	$100,000
Cost to remediate potential environmental impacts	$150,000–$300,000
Likelihood of environmental impacts	20%
Expected value of environmental impacts	$30,000–$60,000
Average expected value of environmental impacts	$45,000
Expected property value	$55,000

TABLE 11.2

Economic Evaluation of a High-value Brownfield Property
That Likely is Contaminated

Property value, no environmental impacts	$250,000
Cost to remediate potential environmental impacts	$150,000–$300,000
Likelihood of environmental impacts	60%
Expected value of environmental impacts	$90,000–$180,000
Average expected value of environmental impacts	$135,000
Expected property value	$115,000

TABLE 11.3

Economic Evaluation of a Brownfield Property with Negative
Worth Due to Real or Perceived Contamination

Property value, no environmental impacts	$100,000
Cost to remediate potential environmental impacts	$150,000–$300,000
Likelihood of environmental impacts	50%
Expected value of environmental impacts	$75,000–$150,000
Average expected value of environmental impacts	$125,000
Expected property value	−$25,000

Because of the perception that the property is likely to be contaminated, it actually has negative worth. That if the property is not contaminated, then the project is economically viable. In such cases, the uncertain potential costs of remediation results in a lack of interest by the business community.

11.3 Objectives of the Brownfields Act

The barriers to expansion, redevelopment, and reuse of brownfields can be grouped into two categories: liability protections and economic incentives.

11.3.1 Liability Protections under the Brownfields Act

11.3.1.1 Relief from Superfund Liability

One of the primary goals of the Brownfields Act was to establish the criteria with which a purchaser of contaminated sites could be exempt from Superfund liability. The prospective purchaser would have to satisfy the Bona Fide Prospective Purchaser provisions, as described in Chapter 6, which include conducting an All Appropriate Inquiries-compliant Phase I environmental site assessment and having no business relationship with any prior property owner or operator who may have been responsible for the contamination on the property. The BUILD Act of 2018 expands the liability protections for local and state governments that seek to take control of contaminated properties within their jurisdiction and expanded the Bona Fide Prospective Purchaser provision to include entities that have

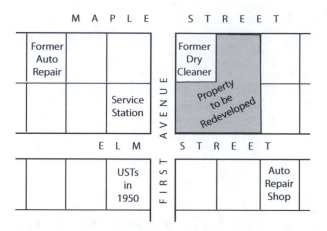

FIGURE 11.1
Environmental concerns in a typical urban area.

tenancy or leaseholds on the property. State brownfields programs have similar stipulations for developers to avoid similar liability under state regulations.

11.3.1.2 Relief from Off-site Contamination Concerns

Brownfields often lie in urban areas that have become blighted through past operations, age, and disuse. Such areas may harbor numerous properties with environmental concerns, as demonstrated on Figure 11.1. Financial organizations may be reluctant to supply capital to a redevelopment project in such an area for fear of being sucked into the cleanups and liabilities associated with several of the neighboring properties. The developer may not show an interest in redeveloping such a property not only for financial concerns but also due to the complexities deriving from multiple contaminated properties contaminating each other and other nearby properties.

In acknowledgement of this issue, several states have passed brownfields legislation that grants the developer immunity from remediating contamination emanating from another site (especially if the same contaminants are found on the subject property). In exchange, the developer is typically asked to commit the requisite resources to the project and to complete the project in a certain timeframe. This liability protection is a win-win: a major barrier has been removed for the developer, and the municipality benefits through the conversion of a blighted property into one that will add to the tax base and possibly attract other developers to the area. It should be noted that state brownfields laws and provisions do not exempt the prospective purchaser from Superfund liability.

11.3.2 Economic Incentives under the Brownfields Act

In addition to avoidance of liability, federal and state brownfields programs offer a wide variety of economic incentives to prospective brownfields redevelopers. Some of these incentives are described below.

11.3.2.1 Brownfields Assessment Grants

Seed money for a redevelopment project often is needed during the feasibility phase of the development project. Federal programs such as the Community Block Grant Program administered by the U.S. Department of Housing and Urban Development (HUD) are designed to provide seed money to developers so that the project does not die before it can be born.

11.3.2.2 Low Interest or No Interest Loans

Many federal and state programs offer low interest or no interest loans to prospective brownfield redevelopers. The objective of the loan programs is to provide seed money upfront so that the planning and assessment portions of the project can begin and make the project economics more attractive. Typically, the loans come out of a *revolving loan fund* that must be paid back within a specified time period so that the same money can be loaned to the next qualifying brownfields project.

11.3.2.3 Tax Abatements and Tax Forgiveness

Certain states, counties, and municipalities have set up zones that are subject to lower taxes to spur development. Such zones may be known as "opportunity zones," "urban enterprise zones," or some similar designation. In addition, governmental agencies can offer lower or no taxes to specific redevelopment projects to enhance their economic viability. This often is done if the project is of sufficient size that by itself it is expected to spur economic revitalization of the surrounding area.

11.3.2.4 Tax Credits

Certain states, counties, and municipalities go beyond tax abatement and tax forgiveness by offering tax credits to brownfield developers. The amount of the tax credit typically is tied to the expected income stream from the redevelopment. Tax credits are particularly popular among developers that are not-for-profit, non-governmental organizations (NGOs). Since they don't pay income taxes, they are allowed to sell these tax credits to non-governmental entities that are subject to taxation. This is a popular method of raising funds for NGOs involved in the construction of affordable and supportive housing and other such below market rate housing projects.

11.3.2.5 Tax Increment Financing

Figure 11.2 is a graph demonstrating how a TIF works. *Tax increment financing* (TIF) is a public financing method that is designed to spur new development, increase property values, and add to the tax base which, by doing so, will allow the investment to pay for itself, at least in theory. Tax increment financing typically is used on large, area-wide redevelopment projects.

For a brownfields redevelopment project, a public agency, a quasi-public agency, or a public-private partnership invests money towards brownfield cleanups in the designated area. Once completed, the project creates jobs and attracts other businesses which in turn create jobs. Property values rise and tax revenues increase from additional property taxes. Part of the tax revenues goes to the basic municipal services that were already being provided, such as schools, police, fire protection, water and sewer services, and so on. The other part of the tax revenue, which represents the "tax increment," or additional taxes collected due to the rise in property values, is used to cover the initial outlay for the redevelopment. The money a city invests in TIF projects is often raised by issuing bonds whose interest is paid over the period of the bond.

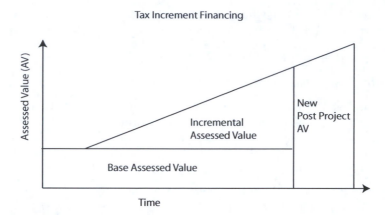

FIGURE 11.2
Tax incremental financing.

11.3.2.6 Insurance Protection

There is available private insurance protection from liability, discussed below, and whose costs may be partially defrayed by some state governments as an incentive for redevelopment.

11.4 Integrating Brownfields Redevelopment with Urban Redevelopment

The 2017 Brownfield Federal Programs Guide, published by the U.S. Environmental Protection Agency (USEPA), lists brownfields programs sponsored by numerous and diverse federal and multi-state entities in the United States. Each of these agencies seeks to address brownfields within the context of its mission. Some examples of agencies, their missions, and the programs they offer to support those missions are provided in Table 11.4.

TABLE 11.4

Examples of U.S. Agencies and Their Missions regarding Brownfields

Name of Agency	Mission regarding Brownfields	Programs Offered
U.S. Department of Agriculture	Revitalize rural communities	(1) Renewable Energy and Energy Efficiency Improvements Program; (2) housing programs; (3) community facilities programs; (4) business programs; (5) cooperative programs; (6) electric programs; (7) telecommunications programs; (8) water and environment programs; and (9) community development programs
U.S. Department of Commerce	Encourage brownfields redevelopment	Provides grants, technical assistance, and loans to states, regions, and communities and to infrastructure projects

(Continued)

TABLE 11.4 (*Continued*)

Examples of U.S. Agencies and their Missions regarding Brownfields

Name of Agency	Mission regarding Brownfields	Programs Offered
U.S. DOC—National Oceanic and Atmospheric Administration	Bolstering the economic vitality of coastal communities	Provides outreach and technical assistance to communities
U.S. Department of Defense		
U.S. Department of Energy	Advance the national, economic, and energy security of the United States	Provide technical and economic assistance to communities; promote energy-saving technologies
U.S. Department of Health and Human Services—Office of Community Services	Address the economic and social services needs of the urban and rural poor at the local level	Provides grant money to community development corporations and community action agencies
U.S. Department of Housing and Urban Development	Create strong, sustainable, inclusive communities and quality affordable homes	Provide grants and loans to state and local governments for integrating brownfields redevelopment planning with transportation and housing planning and to meet safe and affordable housing needs
U.S. Department of the Interior—National Park Service	Preserves natural and cultural resources	Assist state and local governments and community-based organizations in natural resource conservation and outdoor recreation initiatives and in the acquisition of surplus federal lands
U.S. Department of Labor	Advances opportunities for profitable employment	Trains workers so that they can work on brownfields redevelopment projects
U.S. Department of Transportation—Federal Transit Administration	Provides financial and technical assistance to local public transit systems	Encourages public work projects to include brownfield redevelopment and encourages brownfield redevelopers to consider public transportation in their plans
National Endowment for the Arts	Encourage participation in the arts	Using art to improve the esthetics of brownfield redevelopment projects
U.S. Small Business Administration	Aid and protect the interests of small businesses	Provide loans to small businesses interested in locating on revitalized brownfields
Appalachian Regional Commission	Strengthen economic growth in Appalachia	Provides grant money to areas and counties

Source: U.S. Environmental Protection Agency, 2017 Brownfields Federal Programs Guide, 2017.

11.5 Cleanup of a Brownfields Site

On a brownfields redevelopment project, the environmental consultant will be part of a development team. That team will consist of some or all of the following professions and entities:

- Property owner. In certain circumstances, governmental entities may choose to employ eminent domain powers to take a property from its owner. So-called "takings" are done for the greater public good and not for the benefit of a particular non-governmental entity. Public entities taking or coming into ownership of private property are protected from Superfund liability under the Brownfields Act.

- Developer (if different from the property owner). There may be multiple developers on large or complicated projects.

- Financial institution(s). As shown above, there are numerous financial tools available to a brownfields redeveloper. Multiple means of financing may be used on a single project. In addition, different financial institutions can provide liquidity from different sources or to different portions of the project. There can be first-tier lenders, second-tier lenders, and so on. As a result, the financing of brownfields projects can get very complicated, and require the services of a professional well-versed in the various financial instruments to be employed on a project.

- Company providing environmental insurance. In addition to pollution liability insurance, insurance companies may offer "excess of indemnity" insurance on large, complicated projects. Excess of indemnity insurance provides insurance if costs on the overall project exceed an agreed-upon limit.

- Public regulatory agencies. Because brownfield redevelopment covers so many issues, multiple regulatory agencies will get involved. It is helpful to define a lead regulatory agency at the beginning of the project to avoid turf battles as the project progresses.

- Municipal agencies. In addition to the building department, other municipal agencies, such as public works, participate in the planning and execution of a brownfield project.

- Utilities. Public utilities may get involved in the planning stage of a project to ensure that the new development will have adequate electrical and usually natural gas service.

- Community groups. Community groups, some of which may be formally organized and recognized by the local government, may provide input on the proposed development, especially if it may change the character of the area.

- Architect. The design of the new building can be influenced by the existing contamination. In addition, vapor mitigation measures, which often are needed in a building being constructed on a brownfield, must be coordinated with the architecture designs. Vapor mitigation measures are described in Chapter 10 of this book.

- Attorneys. The brownfield developer will need various types of legal advice. A real estate attorney is an integral part of the team on a brownfield redevelopment project, as may be attorneys with expertise in project financing, construction, and environmental regulations, if the investigation and remediation of the real estate will be complicated.

- Project manager. Some brownfield redevelopers hire project managers who specialize in construction projects, especially if they do not have this expertise in-house.

- Geotechnical consultant. The geotechnical consultant will test the subsurface for its ability to withstand the load to be constructed on the property. Geotechnical consultants employ many of the same methods as used by environmental consultants, namely borehole drilling and soil sampling. However, soil samples are tested for their physical properties rather than their chemical properties.

- Asbestos abatement contractor, if there is a building on the property that requires asbestos removal prior to demolition.

- Demolition contractor, if there is a building on the property.

- General construction contractor. The general contractor, or GC, is responsible for the construction of the new building. The GC will hire subcontractors for tasks that it does not have in-house. Typical subcontractors will perform tasks related to steel construction, masonry, roofing, electrical and plumbing systems.

The environmental investigation and remediation of a brownfields site is similar to the investigation and remediation of any contaminated site, with some notable exceptions. As demonstrated above, the environmental consultant must coordinate activities with a host of other participants and stakeholders. In addition, the environmental investigation and remediation must be oriented towards the future usage rather than current usage of the property. If the property is being repurposed for residential usage, then residential remediation criteria must be employed, regardless of the current or former usage of the property. Importantly, the timeliness of the environmental remediation is critical to the success of the overall redevelopment project. Remedies that take years to achieve generally are not desirable on brownfields sites where the developer has a tight schedule, as do its financers and other stakeholders, in many cases. For instance, some public grants and loans have so-called "sunset clauses," mandating that the money be spent within a certain time period. The environmental consultant must take this into account when selecting the remedial action(s) for the property.

In certain circumstances, the environmental consultant's findings can influence the proposed construction. For instance, if the environmental due diligence phase of the project identifies a former filling station or dry cleaner on the portion of the property where the building is to be constructed, or if the subsurface investigation identifies significant contamination on that portion of the property, then the project team may choose to move the future building to an uncontaminated, or presumably uncontaminated, portion of the property and place the parking area, or passive recreational portion of the property, where the contamination is present or suspected to be present.

11.6 Brownfields and Environmental Consultants

Environmental consultants can wear many hats on a brownfield redevelopment project. When the project is being formulated, environmental consultants can help to obtain financing for the project by applying for public or private grants or by assisting others in these endeavors. The consultant may work with the lead developer and the project attorney or project manager in identifying potential sources of financing and completing the applications.

Before a shovel is placed into the ground on a redevelopment site, the environmental consulting is busy performing all or some of the following activities:

- Perform a Phase I environmental site assessment (see Chapter 6), which is a fundamental step in determining the feasibility of the redevelopment project and obtaining the necessary financing for the project. A vapor encroachment survey, as described in Chapter 10, may be conducted in concert with the Phase I ESA.
- Conduct a subsurface investigation (as described in Chapter 7). The subsurface investigation may include testing for the presence of soil vapors within the

footprint of the proposed building (see Chapter 10). If vapor intrusion concerns are identified, then the consultant will work with the project architect to design a vapor mitigation system for the new building.

- Prepare an environmental impact statement (EIS) if federal or in some cases state or local funding is involved (see Chapter 13). The EIS findings may indicate that changes need to be made in the property redevelopment, such as identifying the need for more parking or better site access, or concerns regarding noise generated during construction or concerns regarding shadows that the building would cast onto sensitive properties. The services of an environmental consultant may be needed if there are concerns relating to waterfronts, streams, wetlands, or endangered or threatened species, as described in Chapter 12.

- A radon survey (see Chapter 18) may be warranted if there is an existing building on the property, to assess whether the new building should be outfitted with a radon mitigation system. If no building is present and a radon hazard is suspected, then the development team may authorize the environmental consultant to construct a radon mitigation system prophylactically.

- Perform a hazardous materials inventory (HMI), as described in Chapter 3, if there is an existing building on the property. The HMI may include, at a minimum, an asbestos survey, as described in Chapter 14, a survey for lead-containing paint (see Chapter 15) to comply with OSHA, and a survey for PCB-containing materials, universal wastes, and hazardous wastes, as described in Chapter 3. If asbestos-containing materials are identified, then an environmental consultant will design and monitor the asbestos removal (none of the other abatement options would be feasible since the building would be slated for demolition). The consultant may hire the asbestos abatement contractor directly or solicit bids from qualified contractors so that they can be hired by a third party, usually the lead developer or the demolition contractor.

    The environmental consultant also might be responsible for the removal of lead-containing paint, PCB-containing materials, and the universal and hazardous wastes.

Once the hazardous materials have been removed and the building has been demolished, a myriad of other activities await the environmental consultant. Soil or groundwater remediation, if warranted, typically will be performed after the building has been demolished, although some remediation can be conducted prior to building demolition. Soil remediation often can be performed as part of the proposed construction, such as by removing contaminated soil so that the building basement can be constructed, or remediating groundwater by dewatering the excavation, which would have occurred regardless of environmental conditions on the property. The environmental consultant may perform perimeter air monitoring during the removal of contaminated soils (see Chapter 9) and other contaminated media, and will prepare the necessary paperwork to submit to the lead regulatory agency as applicable.

If vapor mitigation measures are warranted (see Chapter 10), the environmental consultant will monitor their construction and certify their completion. If the vapor mitigation measures include a sub-slab depressurization system (SSDS), the consultant will monitor its startup, and prepare an operations and maintenance manual to ensure proper operation of the system in the future.

If the property hasn't been remediated fully, then it is likely that the applicable regulatory agency has established activity and use limitations (AULs) on all or part of the property.

An environmental consultant will monitor the effectiveness of the AULs as long as they are needed, which could in theory be forever, if the contaminants do not degrade naturally. Environmental consultants also will monitor the effectiveness of the SSDS, if one was installed in the building.

In summary, a brownfields redevelopment project has the potential to require all of the tools in the toolbox of an environmental professional.

Bibliography

Good Jobs First, 2019. Tracking Subsidies, Promoting Accountability in Economic Development. www.goodjobsfirst.org.

Grand River Corridor web site, http://grandriver.fhgov.com/About/Funding.aspx.

Hersh, B., 2018. *Urban Redevelopment—A North American Reader.* Routledge, A Taylor & Francis Group.

Hollander, J.B., Kirkwood, N.G., and Gold, J.L., 2010. *Principles of Brownfield Regeneration.* Island Press.

National Association of Local Government Environmental Professionals, 2006. Superfund Liability—A Continuing Obstacle to Brownfields Redevelopment.

U.S. Environmental Protection Agency, 2006. Anatomy of Brownfields Redevelopment. https://www.epa.gov/brownfields/anatomy-brownfields-redevelopment-october-2006.

U.S. Environmental Protection Agency, 2015. Unlocking Brownfields Redevelopment: Establishing a Local Revolving Loan Fund Program.

U.S. Environmental Protection Agency, 2017. 2017 Brownfields Federal Programs Guide.

U.S. Environmental Protection Agency, 2017. Putting Sites to Work—How Superfund Redevelopment in EPA Region 2 Is Making a Difference in Communities.

U.S. Environmental Protection Agency, 2011. Handbook of the Benefits, Costs, and Impacts of Land Reuse and Development. EPA-240-R-11-001.

12

Ecological Investigation, Protection, and Restoration

Ecological investigations entail the assessment of the health of all or parts of an ecosystem. An *ecosystem* consists of the totality of life, its biodiversity, flora and fauna interacting with abiotic factors—air, soil, sunlight, and water. When an ecosystem is unhindered by humans, the result is a natural engine that provides terrestrial life with a healthy environment, an environment predicated on clean air and water. A healthy ecosystem is characterized by high levels of biodiversity, whereas an unhealthy ecosystem is deficient in biodiversity or is plagued by invasive species, often a result of large uninterrupted, anthropogenically converted landscapes.

Technically ecosystems can be damaged by changing one or more of the following factors:

- The amount of water available to the system (such as through water diversion, river channelization, or natural means, such as flood or drought)
- The configuration of the system, such as through the construction of dams or the infilling of surface waters
- The system chemistry by introducing chemicals into the system or changing the basic water chemistry parameters
- The biology of the system through all of the above, in addition to the introduction of non-native species, or destruction of native species

An ecological investigation can be triggered by a future event that has the potential to disturb an existing ecosystem. The National Environmental Policy Act (NEPA), which is described in Chapter 13, mandates the evaluation of a planned action on the ecology of the area of the planned action, considering a wide range of potential impacts of the proposed action on the area's ecology. An ecological investigation also can be triggered if the ecosystem has been damaged or is suspected to have been damaged by one of these factors. Contamination, whose investigation and remediation are described in Chapters 6 through 9 of this book, triggers an ecological investigation when it is suspected to have altered the chemistry of a nearby surface water body. Of specific interest in a site investigation or remedial investigation, is the impact of contamination on *ecological receptors*, which include surface water, sediments, wetlands, flora, and non-human fauna. Special attention is given to threatened and endangered species of flora and fauna because of the fragility of their existence. The following sections describe the investigation of these ecological receptors and mitigation methods to repair damaged receptors.

12.1 Surface Water and Sediment Investigation and Mitigation

12.1.1 Pollution Sources

Sources of surface water and sediment pollution originate from either point sources or non-point sources. *Point sources*, as the name implies, originate from a particular point or place, such a wastewater pipe outfall (see Figure 12.1). *Non-point sources* (NPSs) result from rainfall or snowmelt that picks up contaminants while moving over and through the ground and depositing them in nearby water bodies. Typical contaminants from non-point sources include fertilizer and pesticide runoff from a farm; chemicals settling out of the air, chemicals transported downstream from a variety of sources, or no discernable source; and diffuse anthropogenic pollution (DAP), such as particulate emissions from motor vehicles and stationary combustion sources, such as power plants. Non-point sources, in the aggregate, may constitute a larger, perhaps regional problem that can be hard to diagnose and difficult to mitigate.

12.1.2 Pollution Pathways in Surface Waters

As with human receptors, the pathways to ecological receptors include inhalation, ingestion, and dermal exposure (injection is not considered an exposure route for non-human animals). However, there is an additional exposure route for non-human animals: bioaccumulation, which represents an indirect exposure route through the consumption of contaminated prey.

Bioaccumulation is the process by which, once consumed, a contaminant can work its way up the food chain. It does so by absorbing into the tissue of the animal that consumed

FIGURE 12.1
A wastewater outfall pipe. (Courtesy of GZA GeoEnvironmental, Inc.)

it at a faster rate than it can be metabolized or excreted. When that animal is consumed by another animal, that animal absorbs the contaminant, and so on. An extreme form of bioaccumulation, known as biomagnification, is when the contaminant increases in concentration within the animal as it moves up the food chain. As with bioaccumulation, biomagnification occurs for persistent chemicals that are not easily metabolized or excreted. The contaminant actually can increase in concentration under certain bioenergetic conditions. Such contaminants are known as *bioaccumulants*.

12.1.3 Surface Water Sampling and Analysis

Evidence of stressed vegetation, sheens on the water or sediments, seeps, or discolored soils or sediment are indicative of present or past spills. Evidence of fish kills, physical abnormalities, or other degradation of flora and fauna populations may also be indicative of present or past spills and trigger surface water and sediment sampling.

Surface water should be sampled if discharges of contaminants into the surface water body have occurred or are expected to have occurred. When point source pollution is suspected, samples should be collected as close to the suspected source in a stationary water body, or the point of deposition in a flowing water body, which is usually the point where the water is moving the slowest. In a flowing water body, the sample collected at the point of deposition should be bracketed by samples upstream and downstream. The upstream sample provides quality control by determining background contaminant concentrations. The various parameters that commonly are used to measure surface water quality are discussed in Section 4.7 of this book.

Figure 12.2 shows a fictitious manufacturing facility with a wastewater pipe emptying into a brook. Some volatile constituents will enter the atmosphere when the wastewater

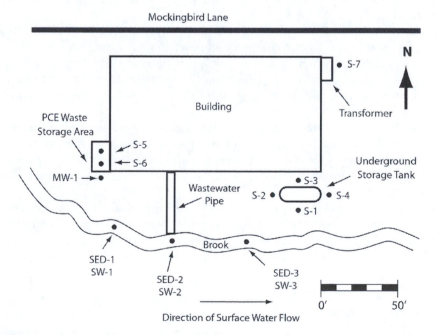

FIGURE 12.2
Surface water and sediment sampling to test for point source pollution emanating from a wastewater pipe at a manufacturing facility.

carrying the volatile compounds discharges from the pipe. Most of the dissolved contaminants and non-aqueous liquids floating on the wastewater, however, will mix with the receiving brook water. These chemicals and liquids will move downstream, although chemicals with high soil-water coefficients or high organic carbon coefficients may adhere to the sediment in the bottom of the stream. Under some hydraulic conditions, chemicals may work their way into the underlying soils beneath the stream.

To assess surface water quality, or the effects of non-point pollution sources on the water body, a more thorough sampling of the water body is warranted. Since there are wide variations in the types and sizes of surface water bodies, there is no "one size fits all" sampling procedure. However, there are certain universal principles that will aid the consultant in the proper design of a sampling and analysis plan.

As a general rule, the quantity of water samples to be collected depends on the size of the water body and the expected chemical and biological variations within the water body. For instance, one sampling point is usually sufficient for narrow/shallow streams that are suspected to have homogeneous conditions along the stream. For deeper and wider streams, more sampling locations across various "transects" (cross-sections of the stream) are necessary because physical conditions can vary horizontally and vertically, which can affect the chemistry of the water.

The amount of interaction with the atmosphere varies with depth in deeper water bodies. Collecting samples from the middle and lower depths of the water body will avoid water that has interacted with the atmosphere and therefore may not be representative of the entire water body. On the other hand, there may not be such a thing as a representative sample in a water body with vertical circulation patterns, or influences from discharges, tributaries, and so forth. Depending on the goals of the investigation, samples from multiple depths may need to be collected. Temperature readings at varied depths in the water body can indicate whether the water chemistry is sufficiently stratified to warrant multiple samples.

When collecting multiple samples from a water body, one should be aware that the collection process itself can affect subsequent samples. For instance, collecting a sample in a river could agitate the water that is heading downstream and thus affect subsequent downstream samples. Collecting samples from downstream to upstream locations would avoid this problem. Similarly, sampling should proceed from the top to the bottom of a water body.

Collecting samples at different times often is needed to characterize the water body. The effects of seasonal weather (cold vs. hot, wet vs. dry), and related biological variations; high flow/low flow stages for rivers and streams; and tidal variations for applicable water bodies, can lead to chemical variations that cannot be measured or understood in one sampling event.

The quality and quantity of water flow are often important parameters in assessing flowing surface waters. High flow rates, or turbulent flow, can affect chemical measurements, especially for volatile organic compounds (VOCs) and sample collection itself, since collecting a water sample in the middle of a fast-flowing river can be difficult or impossible.

12.1.4 Sediment Sampling

Sediments as well as the surface water itself should be sampled when contamination is known or suspected to have occurred since contaminants can sorb onto soil particles. Contaminated sediments can be a repository for contaminants that can be remobilized back into the overlying water body, and a conduit for pollution into the subsurface and pathway for pollutants into the underlying aquifer (see Chapter 5). Sediments also can provide a pathway by which contaminants enter the food chain through flora and bottom-feeding fauna, which can in turn be a source of food for fish, small mammals, and birds.

As with the surface water samples, sediment samples should be collected upstream, downstream and at the location of the point discharge in a flowing water body and near the point of discharge in a stationary water body. The water sample should be analyzed for the contaminants suspected or known to have been released to the surface water body.

Because of the dangers of bioaccumulation to aquatic animals, sediment sampling and analysis should focus upon those chemicals that are persistent in the aquatic environment, have high bioaccumulation potential, are toxic or teratogenic (causing birth defects) to aquatic organisms, and have a high frequency of detection. Analyzing the tissues of fish and other aquatic animals and can assist in identifying bioaccumulative effects in the aquatic food chain.

The goals of sediment sampling are the identification and delineation of contamination, which is similar to soil and groundwater sampling (see Chapters 7 and 8). In the case of sediment contamination, however, background conditions may be difficult to establish, since the point of origin of the contamination may be difficult to identify. In addition, sediment contamination changes with time, since the contamination is located in a dynamic environment.

At a minimum, samples should be collected from the top several inches of sediment, possibly supplemented by deeper samples. Collecting measurements for total organic carbon (TOC), pH, and particle grain size aids in the understanding of contaminant migration through the sediments.

12.1.5 Ecological Risk Assessment

As with soil and groundwater sampling and analysis, generic standards are available by which to compare analytical results to establish whether the contaminant concentrations present an "acceptable" risk to the environment. However, because of the myriad variations in environmental settings, these values can often be misleading. Therefore, where ecological impacts are suspect, an ecological risk assessment is performed.

An ecological risk assessment is analogous to a human health risk assessment as described in Chapter 8 in that its goal is to establish levels of contaminants that will not have a detrimental effect on the subject fauna and flora. The contaminants of concern are generally the same for ecological and human health risk assessments, although the levels of concern signified by certain contaminants differ.

The environmental risk assessment can be thought of containing three phases (NJDEP 2011):

- Phase I: Problem Formulation
- Phase II: Analysis
- Phase III: Risk Characterization

The goal of the first phase of the assessment is to develop an *ecological conceptual site model* (ECSM). The ECSM is similar to the conceptual site model described in Chapter 7. Possible source-to-pathway-to-receptor scenarios are developed for the various contaminants of concern in the study area. The four exposure routes discussed in Section 12.1.2, namely ingestion, direct contact, inhalation, and bioaccumulation, are considered.

Preliminary data is collected to investigate the potential exposures to receptors under the various scenarios. The goal of the data collection is to establish *measurement endpoints*, that is, the results of sampling and analyses that are used to estimate the exposure, effects, and characteristics of the ecosystem. These measurement endpoints are as varied as the ecosystem itself and are difficult to generalize. They include measurements of the parameters discussed earlier in this chapter for surface water, sediment, and wildlife sampling, as well as parameters specific to a given scenario.

In the second phase of the ecological risk assessment, the data collected in Phase I are evaluated and potential contaminant exposures to receptors and their potential effects on the receptors are estimated. The evaluation considers all four of the above-mentioned exposure routes.

The third and final phase of the ecological risk assessment is the *toxicity assessment*, which entails:

- Researching the latest toxicological studies involving the chemicals of concern
- Performing biological surveys to assess actual area conditions
- Performing toxicity tests on actual media

A key component of ecological risk characterization is *food chain modeling*, which predicts how contaminants work their way up the food chain. The pathways considered in food chain modeling include bioaccumulation as well as direct consumption of contaminated water, soil, or sediment. The total potential exposure for the animal in question, known as the *potential dose*, is the sum of these three pathways.

Estimating the contribution of bioaccumulation to the total potential exposure is done by conducting a *bioaccumulation (uptake) study*, which includes a tissue residue study. Alternatively, the investigator can use generic *sediment-soils-to-biota bioaccumulation factors* (BSAFs) and *bioaccumulative factors* (BAFs) that are published by the U.S. Army Corps of Engineers (USACOE) (http://el.erdc.usace.army.mil/basnew), the U.S. Environmental Protection Agency (USEPA) (www.epa.gov/med/Prod_Pubs/bsaf) and other sources.

Once an assessment endpoint has been calculated, it is compared to a *toxicity reference value* (TRV). If the assessment endpoint is greater than the TRV, then an ecological hazard exists and must be remediated, or avoided, in the case of a proposed action.

Because biological systems vary seasonally and over time in other ways, the uncertainty surrounding estimates of ecological risk may be much greater than those associated with a human health risk assessment. The investigator should account for variabilities when evaluating the results of the ecological risk assessment.

12.1.6 Surface Water Restoration

As with soil and groundwater remediation, the goals in surface water restoration typically are based on generic surface water cleanup criteria, which in turn are based on the expected highest and best usage of the surface water body. These uses are defined by exposure scenarios, with protected watersheds meriting the most stringent criteria. Next in line are fresh surface water bodies, which require more protection than salt water bodies. Other surface water categories, in descending order of importance, include drinking water sources, surface waters that sustain fish consumed by humans, surface waters that sustain only fish that are not consumed by humans, surface waters with no human consumption pattern but are consumed by native wildlife, and surface waters with no consumption by or habitat for humans or wildlife.

Even though generic standards are based on exposure scenarios, area-specific remediation standards can be generated by performing an ecological risk assessment, as described earlier in this chapter, or generic.

Step one in restoring a surface water body is to stop or retard the source of the contamination. When the contamination originates from a point source, eliminating that point source is the beginning of the restoration process. If the point source cannot be removed, the surface water body may be restored by using the *total maximum daily load* (TMDL) concept. The goal of the TMDL concept is to limit and/or decrease the aggregate contaminant load being introduced into the water body and allowing natural forces to clean up the water body over a period of time, in some cases years or even decades. Dischargers are apportioned TMDLs based on their operations and usage of the water body. Government often gets involved in forming stakeholder groups, assigning TMDLs, and monitoring and enforcing compliance.

Care must be taken to avoid damaging the water body in the course of fixing it, by inadvertently changing the water chemistry or the physical attributes of the water body in a way detrimental to the resident flora and fauna.

12.1.7 Sediment Remediation

Sediment remediation generally requires source removal since pathway mitigation often is not feasible. Dredging contaminated sediments from the surface water body can remove the main mass of contaminants. Dredging, however, can stir up contaminated sediments which can reintroduce contaminants into the surface body's food chain. *Hydraulic dredging* is a process by which sediments are sucked through a pipe that trails at the bottom of the surface water body, similar to vacuum cleaner. Hydraulic dredging is less likely to stir up and re-suspend contaminated sediments, and is therefore more desirable than traditional physical dredging, although it is the slower and more expensive method of dredging.

12.1.8 Cultural Eutrophication and Mitigation

Of particular concern to the health of surface water bodies are phosphorus and nitrogen. Phosphorus and nitrogen are nutrients used by plant and animal life. High levels of phosphorus and nitrogen accelerate an ecosystem's productivity, resulting in a sharp growth in aquatic plant life, such as the algal bloom shown in Figure 12.3, or the spread of invasive and nuisance aquatic plants. This is known as *cultural eutrophication*, which is an increase in plant life (eutrophication) due to human (cultural) activities. Algal blooms can block sunlight, killing off other aquatic plants. Certain algal species can also produce toxins that can affect other plant and animal life, including humans (e.g., red-tides algae; cyanobacteria). Phosphorus and nitrogen often originate from fertilizers that are transported into a water body by storm water runoff. Once their introduction is brought under control, the algae die and decompose, resulting in an oxygen-deficient environment that can kill aquatic animals.

FIGURE 12.3
Algal bloom in a river in Sichuan, China. (Courtesy of Felix Andrews.)

12.2 Wetland Identification, Delineation, and Mitigation

Section 404 of the Clean Water Act identifies the following six categories of special aquatic sites:

- Sanctuaries and refuges
- Wetlands
- Mudflats
- Vegetated shallows
- Coral reefs
- Riffle and pool complexes

Of these six categories, *wetlands* merits special attention due to its prevalence throughout the United States, its highly productive ecology (when healthy), its biological diversity, and its impact on real estate and development.

12.2.1 Definition of Wetlands

Identifying the extent of a wetland on a property is critical, particularly on properties slated for development, since Section 404 of the Clean Water Act regulates the disturbance of managed wetlands. Section 404 of the Clean Water Act defines wetlands as:

> those areas that are inundated or saturated by surface or ground water at a frequency and duration sufficient to support, and that under normal circumstances do support, a prevalence of vegetation typically adapted for life in saturated soil conditions. Wetlands generally include swamps, marshes, bogs, and similar areas.

This definition contains phrases that need to be clarified.

Under normal circumstances—this does not include areas disturbed by human influences, and conditions at times of extreme droughts or similar events.

A prevalence of vegetation—Vegetation associated with a wetland, hydrophytic vegetation, must dominate an area to be considered a wetland.

For a wetland to be present, positive *wetland indicators* of hydrology, soil type, and vegetation types must be present. Sole reliance on the presence of wetland vegetation can be misleading because many plant species commonly associated with wetlands also can grow successfully on non-wetlands. Furthermore, hydrophytic vegetation and hydric soil conditions could persist for decades after the alteration of the area hydrology rendered the area a non-wetland.

12.2.2 Wetland Hydrology

A wetland need not be wet all of the time and may not often even have visibly standing water. The frequency and duration of wetness depends upon many factors. However, all wetlands have sufficient water so that the surface soils are sufficiently wet to create anaerobic conditions in the soil for a significant portion of the growing season. Either surface water, or groundwater at an elevation no more than 14 inches (4.7 cm) below the soil surface, should be present for at least several weeks of the growing season. Because the growing season is longer in the warmer climates, the process of wetland development generally is faster in the southern U.S. than in the northern U.S.

12.2.3 Hydric Soils

Most surface soils have aerobic conditions due to the ready flow of air through the unsaturated pore spaces of the soil. However, when soils become saturated with water due to rainfall and flooding, the water impedes the flow of gases, including oxygen, through the soil. Microbial activities also act to deplete the soils of oxygen, creating hydric conditions that lead to the creation of hydric soil. *Hydric soil* is "a soil that formed under conditions of saturation, flooding, or ponding long enough during the growing season to develop anaerobic conditions in the upper part" (Federal Register, July 13, 1994). The National Technical Committee for Hydric Soils (NTCHS) defines and develops or accepts criteria for hydric soils.

Hydric soils have been saturated often and over many years, allowing them to develop unique properties that can be recognized in the field. Because of the lack of oxygen in the soils, reducing conditions prevail. Reducing conditions cause the creation of iron and manganese oxides, which give hydric soils their distinctive coloring. *Gleyed soils*, a type of hydric soils, have a distinctive grey coloring, sometimes greenish or bluish grey. Hydric soils often have *mottles* (also called redoximorphic accretions and depletions), which are spots of contrasting color.

Hydric soils support hydrophytic vegetation. If drained by human activity, such as the construction of dams or levees or by some other impediment to natural surface water flow, or by natural conditions such as prolonged drought, they may no longer qualify as hydric soils.

Water is introduced into an area by precipitation, natural drainage patterns, periodic flooding, and high-water tables. A wetland's ability to stay wet depends upon the amount of water introduced into the area and the soil's ability to retain that water. A soil's ability to retain water depends upon the type of soil, its porosity, and its permeability. Beach sand, for instance, may become saturated constantly by wave action. However, the water quickly drains from the sand once a wave recedes, and dries quickly. Clays and hardpan soils are common to wetland environments because of their ability to retain water. However, sandy wetlands do exist, depending upon high groundwater levels present for a significant portion of the growing season.

12.2.4 Wetland Vegetation

Wetland vegetation is defined by the ability of the individual species of plant to actively grow and reproduce in an anaerobic environment. There are various wetland plant communities, which are species that are naturally associated with each other. Vegetation associated with wetlands is known as *hydrophytic vegetation*, that is, vegetation typically adapted for life in saturated, anaerobic soil conditions. The presence of scattered individual plants does not make a community and cannot be used for wetland classification.

12.2.5 Classifying Wetlands

Wetlands are classified by system, which describes the overall environmental setting, and subsystem, which describes the frequency of inundation. There are five wetland systems: marine (oceanic), estuarine (associated with an estuary), riverine (associated with a river), lacustrine (associated with a topographic depression or a dammed river channel) and palustrine (not directly associated with a surface water body). These systems, and their subsystems, are described below.

Marine systems are oceanic wetlands. There are two types of marine wetlands. Marine "subtidal" refers to wetlands in which the substrate, i.e., the portion of the soil that supports life, is constantly submerged. Marine "intertidal" refers to wetlands in which the substrate is exposed by tides.

Estuarine systems are deep water tidal habitats and adjacent tidal wetlands that are connected to the ocean or other large salt water body, although this connection is obstructed to varying degrees by land (see Figure 12.4). The water in an estuarine wetland is at least occasionally diluted by freshwater runoff from the land. Similar to marine systems, estuarine "subtidal" lands are constantly submerged, and estuarine "intertidal" lands are exposed by tides.

Riverine systems include wetlands and deep-water habitats contained within a channel, such as a river, excepting wetlands dominated by trees, shrubs, etc., and habitats with brackish water. There are four riverine subsystems. Riverine "tidal" lands are periodically exposed by tides. There are two riverine perennial subsystems,

FIGURE 12.4
A typical marshy area. (Courtesy of GZA GeoEnvironmental, Inc.)

distinguished by a lack of tidal influence and water flow throughout the year. Riverine perennial lower subsystems have a low flow gradient and a slow water velocity. A riverine upper perennial subsystem has a high flow gradient and fast water velocity. In a riverine intermittent wetland, the water body flows only part of the year.

Lacustrine systems are wetlands and deep-water habitats that are situated in a topographic depression (such as a lake) or a dammed river channel that is larger than 20 acres (80,000 m^2) in area. The water body lacks trees, shrubs, persistent "emergent" plants (i.e., plants that are rooted below the water but extend above the water surface), and emergent mosses or lichens with greater than 30% areal coverage. Lacustrine systems include permanently flooded lakes, reservoirs, intermittent lakes, and tidal lakes with fresh water. Deepwater habitats within the lacustrine system are known as *limnetic* lacustrine systems. Wetland habitats within the lacustrine system are known as *littoral* lacustrine systems. They extend from the shoreward boundary of the system to a depth of 6.6 feet (2 m) below low water or to the maximum extent of non-persistent emergent plants.

Palustrine systems include all non-tidal wetlands that are dominated by trees, shrubs, persistent emergent plants, emergent mosses or lichens, as well as all such wetlands that occur in fresh water tidal areas. Palustrine wetlands range from permanently saturated or flooded land (as in marshes, swamps, and lake shores) to land that is wet only seasonally (as in vernal pools).

12.2.6 Background Research on Wetlands

Before conducting a wetland survey, the consultant should tap into the myriad of available public resources regarding wetlands. The USACOE *Wetlands Delineation Manual* describes the procedures for the identification and delineation of jurisdictional wetlands, that is, wetlands that are regulated under Section 404 of the Clean Water Act. The *National Wetlands Inventory* (NWI) contains maps covering almost 90% of the conterminous United States, portions of Alaska, and all of Hawaii and the U.S. Territories. They are available online at http://www.fws.gov/wetlands. State, county, and local agencies, such as planning and engineering departments, have information regarding wetland locations within their jurisdictions. Geographic information system (GIS) layers that show wetland and hydrologic boundaries and soil classifications also are available online. Other available resources include:

- United States Geological Survey (USGS) topographic maps, which show the general locations of wetlands.

- The Natural Resource Conservation Service (formerly the Soil Conservation Service), a division of the United States Department of Agriculture, has published soil surveys by county for most of the United States (http://soils.usda.gov/survey). The soil surveys show the locations of soil types and provide their general characteristics.

- Ecological information is available in environmental impact statements (see Chapter 13) and similar publications.

- Aerial photographs from public and private sources are also useful in obtaining general information about the presence of wetlands. This is particularly of infrared aerial photography in which red is indicative of live vegetation and the tone of red can be a guide to the density and health of the vegetation and its rate of growth. Dead vegetation usually appears as various shades of tan or green on infrared photographs.

- The U.S. Fish and Wildlife Service (USFWS) has a comprehensive list of the thousands of species of hydrophytic plants and the places where they are present. These plants can be used in the mapping of a wetland.

- Studies performed for public and private entities in support of a proposed development, such as the wetland delineation map shown in Figure 12.5.

As important as these resources are for identifying wetlands, they should never be considered as authoritative regarding an area or certainly a specific property, given the gross scale by which these data sources are provided and their frequent inaccuracies. When it comes down to identifying and delineating wetlands in an area or on a specific property, actual field data is mandatory.

12.2.7 Field Mapping of Wetlands

Wetlands are identified in the field by looking for evidence of inundation, soil saturation, and hydric soils (discussed in Section 12.2.2), and wetlands-related plant communities and dominant species (see Figure 12.6). Evidence of past water inundation includes the presence of living or dead aquatic fauna, the presence of an algal mat or crust, water-stained leaves, moss-covered buttresses, and watermarks on trees. Sediment deposits resulting from water flow, or drift lines, which is an accumulation of debris from a water flow event, represent the maximum level of water from heavy precipitation or flooding. Hydric soils are identified by their distinctive coloration, by the presence of iron deposits or the presence of reduced iron, or by a hydrogen sulfide odor. The field scientist also will analyze drainage patterns, especially near streams, that show the maximum level of water that later receded back into the water body.

Certain intrusive tests can be performed in the field to verify the presence of wetlands. For instance, if water is not at the surface, it may be just under the surface, creating wetland

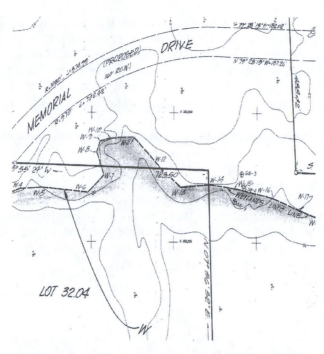

FIGURE 12.5
A portion of a wetland map, showing the location of wetlands and their soil classification.

FIGURE 12.6
A wetland investigator in the field. (Courtesy of GZA GeoEnvironmental, Inc.)

conditions that may not be immediately obvious. To determine the depth that water is present, the field scientist can dig a hole, known as a "soil pit," to a depth of 24 inches (60 cm). Water draining into the hole indicates that the shallow soils are saturated and that the ground surface may be submerged during periods of heavy precipitation or flooding. The field scientist must consider the local variations in water level due to hydrologic cycles, seasonal changes, tidal influences, and so on, to determine whether the saturation depth suggests the presence of wetlands.

In summary, there are three primary factors that define a wetland: ample water, which floods the surface or the near surface of the soils; soils that are characteristic of wetlands; and the presence of biota characteristic of wetlands. These three factors have a causal relationship—they do not appear independent of one another. When one of these factors is missing or in question because of conflicting data (for instance, the presence of vegetation, a nesting bird not characteristic of a wetland, or the lack of hydric soils), the field scientist must use professional judgment to determine the location of the wetland boundary. Knowledge of local conditions and which factors are most diagnostic in identifying wetlands in that area will assist in this determination.

Once the wetland boundary is identified, it is marked in the field, sometimes with flags (see Figure 12.7), which are then surveyed and marked out on a scaled map. The map is then used for planning purposes, which may be subject to regulatory review.

FIGURE 12.7
Field delineation of a wetland using flagging. (Courtesy of GZA GeoEnvironmental, Inc.)

12.2.8 Wetlands Restoration and Creation

Wetland restoration attempts to bring the wetland back to its undamaged condition. *Wetland creation* creates a wetland where none previously existed or expands the size and quality of a currently existing wetland. A successful wetland restoration or creation considers the following factors: proper site selection, hydrology, water quality, substrate augmentation and handling, plant material selection and handling, buffer zones placement, and long-term management. Studying and understanding a healthy, biologically diverse proxy wetland in the vicinity of the restoration site can help in modeling a wetland restoration project.

A healthy wetland must have the proper hydrology. If a wetland has deteriorated because it no longer experiences year-round inundation due to sediment buildup or has not been a wetland because of inadequate available water, then water will need to be introduced to the area through the creation of channels or basins.

Chemicals that are not part of the natural biosystem can overwhelm a wetland and change the characteristics of the site. As with impacted surface water bodies, the first step in restoring a damaged wetland is to stop the processes that damaged it, such as the introduction of chemicals or the prevalence of human activities. If the wetland was a fresh water wetland that was damaged by the encroachment of salt water, then a barrier must be constructed between the wetland and the nearby salt water body. The next step is to remove chemicals that prevent or inhibit plant growth. If the chemicals have worked their way into the soils, as often is the case with salt water encroachment, the damaged soils may need to be removed and replaced.

Soils in a damaged wetland may be starved for nutrients that are critical to plant growth. Increasing the organic content of the soils by augmenting, or mulching, the soils provides a source of needed plant species, microbes, and invertebrates, although care must be taken to avoid introducing non-native species to the area.

The final step in restoring a wetland is reintroducing the proper vegetation to the area. New plants will fare better if they are native to the area and are present in similar, nearby wetlands. Fast-growing plants that can quickly stabilize the substrate and have potential value for fish and wildlife tend to be most successful in a restored wetland. On the other hand, one should avoid plants that will be over-foraged and possibly decimated by animals in the area. Figure 12.8 shows restoration efforts for a property on the Hackensack River in New Jersey.

Once restored, protective measures need to be put into place to prevent damage to the fragile, newly restored area. Construction of a buffer zone can improve the odds of success for the wetland, and is, in fact, a requirement in many jurisdictions. A buffer zone can be an undeveloped, vegetated band around the wetland, or a man-made barrier such as a fence. As a rule, the wider the buffer zone, the better the protection of the wetland. The width of the buffer zone also has to be weighed against the economic impact of loss of that land to development.

The most successful wetland mitigation projects are self-sustaining, requiring no long-term management at all. The progress of the wetland mitigation should be monitored periodically by the field identification of plant communities, dominant species, soil types, hydrology, and so on, until the ecosystem appears to be healthy and stable.

FIGURE 12.8
Restoration of a damaged wetland near the Hackensack River in New Jersey. (Courtesy of Blaine Rothauser.)

12.2.9 Compensatory Mitigation

When the impacts of a proposed action to a wetland, stream, or other aquatic environment are unavoidable, the USACOE requires what is known as *compensatory mitigation*. The USACOE (or designated state authority) usually determines the appropriate form and amount of compensatory mitigation required. There are two types of compensatory mitigation other than wetlands restoration and creation, which are described above:

- *Enhancement*, which alters the wetland to improve a particular function, such as support of a particular species. Wetland enhancement carries the risk that by supporting a particular species, other species are adversely affected.

- *Preservation*, which entails removing a threat to, or preventing the decline of, aquatic resources by an action in or near those aquatic resources. It includes activities commonly associated with the protection and maintenance of aquatic resources through the implementation of appropriate legal and physical mechanisms. Preservation does not result in a gain of aquatic resource area or functions.

Compensatory actions should be undertaken in areas adjacent or continuous to the area to be impacted. If this is not feasible, then compensatory mitigation will occur off-site, ideally in the same geographic area as the area to be impacted. In determining compensatory mitigation, the functional values lost by the resource to be impacted must be considered.

Off-site compensatory mitigation is accomplished through mitigation banks and in-lieu fee mitigation, both of which entail third-party involvement to achieve mitigation goals. The third party, usually a bank or in-lieu fee sponsor, assumes responsibility for the implementation and success of the compensatory mitigation.

A *mitigation bank* contains wetlands, streams and other aquatic resource areas that have been restored, established, enhanced, or preserved. The permit applicant can obtain credits to offset degradation of some other wetlands, stream, or aquatic resource area by buying credits from the bank. The value of the credits assigned to a mitigated area is determined by quantifying the aquatic resource functions restored, established, enhanced, or preserved.

The permit applicant can contribute to an *in-lieu fee mitigation* program, which will result in third-party restoration, creation, enhancement, or preservation activities of the aquatic resource. Governmental agencies or authorized non-profit organizations generally administer these programs.

12.3 Threatened and Endangered Species and Their Habitats

Of special consideration when assessing ecosystems are threatened and endangered species, which are protected under the Endangered Species Act of 1973 (see Chapter 3) and state equivalents. *Endangered species* are plants and animals that have become so rare that they are in danger of becoming extinct. *Threatened species* are plants and animals that are likely to become endangered within the foreseeable future throughout all or a significant portion of its range (see Figure 12.9 for an example of a threatened species). The Endangered Species Act provides all of its protections to endangered species, and many but not all of its protections to threatened species. The USFWS has the authority to determine which protections should apply to each threatened species.

FIGURE 12.9
The bobolink (*Dolichonyx oryzivorous*) is a threatened grassland bird that winters in South America and breeds in the summer in northern United States and southern Canada. (Courtesy of Blaine Rothauser.)

NEPA requires evaluation of a proposed action's potential effects on threatened and endangered species and their habitats. Such studies are required even for development projects that are not regulated by NEPA and its state equivalents. The assessment of the potential effects of a proposed action includes field observations for the endangered or threatened species, but also an assessment of whether the area of the proposed action provides a habitat or could be an appropriate habitat for the species. Understanding the breeding and nesting requirements, or life-cycle features, of a threatened or endangered species is critical to the evaluation process.

12.3.1 Publicly Available Information on Threatened and Endangered Species

Before beginning the field work, the field scientist needs to know what to look for. The USFWS maintains databases of threatened and endangered plant and animal species at http://www.fws.gov/endangered. Most states maintain such databases of threatened and endangered plants and animals within their jurisdictions, which often include a significantly larger list of protected species that are only regionally rare, sometimes at the limits of the natural geographic ranges.

Once the applicable species are identified, the field scientist will investigate the life history requirements of the species to assess whether the study area would be an appropriate habitat for the species. This process begins with knowing the type of lands present in the study area. For instance, if the habitat for the species is a wetland, then the field scientist should review publicly available information to assess whether wetlands are present in the study area. Critical habitats are designated areas believed to be essential to the species' conservation. There are restrictions that range from severe to moderate depending on federal and state mandates. In extreme circumstances, the proposed activity may be prohibited or redesigned to accommodate the species. Most threatened and endangered species regulations buffer the proposed activities from a specific aspect of the species life history to ensure its survival and perpetuation.

12.3.2 Field Surveying

While spotting a threatened or endangered species is sufficient to document its existence, it is not necessary to actually observe the fish in question if it was observed in the river at some time in the recent past. Such observations indicate that the river is an endangered species habitat, providing certain protections that may render the proposed dam construction project dead on arrival, or too expensive to pursue due to the anticipated need for mitigation or modification in the project.

The field scientist should know the species' temperament and preferred environs. If the plant avoids direct sunlight, then it should be looked for in the areas that are shady or dark. If the animal prefers dry areas, then the field scientist should not waste too much time in the damp, low-lying areas or around stream beds.

In addition, knowing an animal's behavior is critical to the timing of the field work. For instance, if the animal of interest is nocturnal, then the field survey should occur at night. An early morning start is best if the animal of interest forages for food when the sun is still low in the sky.

12.4 Invasive Species

Invasive species are species whose introduction into an ecosystem will cause or is likely to cause economic or environmental damage to the ecosystem or to humans. They can supplant native species in their ecological niches, deprive the native animals of their food sources or breeding habitats, or, in the case of invasive plants, crowd out the light, water, and nutrients of the native plants. The presence of invasive species in any environment is the primary metric that restoration ecologists use to characterize natural systems in decline.

The ability of alien species to spread beyond their native habitats has greatly increased in modern times due to the ubiquity of human transportation conveyances by which animals and plants can travel long distances and across wide expanses of water. Because the newly introduced species often lack natural predators, they are able to propagate and spread with impunity. They also typically benefit from naturally fast growth, rapid reproduction, high dispersal ability, and good adaptability to new conditions. Examples of some of the more troublesome and damaging species afflicting the continental United States include the following:

- Kudzu (shown in Figure 12.10), which is a plant that forms dense monocultures that outcompetes native ground cover and forest trees. Kudzu has been documented to grow by up to one foot (0.3 m) per day

- Zebra mussels, which spread via oceangoing vessels into the Great Lakes. Zebra mussels disrupt fresh water ecosystems, damage harbors and waterways, and clog water treatment plants and wastewater pipes

- Red imported fire ants, or *Solenopsis invicta*, sting humans, injecting a venom that can cause swelling, and, in rare cases, a severe allergic reaction known as anaphylaxis. This species has spread throughout the Southeastern United States. Its spread northward is limited due to its inability to live in cold climates

Controlling the spread of invasive species can be broadly defined into six categories:

- Conducting controlled burns of affected areas
- Physically removing the species, through mowing, uprooting, or harvesting in the case of invasive flora, or by trapping and culling in the case of invasive fauna
- Constructing physical barriers to prevent the spread of the invasive species
- Releasing predators, parasites, or diseases that will control an invasive species without harming native species
- For invasive fauna releasing pheromones designed to disrupt the mating of invasive species, or introducing sterile males into the population
- Employ pesticides or herbicides designed to destroy the invasive fauna species. Care must be taken to avoid introducing chemicals that can damage other species in the area or the ecosystem itself. The integrated pest management (IPM) approach can help identify the best control strategy for a given invasive species in a given area

FIGURE 12.10
Kudzu (*Pueraria montana*) overtaking a forest near a highway in Kentucky. Kudzu, which originated from Japan, was initially used to enhance soil fertility on eroded slopes, but its release into our environment has left it with no natural herbivores to control its spread, as it skeletonizes every native shrub and tree in its wake. (Courtesy of Blaine Rothauser.)

12.5 Ecological Management and Environmental Consultants

Environmental consultants provide the brains and the brawn for many portions of ecological management. They supply the boots on the ground, delineating wetlands, identifying threatened and endangered species, assessing the health of water bodies, and so on. Because many of the areas of ecological interest are difficult for mechanized equipment to reach, environmental consultants do much of the hole digging, sample collecting, wading, and bushwhacking through inaccessible areas to conduct their studies.

Because of the diversity of activities that an ecological evaluation entails, environmental consultants often have sub-specialties in disciplines such as wetlands, wetlands restoration, threatened and endangered species, and so on. Various professional licenses are available to environmental consultants, such as the Certified Natural Resources Professional (CNRP) license issued by the National Registry of Environmental Professionals (NREP) and the Professional Wetland Scientist (PWS) certification from the Society of Wetland Scientists (SWS).

Contract laboratories will perform the chemical, biological, and tissue analyses, but much of the data used in ecological studies, such as water temperature, chemical composition, and so on, are collected by consultants in the field. Contractors use heavy earth-moving equipment to perform restoration work, such as the building of barriers and channels, removing damaged soils, and placing healthy soils in their place or on top of the damaged soils. Environmental consultants will prepare contractor specifications for ecological restoration work to verify that the restoration work is being performed in accordance with the specifications. Consultants then will monitor the progress of the newly planted flora, the stability of the restored slopes, wetlands, and so on, and suggest additional restoration or mitigation work if problems arise.

Consultants also negotiate settlements with regulatory authorities for compensatory mitigation and work with regulatory authorities and mitigation banks so that their client's development project that may damage all or part of a wetland area could proceed.

Problems and Exercises

1. Identify two types of wetlands (marine, estuarine, riverine, lacustrine, or palustrine) in your state or territory. Identify two fauna and two flora associated with each of the wetlands.

2. Using the internet, identify a threatened or endangered species in your area. What ecological areas is the species likely to be found?

3. After a wetland has been restored, what parameters might an environmental consultant measure to determine the success of the restoration?

Bibliography

Barbour, M.T., Gerritsen, J., Snyder, B.D., and Stribling, J.B., 1999. Rapid Bioassessment Protocols for Use in Streams and Wadeable Rivers: Periphyton, Benthic Macroinvertebrates, and Fish, 2nd ed. EPA 841-B-99-0002. USEPA Office of Water.

Canter, L.W., 1996. *Environmental Impact Assessment*, 2nd ed. McGraw-Hill.

Federal Geographic Data Committee, 2009. Wetlands Mapping Standard. FGDC-STD-015-2009.

Interagency Workgroup on Wetland Restoration, 2003. Wetland Restoration, Creation, and Enhancement. National Resources Conservation Service.

Kentula, M.E., 1996. "Wetland Restoration and Creation," In *The National Water Summary on Wetland Resources*. J.D. Fretwell, J.S. Williams, and P.J. Redman, Eds. pp. 87–92. Water-Supply Paper 2425. U.S. Geological Survey, Washington, DC.

Lockheed-Martin Energy System, Inc., 1997. Preliminary Remediation Goals for Ecological Endpoints. U.S. Department of Energy, Office of Environmental Management.

New Jersey Department of Environmental Protection, 2005. Field Sampling Procedures Manual.

New Jersey Department of Environmental Protection, 2013. Technical Requirements for Site Remediation, N.J.A.C. 7:26E.

New Jersey Department of Environmental Protection, 2018. Ecological Evaluation Technical Guidance. Version 2.0.

Suter, G.W., II, Efroymson, R.A., Sample, B.E., and Jones, D.S. 2000. *Ecological Risk Assessment for Contaminated Sites*, Lewis Publishers, Boca Raton, FL.

U.S. Army Corps of Engineers, 1987. Corps of Engineers Wetlands Delineation Manual. Technical Report Y-87-1.

U.S. Environmental Protection Agency, 1987. A Compendium of Superfund Field Operations Methods. EPA 540/P-87/001. Office of Emergency and Remedial Response.

U.S. Environmental Protection Agency, 1997. Ecological Risk Assessment Guidance for Superfund, Process for Designing and Conducting Ecological Risk Assessments. EPA 540-R-97-006. Office of Solid Waste and Emergency Response.

U.S. Environmental Protection Agency, 1998. Risk Assessment Guidance for Superfund, Volume II, Environmental Evaluation Manual. EPA/540/1-89/001.

U.S. Environmental Protection Agency, 2000. Bioaccumulation Testing and Interpretation for the Purpose of Sediment Quality Assessment, Status and Needs. EPA 823-R-00-001. USEPA, Office of Water, Washington, DC.

U.S. Environmental Protection Agency, 2008. Compensatory Mitigation for Losses of Aquatic Resources; Final Rule. 40 CFR Part 230.

U.S. Environmental Protection Agency web site: www.epa.gov/owow_keep/NPS/index.html.

U.S. Fish and Wildlife Service, 1979. Classification of Wetlands and Deep Water Habitats in the United States. FWS/OBS-79/31.

U.S. Fish and Wildlife Service, 1989. Federal Manual for Identifying and Delineation Jurisdictional Wetlands.

U.S. Fish and Wildlife Service. National Wetlands Inventory. www.fws.gov/wetlands.

Section IV

Indoor Environmental Concerns

13

Environmental Impact Assessment and Mitigation

The *National Environmental Policy Act* (NEPA) of 1969 is a federal law designed to protect the environment. The Act requires that major federal actions (such as construction projects) be assessed for their potential to significantly impact the environment. It acts as an "umbrella" that covers a variety of federal, state, and local environmental laws to which federal projects must adhere. Environmental reviews have varying levels of intensity depending upon the size and type of the proposed federal action. NEPA review may be required for small projects such as the construction of new parking stalls, a new wind turbine, or new railroad sidings, up to large projects such as new bridges, new airports, or new military bases. An NEPA review may be required for new construction as well as retrofit, reconstruction, or realignment projects. NEPA review occurs prior to major project decision points like final design and construction so ways to avoid, minimize, or mitigate environmental impacts may be incorporated into the project as early as possible. Table 13.1 list federal agencies that have regulations regarding the NEPA process and an example of a project that would require NEPA review.

NEPA review is required if these departments initiate an eligible project or provide funds to the project. There are also state and sometimes local NEPA equivalents of this process that are triggered by public and even private actions that exceed certain impact thresholds.

NEPA review evaluates if the anticipated environmental effects of a federal undertaking (the proposed action) and its alternatives are significant. The NEPA process considers context and intensity of potential impacts when determining if an anticipated impact is significant. Context could include looking at the project through the points of view of other groups such as national, regional, or local groups, or special interest groups. An environmental impact may be beneficial or adverse, and a project may be found to be significantly beneficial on balance even if the project has some adverse environmental impacts.

The NEPA process reviews a wide range of human/community and natural resources that could be affected by a project. This chapter outlines the NEPA process, and provides brief descriptions of the technical elements that are considered in the evaluation process.

TABLE 13.1

Federal Departments Subject to NEPA

Federal Department	Example of Projects Subject to NEPA
Department of Agriculture	Conversion of a large farm or ranch to non-agricultural usage
Department of Commerce	Restoration of an estuary or port
Department of Defense	Conversion of a military base to non-military usage
Department of Energy	Construction of a transmission line, pipeline, or a power plant
Department of Health and Human Services	Construction or expansion of a public park
Department of Housing and Urban Development	Construction of affordable or supportive housing
Department of the Interior	Expansion or construction of an existing mine
Department of Justice	Research and development of DNA analyzers
Department of Transportation	New bridge, tunnel, or highway

13.1 The National Environmental Policy Act Project Scoping and Agency Participation

The beginning of the NEPA process is known as scoping. The public agency with jurisdiction will meet and scope what it believes the project will include, where it will be located, and other basic information to guide the start of project planning. One public agency involved is designated as the lead agency, through which all project information flows. The lead agency defines the needs and objectives of a project, makes the decisions, and manages the project from beginning to end. The other agencies routinely participate in the decision making and coordinate with the lead agency and the other agencies involved in the process, especially in the early stages of the NEPA process.

Project scoping should take into account as many feasible and potential solutions as practicable, but should also clearly define the edges (termini) of the project. The lead agency uses one of three levels to classify each project that requires NEPA:

- Categorical exclusion
- Environmental assessment
- Environmental impact statement

After the project classification is determined, NEPA documents may take varied forms. Commonly, one cohesive environmental document is developed to summarize the NEPA process per project. Less commonly, the NEPA document is divided into two or more "tiers" to reduce repetitive analyses. Tiered documents start with a planning-level broad analysis and continue into subsequently narrower analyses. Supplemental NEPA documents may be developed to amend original documentation, if required.

The NEPA process may result in scope changes as the process unfolds, as discussed later.

13.1.1 Categorical Exclusion

In some cases, the government may issue a *categorical exclusion* (CATEX or CE), which exempts the proposed action from the NEPA process. The CATEX process differs by federal agency, but can range from not requiring additional environmental review, to completing a checklist to broadly document projects impacts, to developing a very short and streamlined version of a NEPA document. A CATEX is usually issued when:

- A course of action is identical or very similar to a past course of action and the impacts on the environment from the previous action can be assumed for the proposed action
- The federal agency with jurisdiction determines that the proposed action will not individually or cumulatively have a significant environmental effect
- When a proposed structure is within the footprint of an existing, larger facility or complex

13.1.2 Environmental Impact Statement

NEPA requires that an *environmental impact statement* (EIS) be prepared to disclose impacts for major federal projects that can significantly affect the environment or will result in significant public concern. In the EIS, potential impacts are fully disclosed and alternative actions that may avoid, minimize, or mitigate the potential impacts are evaluated. NEPA does not require that the planned actions cause no environmental impact at all, but rather that it be the *least environmentally damaging practicable alternative* (LEDPA) that complies with applicable federal, state, and local laws. Because federal agencies typically tailor specific EIS content to their requirements, EIS content may vary across federal agencies and project types.

The major steps in the EIS development process include:

- Notice of intent (NOI)
- Draft environmental impact statement (DEIS)
- Final environmental impact statement (FEIS)
- Record of decision (ROD)

13.1.2.1 Notice of Intent

The notice of intent (NOI) is a public notice stating that an EIS will be prepared and considered. The NOI includes a description of the proposed action, its possible alternatives, project scoping details, and contact information for the project representative from the lead agency.

13.1.2.2 Draft Environmental Impact Statement

A DEIS typically begins with a statement of the purpose and need of the proposed action, followed by a listing of alternatives to the proposed action (as well as a description of the proposed action). For each environmental issue in each DEIS, a *no action alternative* is evaluated. The no action alternative, in which the proposed action does not occur, is the baseline to which the other alternatives are compared. The DEIS describes the project's "affected environment," which typically describes current conditions and provides a comprehensive analysis of the environmental impacts of each of the considered alternatives. In addition to the environmental impact determinations, the DEIS measures how well each of the considered alternatives meets project goals (as listed in the purpose and need statement), and identifies fatal flaws of any alternative.

The DEIS lists all of the permits, licenses, etc. needed to implement the project. Stakeholders and the public are given the opportunity to comment on the DEIS, using the federal agency's preferred public involvement methods, which includes one or more public hearings. The lead agency may respond to public comments by modifying the proposed action, developing new alternatives, modifying the analyses presented in the DEIS, or refuting stakeholder objections.

13.1.2.3 Final Environmental Impact Statement

After the lead agency addresses the comments on the DEIS, a final EIS (FEIS) is issued. The FEIS describes the proposed action and its expected impacts on the environment. If stakeholders disagree with the FEIS, they can protest the decision with the lead agency. Any resulting significant change in scope or alternatives will result in the preparation of a supplemental EIS (SEIS), beginning the NEPA process all over again, or cancelling the proposed action outright.

Once all comments and disputes are resolved, the lead agency issues a *record of decision* (ROD), which is its final formal issuance prior to implementing the proposed action. Similar in effect to a ROD under CERCLA (Comprehensive Environmental Response, Compensation, and Liability Act; see Chapter 8), the ROD informs the public of the approved proposed action and discusses the findings of the EIS, demonstrating how the lead agency's consideration of alternatives identified the preferred alternative. If members of the public are still dissatisfied with the outcome after the issuance of the ROD, they would have to sue the agency in federal court to obtain changes to the ROD.

13.1.2.4 Environmental Mitigation Plan

If the proposed action is expected to cause significant environmental impacts, an *environmental mitigation plan* is prepared and incorporated into the EIS documents and the ROD. Mitigation plans typically are written per environmental resource type, as each environmental resource analysis in NEPA adheres to a different federal law or procedure, with the federal agency with jurisdiction coordinating with the lead agency. For example, a wetland mitigation plan would be written in coordination with the U.S. Army Corps of Engineers, which has jurisdiction over wetlands through Section 404 of the Clean Water Act. Mitigation could include one or more of the following actions:

- Avoid the impact by changing all or part of a proposed action
- Limit the degree or magnitude of the proposed action
- Develop a plan to restore the impacted portion of the environment following implementation of the proposed action
- Develop a plan to restore the impacted portion of the environmental over time
- Compensate for the environmental impact by replacing or providing similar environmental conditions elsewhere

13.1.3 Environmental Assessment

Sometimes it isn't clear whether a proposed federal action will result in significant impacts or significant public concern. An *environmental assessment* (EA) is prepared for projects requiring NEPA that do not meet the agency with jurisdiction's criteria to be a CATEX, and it is not immediately clear if the project will result in significant impacts and warrant an EIS. The EA process is similar to that of the EIS, although it typically results in a more concise document. The conclusion of the EA will determine if there will be or will not be significant environmental impacts as a result of the proposed action.

If the EA indicates that no significant impacts are likely, then the lead agency can release a *finding of no significant impact* (FONSI) to conclude the NEPA process and proceed with the project. The FONSI may be accompanied by a request for minor changes in the proposed project.

13.1.4 Public Participation

The NEPA process is designed to involve the public and other stakeholders so that the current and useful information is available to the decision makers, and conflicts with stakeholders can be avoided. The system is designed so that citizens can be partners in the process, and public participation is encouraged to occur early and often through the NEPA process.

During the NEPA process, the lead agency often will coordinate with stakeholders to try and resolve conflicts that may arise. The lead agency will develop a public participation program to inform the public and solicit participation in the decision-making process. The program must include notices in the *Federal Register* that may include information, meeting, or hearing notices in local media, public hearings, project-specific web pages, or local media. Citizens and citizen groups are asked to provide comments to the lead agency, which then addresses the comments and incorporates them as appropriate into the NEPA document. NEPA development team provides technical support to the public participation program.

13.2 Technical Evaluation for the National Environmental Policy Act

Extensive research goes into the preparation of a NEPA document. It can take years to prepare, and may eventually consist of multiple volumes that can be thousands of pages in length. NEPA document must be written for public consumption and comprehension, while including the proper level of detail needed by regulatory technical experts. It should be written plainly and clearly so that it could be easily understood by laypeople. More detailed technical information should be provided in the appendices.

The next sections describe the factors that are typically evaluated in NEPA documentation. Other factors that may need to be evaluated, depending on the federal agency with jurisdiction of the project, the type of project, and the comments receiving during the scoping process, including: light, energy, utilities, public health and safety, and solid waste and recycling.

13.2.1 Physiography, Geology, and Seismicity

Physiographic data are collected and evaluated, including topography, surface slopes, drainage patterns, rates of erosion, etc. Data are obtained regarding soil types and their physical characteristics. The underlying bedrock must be evaluated for its suitability to support the structures to be constructed. Seismicity is usually evaluated as well, even in areas that are not prone to earthquakes.

13.2.2 Groundwater

The presence of soil or groundwater contamination can delay project construction and drive up its costs. NEPA requires environmental assessments of the area of the proposed action. A Phase I environmental site assessment, which is described in Chapter 6 of this book, is a useful first step in assessing the potential for subsurface contamination (although NEPA does not require that the Phase I ESA comply with the ASTM E1527 standard). If subsurface contamination may be present in the area of the proposed action, soil and groundwater sampling, as described in Chapter 7 of this book, may be conducted. Remediation, as described in Chapter 9 of this book, may be warranted before the proposed action can be implemented.

Special attention is paid to contamination issues if the project area is located over a sole-source aquifer, that is, an aquifer that is the only major source of drinking water to the area. Also of concern is the impact of the proposed construction on groundwater recharge areas, which are the surface areas through which rain will percolate into the subsurface and replenish the drinking water aquifer.

13.2.3 Water Supply

The addition of residents, workers, or power plants (depending on the type of energy produced) to the project area will result in increased water usage in the area. Withdrawing water from a water body or from an aquifer could reduce its availability to other users and may increase the relative concentration of pollutants in that portion of the water body. NEPA requires an evaluation of the expected incremental usage of water anticipated and whether the additional usage will result in water shortages during droughts.

13.2.4 Surface Waters

A large-scale construction project can impact water quality by generating dirt and other run-off that is introduced into surface waters directly or indirectly through the storm sewer system. A baseline evaluation of the surface water environment in the project area includes a detailed description of the surface water bodies; their relationship to drinking water supplies; a description of water flow, including waste water effluent contributions to the surface waters; drainage patterns; the expected path of dirt and other run-off; and the potential for erosion during and after the construction phase of the project. Historical water flow data combined with computer simulations of water usage in the various alternatives under consideration aids in the identification of potentially significant environmental impacts to surface waters. Computer simulations also assess the potential changes in water quality as well from increased load of pollutants and sediments into the water body.

Field sampling is used to fill in data gaps that may be identified in the baseline study. Sampling parameters and methodologies are discussed in Chapter 12.

13.2.5 Wild and Scenic Rivers

The National Wild and Scenic Rivers System Act in 1968 is designed to preserve certain rivers with outstanding natural, cultural, and recreational values in a free-flowing condition for the enjoyment of present and future generations. Three types of rivers are covered by the Act:

- *Wild river areas.* Rivers or sections of rivers that are free of impoundments and generally inaccessible except by trail, with unpolluted and essentially primitive watersheds or shorelines

- *Scenic river areas.* Rivers or sections of rivers that are free of impoundments, with mostly primitive and undeveloped shorelines or watersheds, with some road access

- *Recreational river areas.* Rivers or sections of rivers that are readily accessible by road or railroad, that may have some development along their shorelines, and that may have been impounded diverted in the past.

The Act is designed to safeguard the special character of these rivers, while also recognizing the potential for their appropriate use and development. The National Parks Service, a division of the U.S. Department of the Interior, enforces the provisions of the Act and maintains a database of rivers that are classified as wild or scenic.

13.2.6 Wetlands

Wetlands are described in Chapter 12 of this book. The U.S. Fish and Wildlife Service maintains the National Wetlands Inventory (NWI), and various states, counties, and municipalities often have wetlands maps that have a finer scale than the NWI. Wetlands delineation activities may be required by the federal agency with jurisdiction if the extent of jurisdictional wetlands in the project area is not known.

13.2.7 Flood Plains

Flood plains are lowlands that are along coastlines or near surface water bodies that are prone to flooding. Flood-prone areas are defined by the likelihood that the land will be flooded any given year. Land within a "100-year flood plain" has a 1% chance of being flooded in a given year and land within a "500-year flood plain" has a 0.2% chance of being flooded in a given year. The Federal Emergency Management Agency (FEMA) publishes maps that indicate the locations of flood plains. It also has a category regarding the chance of a property being flooded over a 30-year period, which is useful to agencies that provide 30-year mortgages to property owners.

Addition of a new discharge into a water body as a result of the construction project may increase the flood potential of the receiving water body, especially if there are floodplains in the study area. NEPA considers the potential for actions to impact flood plains or add water loads to flood plains can increase the potential and severity of flood to the extent practicable.

13.2.8 Coastal Barrier Resources

Undeveloped coastal barriers along the Atlantic and Gulf coasts have been protected areas since the passage of the Coastal Barriers Resource Act of 1982. The Coastal Barrier

Improvement Act of 1990 expanded the protected areas to include undeveloped coastal barriers along the Florida Keys, Great Lakes, Puerto Rico, and the U.S. Virgin Islands. The NEPA process requires the identification of these protected areas and, if present, an evaluation of the potential impact on these protected areas by the proposed project.

13.2.9 Natural Resources

Evaluation of the biological setting of the project entails many of the activities described in Chapter 12. Of concern is the potential impact on flora (both upland and wetland) and fauna (both land animals and aquatic organisms). The NEPA process requires evaluations of wildlife, wildlife habitat, endangered species, and vegetation, as described below. Since the presence of flora and fauna vary seasonally in most parts of the country, field data may need to be collected in different seasons, which should be taken into account when scheduling field work for the NEPA documentation.

13.2.10 Wildlife, Wildlife Habitat, and Threatened and Endangered Species

Understanding the wildlife in an area entails identifying the presence of specific species, their relative abundance, and their habitats. This is especially important if endangered species or critical habitats may be present in the project area (see Chapter 12). Endangered species and habitats that could be impacted by the proposed action are identified as part of the NEPA process. Table 13.2 is an example of such a table for an EIS, prepared because of the proposed construction of a roadway in Seattle, Washington. That EIS also included similar tables for birds marine mammals, and waterfowl found in urban Seattle.

Groups of aquatic organisms in or adjacent to the project area that may require study in the EIS may include: benthic organisms (organisms that live at the bottom of the water body) that are the primary food source for most fish species; algae; aquatic plants; zooplankton; fish; and mammalian aquatic animals. Field activities include collecting samples from each aquatic community, from different depths in the water body in the case of zooplankton and fish. In most cases, a four-season study is warranted due to the variations in aquatic communities and life stages in the course of a calendar year.

If endangered species or their habitats are identified, the potential short-term and long-term impacts of the proposed project on wildlife in the area using a variety of descriptive and statistical techniques must be assessed. Collection of field data will help evaluate the potential impacts.

13.2.11 Vegetation

This category includes flora that are not endangered or threatened but are considered important to the area. Upland plants (i.e., plants not associated with wetlands or water bodies, which are discussed above) may be mapped by remote means such as aerial photographs, or by direct field inspections. Dominant plant species are identified in each plant community, and the potential impacts to them by the proposed action are assessed.

13.2.12 Air Quality

To understand the impact of the proposed project on air quality, a thorough understanding of the climate in the project area is needed. Such data are readily available from the National Oceanic and Atmospheric Administration (NOAA), the National Climatic

TABLE 13.2

Land Mammals That May Be Found within Urban Habitat along the Alaskan Way Viaduct Corridor

Common Name	Scientific Name
Common opossum	*Didelphis marsupialis*
Little brown myotis	*Myotis lucifugus*
Yuma myotis	*Myotis yumanensis*
California myotis	*Myotis californicus*
Silver-haired bat	*Lasionycteris noctivagans*
Big brown bat	*Eptesicus fuscus*
Hoary bat	*Lasiurus cinereus*
Townsend's big-eared bat	*Plecotus townsendii*
Long-eared myotis	*Myotis evotis*
Domestic rabbit	*Oryctolagus cuniculus*
Eastern gray squirrel	*Sciurus carolinensis*
Deer mouse	*Peromyscus maniculatus*
Muskrat	*Ondatra zibethicus*
House-mouse	*Mus musculus*
Pacific jumping mouse	*Zapus trimtatus*
Norway rat	*Rattus norvegicus*
Black rat	*Rattus rattus*
Coyote	*Canis latrans*
Raccoon	*Procyon lotor*
Ermine	*Mustela erminea*
Mink	*Mustela vison*
River otter	*Lutra canadensis*
Domestic dog	*Canis familiaris*
Domestic cat	*Felis domesticus*

Data Center, U.S. Environmental Protection Agency (USEPA), and state and local sources. Climate data that contribute to the EIS include monthly precipitation, wind speed and direction; mean monthly temperatures, daily temperature range, mean snowfall (as applicable), and heating/cooling degree days.

In addition, at a minimum, the changes in concentration in ambient air for the following air pollutants that are subject to the USEPA's National Ambient Air Quality Standards (NAAQS; see Chapter 3) are considered in the NEPA process:

- Carbon monoxide and ozone precursors (NO_x and VOC), both which primarily are emitted by motor vehicle emissions
- Particulate matter (PM_{10} or $PM_{2.5}$), which associated with stationary energy sources (such as power plants) and diesel combustion engines (usually buses and heavy trucks)
- Total suspended particulates
- Sulfur dioxide, which is emitted primarily by stationary sources

Once baseline conditions are understood for these pollutants, impact calculations using computer modeling software may be performed to understand how they will change as a result of the project. The goal is the *prevention of significant deterioration* (PSD), as required under the Clean Air Act. Computer models incorporate the meteorological data, the baseline pollutant data, and the predicted changes in stationary and motor vehicle pollution sources, from which the computer model can predicts future increases in pollutants. They can estimate pollutant levels under normal conditions as well as worst-case conditions for the range of alternatives under consideration. If the computer model indicates that the project would not comply with Clean Air Act requirements, then an optimization analysis would be performed, in which certain parts of the project would be changed to assess whether such changes would bring the project into compliance with the Clean Air Act. If the project causes significant increases in air pollution, the project may be cancelled, especially if the project is located in a non-attainment area for that particular pollutant.

Separate from the study of NAAQS pollutants is the study of fugitive dust emissions during construction. Here as well, computer modeling can be used to calculate expected and worst-case scenario dust concentrations downwind of the proposed construction. The results may indicate that anticipated construction methods may need to be changed. Minor changes might entail the implementation of dust suppression techniques, such as air misting. More substantial procedural changes could result in a significant increase in the cost of the project. Air monitoring during construction is described in Chapter 9 of this book.

Due to climate change concerns, many federal and state regulatory agencies are now requiring an analysis of greenhouse gas (GHG) emissions during the NEPA process. The primary GHG pollutants are carbon dioxide (CO_2), methane, nitrous oxide, and fluorinated gases, with CO_2 typically being of primary focus. A GHG evaluation consists of an estimate of the total GHG emissions from the construction and operation of the proposed action. If there is a predicted net increase in GHG emissions, then the NEPA process must consider measures that would offset this increase.

13.2.13 Farmland Protection

The Agriculture and Food Act of 1981 contained the Farmland Protection Policy Act (FPPA) that went into effect in 1994. The FPPA is intended to minimize the impact of federal programs on converting farmland to non-agricultural uses. The NEPA process must evaluate the potential effect of the proposed action on existing farmland in the area. The FPPA defines farmland as including the following three designations:

- *Prime farmland* is farmland which is of major importance in meeting the United States' short- and long-range needs for food and fiber.

- *Unique farmland* is land other than prime farmland that is used to produce high-value food and fiber crops such as citrus, tree nuts, olives, cranberries, and other fruits and vegetables.

- *Land of statewide or local importance* is land used for the production of food, feed, fiber, forage, and oilseed crops that the state or local area considers important to its economy.

Farmland subject to FPPA requirements does not have to be currently used for cropland. It can be forest land, pastureland, cropland, or other land, but not water or urban built-up land.

13.2.14 Noise

Within the context of NEPA, noise includes sounds that are undesirable because it intrudes on people's lives or is intense enough to damage a person's hearing. Whereas the decibel (dB) is the unit of measurement for noise, noise studies measure *A-weighted noise decibel* (dBA) levels. A dBA measurement puts different weights on high-pitched and low-pitched noises since the human ear hears these sounds differently. Table 13.3 provides typical outdoor noise levels in urban, suburban, and rural environments. A sound level of 70 dBA is perceived as twice as loud as a sound level of 60 dBA.

Federal agencies have differing procedures and metrics for noise analysis that are related to their jurisdiction. For example, the Federal Highway Administration (FHWA) studies highway noise, the Federal Railroad Administration (FRA) studies rail noise, and Federal Aviation Administration (FAA) studies air traffic noise. Typically, existing and future year noise levels, with and without the proposed project, are studied to determine where noise impacts are predicted to occur. Noise metrics differ by federal agency with jurisdiction and the noise source being studied, just as each federal agency has differing requirements and regulations for noise analyses. For example, highway traffic noise (FHWA) typically uses L_{eq}, which is an equivalent continuous sound level over a period of 1 hour, and highway traffic noise impacts are identified using noise abatement criteria that are based on (typically) exterior noise levels that interfere with outdoor conversations. Conversely, aviation noise (FAA) typically uses L_{dn}, which is a day-night averaged sound level over a 24-hour period, and aviation noise impacts are identified using noise thresholds for compatible land uses, and can be interior or exterior.

Measuring noise in the field is accomplished using a simple hand-held instrument that meets the requirements of the federal agency with jurisdiction over the project. Once field measurements are collected, predicted noise levels of the proposed action are obtained through the methods typically used by the federal agency with jurisdiction. This may include developing computer noise models to predict noise levels that will be created by the project. The future year noise levels with the proposed action included are then compared to regulatory noise criteria to identify noise impacts.

Noise mitigation can be accomplished by requiring improved sound-proofing for the proposed project or restrictions on its usage.

TABLE 13.3

Typical Outdoor Noise Levels in Various Environments

Setting	L_{dn} (dBA)
City noise (major downtown metropolis)	75–85
Very noisy urban	70
Noisy urban	65
Urban	60
Suburban	55
Small town and quiet suburban	40–50

L_{dn} = day/night sound level.
dBA = A-weighted decibels.

13.2.15 Cultural Resources

Historic, archaeological, and cultural sites, collectively known as "cultural resources," are properties or buildings with national or local significance. Historic sites are buildings, structures, and sites that date back to colonial times. Archaeological sites are usually Native American sites that predate the colonial era. Cultural resources include cemeteries, parks, trails, and other similar features that may or may not have historical or archaeological significance but contribute to the local culture.

General information on historic, archaeological and cultural resources is available on the National Register of Historic Places, and various local equivalents. "Windshield surveys" (surveys conducted while driving) are performed in the area of interest. Structures are judged on their date of construction, their architectural merit, and their importance to local, state, or national history.

Sometimes areas that may contain historical or archaeological artifacts can be identified using infrared aerial photography, the field identification of areas that appear to have been disturbed by humans, or where there is evidence of historic remains. If research indicates that a site may contain buried historical artifacts or structures, an archaeological investigation is performed. The area of interest is gridded, and test excavations or test cores are installed at set locations in the grid. The archaeologist analyzes the excavated soils in an effort to identify historic artifacts. Significant archaeological findings could delay a project while an archaeological dig is conducted, or force the project to be moved or cancelled.

If research indicates that a property to be affected by the proposed action is or should be considered an historic or cultural resource, then the lead agency notifies the state historic preservation officer (SHPO), who, based on the evidence presented by the lead agency, either denies the historic or cultural significance of the property or proposes its inclusion onto the National Register. If the SHPO determines that the proposed action will have an adverse effect on the historically significant property, either the proposed action will need to be modified or the artifacts or structures recovered or moved from their original location and preserved elsewhere.

13.2.16 Transportation

Proposed actions almost always change existing transportation patterns. Transportation studies may involve evaluations of motor vehicle, pedestrian, rail and airport traffic; however, motor vehicles are the most commonly studied mode of transportation. The evaluation begins with a baseline study of the traffic (see Figure 13.1) and parking in and around the project area. Traffic engineers study the usage, capacity, and adequacy of the existing infrastructure, and compare it to accepted federal and state standards.

Computer models are used to estimate traffic demand after construction of the proposed project, as well as the impact on transit access and pedestrian routes from primary and secondary impacts. Models are run for the various alternatives, for average traffic as well as peak times. Proposed traffic scenarios must agree with regional or local transportation master plans (and may need to be approved by local planning agencies). They also interact with air quality because of the involvement of motor vehicles in the analysis.

Efforts to relieve significant impacts on traffic may result in adding, eliminating, or moving proposed entrance and exit ramps; changing the proposed design of arterial intersections; or adding, eliminating, or changing proposed traffic control measures, such as traffic lights, speed limits, and stop signs. Providing alternative modes of transportation may also be considered.

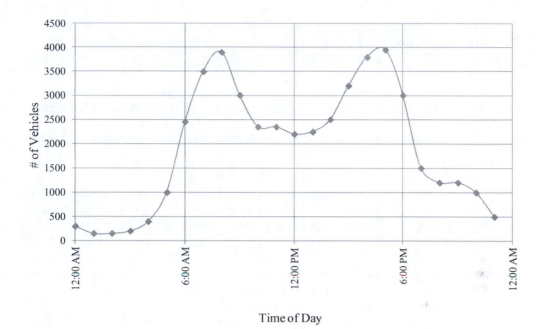

FIGURE 13.1

A typical highway traffic pattern over a 24-hour period. (Adapted from U.S. Department of Transportation, Federal Highway Division, SR 99: Alaskan Way Viaduct Replacement Project: Final Environmental Impact Statement and Section 4(f) Evaluation, July 2011.)

13.2.17 Socioeconomics

A major project can affect the socioeconomic conditions in the nearby local communities. Social impacts include the consequences of the proposed action that can alter the ways in which people live, work, play, and participate in civic life. The following socioeconomic categories are considered in the NEPA process:

- *Demographics.* The proposed project can change the age, ethnicity, income profile, and level of poverty in the study area. Projected changes in demographics are estimated using statistical techniques.

- *Economic Base.* The effect of the proposed action on the local economy needs to be understood. Current economic data that can provide baseline data for a study are available from federal and local economic development agencies, chambers of commerce, and so forth. Direct changes in the economy from the proposed project include the creation of construction jobs. Improved infrastructure may lead to more opportunities for local businesses, but businesses may also move in from outside the area, which could result in the closing or dislocation of local businesses.

- *Housing.* Improvements in the local economy may result in an increase in housing prices, which could have a deleterious effect on middle-income or low-income residents.

- *Local government finances.* Proposed actions generally are designed to increase the economic base of the local government, but they also can increase the long-term costs of governmental services needed to maintain the project. The resulting strain on public services such as police, fire, schools, and health services need to be considered when judging the long-term impacts of the proposed action on the local community.

- *Land use.* The effect of the project on land use needs to be understood, especially its potential effects on open space and recreational properties, such as parks. The proposed project is evaluated for its consistency with local or regional planning documents.

- *Aesthetics.* The aesthetics of the proposed action may have a direct effect on the local communities. Project designers often are willing to spend time and money on the visual quality of the proposed action. Renderings are generally included in the EIS. There are usually numerous comments on the aesthetics of the proposed action, and many changes are made in response to public comments.

Environmental justice (EJ) deals with the concept that the impacts of construction and industry often have a disproportionately high impact on low income and minority communities. Environmental justice considerations run through the entire NEPA process. Outreach to these communities within and around the proposed action is conducted, and both primary and secondary effects of the proposed action on these communities is taken into account. Different agencies handle EJ concerns differently and in accordance with their jurisdiction.

13.3 Cumulative and Indirect Impacts of the Project

Some of the most significant environmental effects caused by a project may result not from the direct effects of a particular action, but from the combination of individually minor effects of multiple actions over time. This phenomenon, known as "cumulative effects," is the impact on the environment which results from the incremental impact of the action when added to other past, present, and reasonably foreseeable future actions regardless of what agency (federal or non-federal) or person undertakes such other actions.

Cumulative effects must be evaluated along with the direct effects and indirect effects (those that occur later in time or further removed in distance) of each alternative. The range of alternatives considered must include the "no action alternative" as a baseline against which to evaluate cumulative effects. The range of actions that must be considered includes not only the project proposal but all connected and similar actions that could contribute to cumulative effects. All related actions can be addressed in the same analysis. For example, the expansion of an airport runway that will increase the number of passengers traveling must address not only the effects of the runway itself, but also the expansion of the terminal and the extension of roadways to provide access to the expanded terminal. The selection of actions to include in the cumulative effects analysis, like any environmental impact assessment, depends on whether they affect the human environment.

13.4 The National Environmental Policy Act Process and Environmental Consultants

Because the preparation of a NEPA document usually requires resources and availability beyond the ability of the lead agency, as well as specialized technical expertise, it is usually prepared by a private-sector consulting firm. Since its preparation involves a number of disciplines and forms, a team of consultants usually is needed to complete the NEPA process. The team may come from one firm or from several firms, each performing one or more parts of the NEPA process.

Many environmental concerns discussed in other chapters of this book, such as environmental due diligence (Chapter 6), site investigation (Chapter 7), remedial investigation (Chapter 8), remediation (Chapter 9), and ecological assessment (Chapter 12) come into play in the NEPA process. In addition, brownfield redevelopment projects (Chapter 11) that use federal funds must go through the NEPA process. Evaluation of other environmental issues such as water supply and noise, also are performed by environmental consultants. Other evaluations, such as cultural resources and transportation, typically are performed by specialists outside of the mainstream of environmental consulting.

Environmental consultants' contributions to the NEPA process include preliminary research that is used to establish baseline conditions, and the collection of field data to fill in data gaps in the baseline data, or to assess the potential impacts of the various alternatives considered. Environmental consultants typically perform computer modeling of air impacts, noise impacts, and predicted water quality. They may prepare fact sheets for stakeholders, attend public meetings as part of the evaluation team, coordinate the various participating public agencies, and assist in the evaluation of alternatives and the redesign of the proposed project as warranted.

Problems and Exercises

1. Go to the USEPA web site and download a Final EIS. Describe how is the project consistent with the mission of the lead agency.

2. Discuss the results of a data set that was collected for a Final EIS on the USEPA website. Discuss why the data changed or didn't change the proposed project.

3. Download USEPA comments letters for a Final EIS. Discuss the objective of the comments and summarize USEPA's responses to the comments.

4. Review an EIS for a project that was cancelled. Discuss whether the project was cancelled due to the findings of the NEPA process.

Bibliography

Bregman, J.I., 1999. *Environmental Impact Statements*, 2nd ed. Lewis Publishers.
Council on Environmental Quality, 2007. A Citizen's Guide to the NEPA: Having Your Voice Heard. Executive Office of the President.

Lawrence, D.P., 2003. Environmental Impact Assessment: Practical Solutions to Recurrent Problems. Wiley-Interscience.

Mangi Environmental Group, Inc., 2010. Final Programmatic Environmental Assessment for the National Institute of Justice Grants Program.

U.S. Department of Transportation, Federal Highway Division, 2011. SR 99: Alaskan Way Viaduct Replacement Project: Final Environmental Impact Statement and Section 4(f) Evaluation.

U.S. General Services Administration web site: http://www.gsa.gov/portal/category/21006.

U.S. Environmental Protection Agency, www.epa.gov.

U.S. Environmental Protection Agency, 1974. Information on Levels of Environmental Noise Requisite to Protect Public Health and Welfare with an Adequate Margin of Safety. Report Number 550/9-74-004. Washington, DC.

14

Drinking Water Testing and Mitigation

14.1 Introduction

In large part because of the Safe Drinking Water Act (SDWA), the water that runs from the tap in the homes, schools, and places of business in the United States is among the cleanest water in the world. The SDWA requires *public water systems*, defined as those systems that provide piped drinking water to at least 25 persons or 15 service connections for at least 60 days per year, to test the drinking water for a variety of contaminants at least annually. The contaminants to be tested for, as described in Chapter 4, include various chemical, biological, and even nuclear parameters, as well as physical properties such as color, temperature, and hardness. Smaller drinking water systems may be regulated under state and local laws.

Since its promulgation, the U.S. Environmental Protection Agency (USEPA) enacted several rules under the SDWA. This chapter discusses the Lead and Copper Rule, which has the most implications for environmental investigations and remediations among the rules promulgated under the SDWA.

14.2 History of Lead in Drinking Water

Lead in drinking water dates back to ancient times. The chemical symbol for lead, Pb, is an abbreviation of the Latin word for lead, *plumbum*. Lead was used in the conveyance system that supplied drinking water to ancient Rome (see Figure 14.1). The English word "plumbing" derives from this Latin root. By its very origins, plumbing has been associated with lead.

Just as with asbestos (Chapter 15) and lead in paint (Chapter 16), lead was utilized in drinking water systems because of its beneficial properties. Lead makes products more durable because of its resistance to physical as well as chemical breakdown, and it is relatively inexpensive to mine and manipulate. As demonstrated in Figure 14.1, lead pipes can remain in service for a very long time.

Beginning in the 1800s and well into the twentieth century, lead was commonly used in water piping, solder, and plumbing fixtures (e.g., faucets, taps, elbows, and valves). Buildings constructed before 1930 most likely were serviced by lead-based piping, after which they were replaced by copper pipes and galvanized steel pipes. The phasing out of lead in piping, however, did not extend to solder, which remained predominantly lead-based into the 1980s. In January 2011, Congress passed the Reduction of Lead in Drinking

FIGURE 14.1
Lead piping at an ancient Roman bath.

Water Act, which required that pipes, pipe fittings, plumbing fittings, and plumbing fix-tures used for drinking water be lead-free. The Act defined "lead free" as a weighted aver-age of 0.25% lead calculated across the wetted surfaces of the above-mentioned plumbing equipment.

14.3 Health Effects of Lead in Drinking Water

Lead is one of the most prevalent drinking water problems in the western world. In the context of environmental investigations and remediations, it is important to note that lead is often a building-specific problem, even for buildings serviced by public water systems, but especially for those that have systems that aren't monitored under the SDWA. Because the problem can be building-specific (and therefore property-specific), it can affect the sale and development of properties, which is why lead is the only drinking water concern men-tioned explicitly in the ASTM Phase I standard (see Chapter 6).

Lead, if present in the plumbing of a water delivery system, can leach into the water due to *corrosion*, which is "the deterioration of a material, usually a metal, that results from a reaction with its environment" (NACE International, 2000). Although corrosion occurs more readily when the water is acidic (low pH), or "soft" (low calcium), some lead will leach into the water regardless of the water's physical properties or chemical composition. Drinking lead-contaminated water, even in small doses, can impact the health of the people drinking the water, particularly young children and pregnant women. Lead can cause neurological damage and even death if ingested in sufficiently large quantities. The reader may want to review the case of the Flint, Michigan, municipal water system, in which thousands of residents became sickened by the introduction of lead-contaminated water into the public water system. (The USEPA web site for the Flint drinking water crisis is available at www.epa.gov/flint/flint-drinking-water-documents.)

14.4 Health Effects of Copper in Drinking Water

Whereas lead is a contaminant that has well-documented health effects, the health effects of copper are less severe. Ingestion of copper can cause gastrointestinal distress, but there are no documented long-term health effects associated with the ingestion of copper. Copper is regulated under the SWDA not because of its health effects, but because its presence in drinking water can be used as a surrogate to indicate the presence of corrosion in the drinking water delivery system.

14.5 Lead and Copper Rule

The SDWA established a maximum contaminant level (MCL) of 15 micrograms per liter (µg/L) for lead in drinking water and 1,300 µg/L for copper in drinking water. Some states have lower MCLs for lead and copper in drinking water. The SDWA also established an MCL "goal" (MCLG) of "non-detect" for lead, the level of detection being subject to the analytical method employed. The MCL is enforceable; the MCLG is not. If the lead or copper level in drinking water exceeds the MCL (or its state standard), then the situation must be mitigated to bring the drinking water into compliance with the regulations.

The 1986 amendments to the SDWA included a total ban of the usage of lead in piping (known as the *Lead Ban*), established a limit of less than 0.2% lead in piping solder and flux (the material used in association with the solder) and banned the usage of more than 8.0% lead in faucets and other plumbing fixtures. Prior to 1986, solder often contained as much as 50% lead. The Lead and Copper Rule, promulgated in 1991, did not ban the use of copper in drinking water systems; in fact, copper pipes are commonly used to this day in delivering drinking water to the user.

Even though the Lead Ban was passed in 1986, states were not required to implement the ban until 1988, and in many states the ban wasn't enforced until 1991. Even that date is risky to rely on, since plumbers commonly used old materials stock, or ignored the ban altogether, for several years afterward. In fact, it was still permissible to sell piping and fittings containing lead in the United States until August 6, 1998. Therefore, one cannot rule out the presence of lead in drinking water systems of buildings or portions of buildings that were constructed well into the 1990s.

14.6 Investigating Sources of Lead in Drinking Water

To conduct a lead-in-drinking-water investigation, one must first understand the potential sources of lead in a drinking water system. It is helpful to picture a drinking water system as having three major components: water withdrawal points, a delivery system, and access equipment.

Drinking water is withdrawn from surface water or groundwater (see Chapter 5 to learn more about groundwater), or both. Surface waters that provide drinking water are generally fresh water lakes, man-made reservoirs, and rivers, although many countries are now drawing much of their drinking water from salt water bodies and making the water potable using desalination plants.

Figure 14.2 is a schematic drawing of a water system for which groundwater is the drinking water source. Because the water source is below ground, drinking water wells must be installed to access the groundwater. The well is drilled into an aquifer and outfitted with a screen within the aquifer, which allows the water to flow into the wellbore. A pump installed in the well draws the water up to the surface and into the water treatment plant, after which the water enters the delivery system. The delivery system consists of underground piping, first through mains and then through a series of progressively smaller diameter pipes, until it reaches the buildings that will use the water. There is a service connection, typically on public property such as a roadside curb, that allows water to flow from the public water system into the building and makes it available for use by the building occupants. Within the building are various pipes from the *point of entry*, as described later in this chapter.

Figure 14.3 is a schematic drawing of a water treatment plant that draws its water from a river. The water enters the plant through underground piping; is treated by various

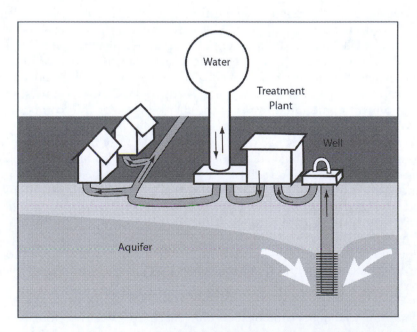

FIGURE 14.2
Groundwater source of drinking water. (From Ohio EPA. With permission.)

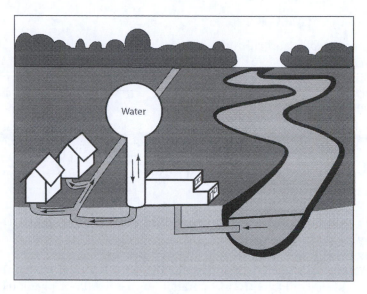

FIGURE 14.3
Surface water source of drinking water. (From Ohio EPA. With permission.)

processes within the plant to render the water potable; and then piped to a storage facility, typically a water tower. The water tower releases the water on an as-needed basis into the delivery system, as described above.

Lead can be present in the equipment used to extract and pump the water from its source into the water delivery system. In the case of systems that utilize groundwater, the equipment would include the water productions wells and the pumps that bring the water to the surface. Lead also could be present in any and all of the various piping that comprises the delivery system, as well as the rock formation itself that stores the water. Solder, which is used to fuse pipes together, contained lead until its usage in solder was banned the mid-1980s. Solder also was used in plumbing fixtures, such as faucets, and often the fixtures themselves were partially comprised of lead.

Homeowners and businesses with their own private potable water systems typically get their drinking water and water to be used for irrigation and other non-potable purposes from private wells that draw groundwater from aquifers. The USEPA estimates that 15% of all homeowners have private wells on their property. These systems are more likely to have lead issues than public water systems since the smaller systems are not regulated under the SDWA.

14.7 Investigating for the Presence of Lead and Copper in Drinking Water

Assessing the presence of lead and copper in a drinking water system follows the same general guidelines as other environmental investigations described in this book. The first step is thorough research into the history of the equipment to be assessed. The information collected is reduced into a format, in this case known as a plumbing profile, that allows the investigator to develop a sampling and analysis plan. These steps are described below.

14.7.1 Developing a "Plumbing Profile"

Before collecting samples and running analyses for the presence of lead, it is prudent to first develop a *plumbing profile* for the subject building or buildings. The objective of the plumbing profile is to identify all potential sources of lead and copper in the drinking water supplied to the building. The investigator will collect information regarding all components of the plumbing system, from the water extraction point to the point of entry into the building, through the piping, concluding with an inventory of water outlets inside the building. Information often will be available from multiple sources, including the public water system owner and operator, as applicable (they could be separate entities), and the owner of the subject building. The local regulatory authorities, such as the town and county health departments, also may have records regarding the drinking water equipment serving the building.

The plumbing profile will review the history of the water extraction equipment, conveyance piping, and access points. This history would entail both the time of installation of the equipment as well as the records of any upgrades, repairs, or replacements that may have occurred subsequent to their installation. The plumbing profile also will include water treatment systems currently in operation, and records of previous testing of the drinking water at any point in the distribution system.

If the public water system has had problems in the past with elevated concentrations of lead, then the end users are likely to have lead problems as well. In some jurisdictions, the use of lead in service connectors was required prior to the Lead Ban. If the building was constructed prior to the Lead Ban and has its own water source (known as a *Non-Transient, Non-Community Water System*, or NTNCWS), then the investigator should assess possibility of the presence of lead in the drinking water.

Figure 14.4 shows the piping system on a floor of a typical commercial building. There are pipes carrying hot water and pipes carrying cold water. Both systems have vertical pipes, called "risers," that distribute the water from the piping entering the building, typically located below ground, to the higher floors. In multi-story buildings, bathrooms and other water withdrawal points tend to be in the same relative place on the floor because they connect to the same risers. The wall behind which these pipes run is colloquially known as the "wet wall." Connecting to the risers are horizontal pipes that bring the hot and cold water to the water outlets for consumption.

There are three areas of water withdrawal in Figure 14.4: the men's bathroom, the women's bathroom, and the janitor's closet. There may be multiple withdrawal points in the bathrooms, the withdrawal points of interest being the sinks. Inspection of the hardware, study of the systems and their installation and maintenance history will determine how many withdrawal points to sample within the building. A visual inspection of the plumbing equipment can be useful. Lead pipes typically are dull gray in color and easily scratched by a steel object. If the equipment meets these criteria, one may assume that the equipment contains lead and bypass the collection of a water sample for analysis, as described below. That said, lead analyses are relatively inexpensive, so it is usually worthwhile to verify the visual clues and collect a water sample for analysis.

14.7.2 Development of a Lead Content Sampling Plan

Once the water outlets whose drinking water may contain lead have been identified, the environmental consultant will design a *lead content sampling plan* (LCSP). The LCSP should be based on the plumbing profile. The plan is to thoroughly assess the drinking water system for lead hazards to building occupants.

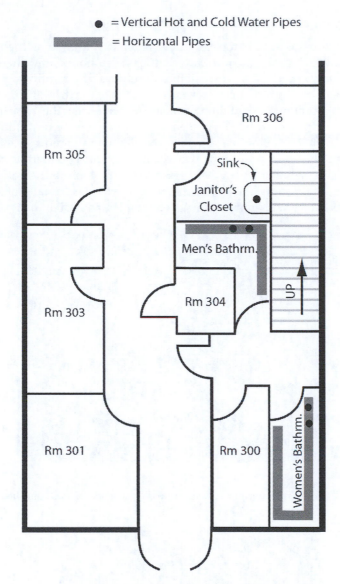

= Vertical Hot and Cold Water Pipes

= Horizontal Pipes

FIGURE 14.4
A schematic diagram of hot and cold water piping in a typical office building.

The LCSP should include the who, what, where, why, and how of the proposed sampling program. Included in the plan should be floor plans that show the approximate locations of the point of entry of the drinking water, the distribution pipes, and the proposed sampling locations. It should include the names of the people who will be conducting the sampling and the site contacts. It also should contain quality control instructions, including directions to flush the proposed sampling locations less than 6–8 hours before the sampling is to be conducted, as recommended by the USEPA.

For large distribution systems, it is very costly and unnecessary to sample every access point for lead and copper content. The LCSP can forego sampling of water access points that are not typically used for consumption, such as toilets and outdoor hose connections.

14.7.3 Sampling for Lead and Copper in Drinking Water

The first sample to be collected from an outlet is what is known as the *first draw sample*, (see Figure 14.5), which typically is collected at least 6–8 hours after the last usage of the outlet. This time lag would allow corrosion to take place if conditions allow, making the first draw water sample the worst-case scenario for lead in the drinking water at that outlet. The closer system components containing lead are to that outlet, the more likely lead will be detected in the first draw sample. The presence of copper at a concentration above its MCL indicates that corrosion is taking place, which should be mitigated, as described below.

The containers to be used to store the water sample should be inert, that is, non-reactive with the chemicals in the water with an acid preservative to stabilize the lead-containing compounds in the water. Typically, the containers are made of glass or laboratory-grade plastic. A fixed-based laboratory will analyze the water samples, as shown in Figure 14.6, although

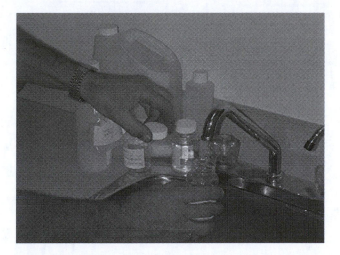

FIGURE 14.5
Collecting samples of tap water. (Copyright Agriculture and Agri-Food Canada, published by the Government of Canada.)

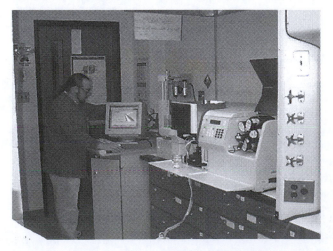

FIGURE 14.6
Laboratory analysis of drinking water sample. (From www.water-research.net. With permission.)

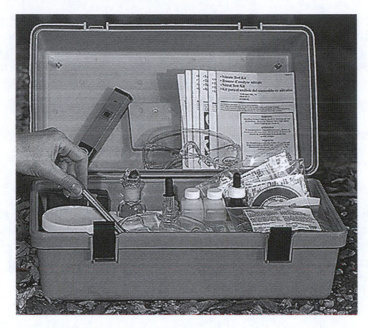

FIGURE 14.7
Portable laboratory for water analysis. (Courtesy of Lab Safety Supply, (now Grainger) Janesville, WI, www. labsafety.com.)

there are test kits readily available at various retail outlets or through online vendors. The results obtained from test kits are not as accurate as those obtained in a laboratory, but the test kits are inexpensive and provide results instantaneously. Portable laboratories occupy a middle ground (Figure 14.7) due to their ability to analyze water samples almost as quickly as a test kit, but with much higher accuracy and precision.

If elevated concentrations of lead are detected in this sample, the next task is to locate the source of the lead, be it in the water delivery system to the point of entry, the delivery system inside the building, the plumbing fixtures, or a combination of the three. There are several ways of doing this. One way is to collect a sample at the water's point of entry into the building. If lead is detected above acceptable limits in the point of entry sample, then at least part of the problem lies with the water purveyor.

To determine if there is a lead problem at the plumbing fixture tested, one must discount the lead contribution from outside sources. For example, if 40 μg/L of lead was detected in the water obtained from a kitchen sink and 15 μg/L was detected at the point of entry, then one can conclude that the sink fixtures contributed approximately 25 μg/L of lead.

A more complicated example would be if the water emanating from the kitchen sink contained 17 μg/L of lead and the water at the point of entry contained 9 μg/L of lead. In this example, it would appear that the kitchen fixture contributed 8 μg/L of lead. Although the drinking water exceeds the limit, neither the source piping nor the fixture is by itself contributing lead over the regulatory limit. The situation still would require mitigation, since the building occupants would be exposed to lead-contaminated water.

Commonly, first draw water sampling is supplemented by what is known as a *flush sample*, which is collected from the same outlet from which the first draw sample was obtained. The objective of the flush sample is to purge the standing water from the outlet piping and allow fresh water from outside the fixture to pour through the faucet and be sampled. In smaller buildings, an adequate flush sample can be collected after letting the water run for 15–30 seconds. In larger buildings, one should let the water run until the pipes have been purged of standing water. It may be necessary to let the water run for 1–3 minutes prior to sampling, or even longer in buildings with complex water delivery systems.

Sampling results can be affected by numerous factors, so it is advisable to test multiple fixtures, in smaller buildings all fixtures, before drawing conclusions about the sources of lead detected in the drinking water at one location.

14.8 Lead and Copper Mitigation

Once the lead or copper problem has been identified, the next step is to mitigate (remediate) the problem. If the problem lies outside the building, then the solution may mean requesting that the water purveyor fix the problem. If the problem involves the water delivery system inside the building, then the building owner will have to fix the problem to protect the health of the building occupants.

Lead mitigation falls into two basic categories: passive mitigation and active mitigation. *Passive mitigation* refers to a solution that doesn't involve engineering design and construction. For instance, allowing the water to run so that water that had been standing in the interior piping has been flushed is a remedy, although it is difficult to enforce since small children would have ready access to toxic drinking water. If this remedy is selected, samples should be collected after different flush times and from all of the lead-containing fixtures to determine how long the water would need to run before it can be consumed safely.

Another passive mitigation method is to not consume the water altogether and to rely on bottled water. Bottled water, however, can be expensive, especially if used for cooking as well as direct consumption. Because of the expense and inconvenience involved, bottled water generally is considered a temporary solution until a permanent solution can be implemented, except in situations where low consumption of water is expected, such as in an office building.

The most reliable solutions to the lead problem involve *active mitigation*, which entails the design and construction of a remediation system. The simplest but often the costliest solution is to replace the lead-containing delivery equipment with lead-free equipment. For a sink or a water fountain, this might mean simply replacing the faucet or the piping leading to the outlet. If, however, it involves removing pipes, which are typically concealed behind walls or beneath floors, then the mitigation could be very expensive and highly disruptive to building occupants.

If the water source is a private water system, then the building owner may be responsible for the entire system, not just the equipment inside the building. In such cases, it may be advisable to install a treatment system between the lead-containing equipment outside the building and the piping and outlets inside the building. Such devices are known as POET (*point-of-entry treatment*) devices.

One such POET system is a water purification unit. A typical water purification unit involves the process of *reverse osmosis*, shown on Figure 14.8. In such a system, lead is forced to move across a semi-permeable membrane, going from an area of higher concentration to an area of lower concentration. Once on the other side of the semi-permeable membrane, the lead can then be removed by flocculation or the lead-contaminated water could be discharged to the sanitary sewer system. A reverse osmosis system can also be installed on an individual fixture, which would avoid the

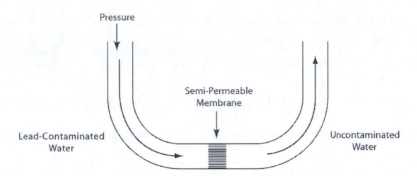

FIGURE 14.8
A schematic diagram showing the process of reverse osmosis.

costs of replacing the plumbing hardware for that water outlet. Such systems at the tap are known as *point-of-usage* (POU) devices.

As indicated above, high concentrations of copper in drinking water are indicative of corrosion in the delivery system. Therefore, the remedy needs to address the corrosion as well as the associated lead issue. This is best accomplished by adjusting the pH of the water coming into the building by treating it with devices such as calcite filters, soda ash or phosphate solution tanks. Another way to address corrosion in the piping system is the equivalent of encapsulation in asbestos abatement (see Chapter 15), namely either coating the pipes with a sealant that will prevent the water from interacting with the lead in the piping, or soldering or running a non-metallic sleeve through the piping so that the drinking water does not come into contact with the corroded metal.

The effectiveness of the engineered solutions must be verified by confirmation testing of the water at the locations where the lead problem was detected. Periodic testing and maintenance of engineered systems such as POET devices is also advisable to ensure that the lead problem, once solved, does not return.

14.9 Drinking Water Testing and Environmental Consultants

Environmental consultants typically test drinking water on private properties as part of due diligence in support of a property transaction or financing or in response to a potential health concern. Consultants also will conduct such testing for the owners of public water systems, although the larger purveyors of drinking water typically will keep such testing in-house. Contract laboratories will perform the analyses of the drinking water samples.

Problems and Discussion Questions

1. Lead is detected in the drinking water of a private residence. A water sample collected from the point-of-entry into the building did not contain detectable concentrations of lead. What are the potential sources of the lead?

2. The following questions refer to the table below.

Sample No.	Sample Location	Lead Concentration, First Draw (µg/L)	Lead Concentration, Flush Sample (µg/L)
1	Point-of-entry into house	4	4
2	2nd floor bathroom sink	20	5
3	Kitchen sink	12	5
4	Outdoor tap	5	5

a. Interpret the analytical results shown in the table.

b. Describe potential steps to be taken to remediate the hazard presented by lead in the drinking water.

3. The following questions refer to the table below.

Sample No.	Sample Location	Lead Concentration, First Draw (µg/L)	Lead Concentration, Flush Sample (µg/L)
1	Point-of-entry into house	0	10
2	2nd floor bathroom sink	8	18
3	Kitchen sink	0	10
4	Outdoor tap	20	20

a. Interpret the analytical results shown in the table.

b. Describe potential steps to be taken to remediate the hazard presented by lead in the drinking water.

Bibliography

NACE International, 2000. NACE International Glossary of Corrosion-related Terms. (Item No. 26012).

New Jersey Department of Environmental Protection, 2017. Lead in Drinking Water at Schools Facilities. Version 3.0.

New Jersey Department of Environmental Protection, 2017. Guidance for Developing a Lead and Copper Sampling Plan.

U.S. Environmental Protection Agency, 1993. Lead in Your Drinking Water: Actions You Can Take to Reduce Lead in Drinking Water. EPA 810-F-93-001.

U.S. Environmental Protection Agency, 1994. Lead in Drinking Water in Schools and Non-Residential Buildings. EPA 812-B-94-002.

U.S. Environmental Protection Agency, 1998. Commonly Asked Questions: Section 1417 of the Safe Drinking Water Act and the NSF Standard.

U.S. Environmental Protection Agency, 2001. Drinking Water Standard for Arsenic.

U.S. Environmental Protection Agency, 2001. Technical Fact Sheet: Final Rule for Arsenic in Drinking Water.

U.S. Environmental Protection Agency, 2001. Radionuclides Rule: A Quick Reference Guide.

U.S. Environmental Protection Agency, 2002. Drinking Water from Household Wells. EPA-816 -K-02-003.

U.S. Environmental Protection Agency, 2008. Ground Water Rule: A Quick Reference Guide.

U.S. Environmental Protection Agency, 2008. Ground Water Rule Compliance Monitoring: A Quick Reference Guide.

U.S. Geologic Survey, 2018. General Introduction for the National Field Manual for the Collection of Water-Quality Data. Techniques and Methods 9-A0, Version 1.1.

15

Asbestos Identification and Abatement

15.1 Introduction

It was known as the magic mineral. Asbestos is a naturally occurring mineral that was valued since ancient times for its resistance to heat and flame. The ancient Greeks gave the mineral its name, which means "inextinguishable." Its modern history began in the mid-1800s, when the construction industry began incorporating asbestos fibers in its roofing materials to prevent fires from spreading in urban areas. Eventually, asbestos was used not only for its fire-resistance abilities, but also its ability to provide insulation, sound-dampening, its tensile strength and general abundance.

15.1.1 Types of Asbestiforms

There are six known asbestos minerals, as follows:

- *Chrysotile*, a white or greenish-colored mineral, was the most commonly used asbestiform in the United States and the only serpentine asbestiform (see Figure 15.1).
- *Amosite*, also known as "brown asbestos," is an amphibole asbestiform that derives its name from the asbestos mines of South Africa.
- *Crocidolite*, also known as "blue asbestos," is an amphibole asbestiform that was mainly mined in South Africa and Australia, and had limited usage in the United States.
- *Anthophyllite* is an amphibole asbestiform that has excellent resistance to chemicals and heat. It was primarily used in the United States in decorative and acoustical material.
- *Tremolite* is a white to yellow-colored amphibole asbestiform that has a major ingredient in industrial and commercial talc. There is also a non-asbestiform type of tremolite.
- *Actinolite*, an amphibole asbestiform that is greenish to white in color, has poor resistance to chemicals and had limited commercial usage. It may be found in commercial and industrial talcs.

Except for chrysotile, all asbestiforms fall into the mineralogical category of amphiboles, which form hard, needle-like particles that will stick to and damage soft tissues in the lungs and other internal parts of the body. Chrysotile is a serpentine which, as its name

FIGURE 15.1
Naturally occurring chrysotile asbestos. (Courtesy of the Natural History Museum, London, UK.)

implies, has a wavy morphology. Serpentine asbestiforms tend to get lodged predominantly in the upper respiratory pathways. Amphibole asbestiforms are generally recognized as more hazardous to human health than chrysotile.

15.1.2 Health Problems Related to Asbestos

There are three known asbestos-related diseases: asbestosis, lung cancer, and mesothelioma.

Asbestosis is a scarring (fibrosis) of the lung that impairs the elasticity of the lung tissue and restricts breathing. Asbestosis has a latency period of 10 to 30 years and generally is associated with a long, heavy exposure to airborne asbestos fibers.

Lung cancer is a malignant tumor of the bronchi covering. Although there are many causes of lung cancer, there is a clear increase in risk among people who worked with asbestos. There is no threshold or level of exposure below which the risk of lung cancer is not increased. The typical latency period for lung cancer resulting from asbestos exposure is 20 to 30 years.

Mesothelioma is a cancer of the mesothelium—the lining of the chest or the lining of the abdominal wall. It is the only known type of cancer directly attributed to asbestos exposure. By the time it is diagnosed, it is almost always fatal. There is no exposure threshold for mesothelioma, and the disease may not manifest itself until up to 40 years after the time of exposure to asbestos.

15.1.3 Regulatory Framework for Asbestos

In 1973, the U.S. Environmental Protection Agency (USEPA) banned the application of asbestos-containing spray-on surfacing material and visible emissions of asbestos under the National Emission Standards for Hazardous Air Pollutants (NESHAPs) section of the Clean Air Act. Other forms of asbestos-containing materials (ACMs), which the USEPA defines as having greater than 1% asbestos by weight, gradually became less common over the years, although some ACMs are still being sold in the United States.

In 1986, the USEPA amended the Toxic Substances Control Act (TSCA) with the Asbestos Hazard Emergency Response Act (AHERA). AHERA, which was designed to minimize asbestos hazards in schools for grades K through 12, set up elaborate systems for the testing, reporting, training, and maintenance of asbestos-containing building materials (ACBMs). The ACBMs include materials that were part of the building interior and excludes exterior materials, such as roofing materials and exterior wall coverings, as well as temporary equipment, such as blackboards, laboratory hoods, and other laboratory equipment. ASHARA, or the Asbestos School Hazard Abatement Reauthorization Act, which was enacted in 1990 and took effect in 1994, extended the AHERA rule to non-school buildings that are not private residences with less than ten dwelling units. Through ASHARA, the AHERA protocols became the norm for virtually all commercial properties. In addition, many states and some cities have their own asbestos abatement regulations, which, by law, must be at least as stringent or more stringent than the federal laws.

The federal Occupational Safety and Health Administration (OSHA) regulates worker exposure to asbestos. The OSHA regulations are designed to prevent airborne asbestos in the workplace from exceeding the current permissible exposure limit (PEL) of 0.1 fibers per cubic centimeter (f/cc). In addition, OSHA's Asbestos Standard for Construction, codified in 29 CFR 1926.1101, is designed to protect workers who may become exposed to asbestos in the workplace. It plays a role in the selection of personal protective equipment for asbestos abatement workers, as described later in this chapter.

15.1.4 Types of Asbestos-Containing Materials

Asbestos was used in literally thousands of products. A summary of building products in which asbestos was used is provided in Table 15.1.

Certain materials can be designated as non-ACBM *per sé* (without the need for sampling or obtaining corroborating documentation), including the following:

- Certain types of poured concrete,
- Concrete blocks, bricks, and terra cotta materials,
- Glass products,
- Unpainted wood,
- Fiberglass insulation. Fiberglass is thermal insulation that has generally replaced asbestos. It is typically yellow in color, and easier to squeeze than ACM thermal insulation.
- Rubberized or plastic insulation materials,
- Metallic objects and sheet metal

TABLE 15.1

Products That May Contain Asbestos

Cement Asbestos Insulating Panels	Insulation, Thermal Sprayed-On	Cooling Tower, Fill
Cement Asbestos Wallboard	Blown-In Insulation	Cooling Tower, Baffles or Louvers
Cement Asbestos Siding	Insulation, Fireproofing	Valve Packing
Roofing, Asphalt Siding	Taping Compounds	Waterproofing, Asbestos Base Felt
Roofing, Asphalt Saturated Asbestos Felt	Packing or Rope (at penetrations through floors or walls)	Waterproofing, Asbestos Finishing Felt
Roof, Paint	Paints	Waterproofing, Flashing
Roofing, Flashing (tar and felt)	Textured Coatings	Dampproofing
Roofing, Flashing (plastic cement for sheet metal work)	Flexible Fabric Joints (vibration dampening cloth)	Plumbing, Piping Insulation
Laboratory Hoods	Fire Curtains	Plumbing, Pipe Gaskets
Laboratory Oven Gaskets	Elevators, Equipment Panels	Plumbing, Equipment Insulation
Laboratory Gloves	Elevators, Brake Shoes	Electrical Ducts (cable chases)
Laboratory Bench Tops	Elevators, Vinyl Asbestos Tile	Electrical Panel Partitions
Putty and/or Caulk	HVAC Piping Insulation	Electrical Cloth
Door Insulation	HVAC Gaskets	Insulation, Wiring
Flooring, Asphalt Tile	Boiler Block or Wearing Surface	Stage Lighting
Flooring, Vinyl Asbestos Tile	Breeching Insulation	Incandescent Recessed Fixtures
Flooring Vinyl Sheet	Fire Damper	Chalkboards
Flooring, Backing	Duct Insulation	Ceiling Tile
Plaster, Acoustical and Decorative	Ductwork Taping	Flue, Seam Taping

15.1.5 Components of Buildings

To identify suspect ACMs in a building, it helps to understand the construction of buildings.

A building superstructure consists of a foundation, floors, a roof, and load-supporting walls (i.e., walls designed to support all or part of a building). Building foundations usually are comprised of poured concrete. Exterior walls and interior load-bearing walls are typically composed of concrete block, brick, or, in taller structures, a steel I-beam frame. None of these structural materials, excepting the mortar used in the wall construction, typically are suspect ACM. In residences and smaller commercial buildings, the basic structure consists of 2′ × 4′ wood planks, which also aren't suspect ACM.

On the other hand, materials that are used to coat the structural building components and building "improvements" are often composed of suspect materials. Foundations and floors are often covered by resilient floorings or carpeting, whose adhesive is a suspect ACM. Fireproofing applied to steel beams are suspect ACM. Non-load bearing wallboard panels and the joint compound used to smooth out the seam between the panels are suspect ACM. Interior walls in older buildings are composed of plaster, which is a suspect ACM. Before being banned, exterior walls were often covered with cement panels usually known by their trade name: Transite™.

Office buildings often have suspended ceiling grids with "lay-in," "drop," or "dropped" ceilings. They earned this name because are situated within an aluminum lattice that is hung from the solid ceiling. Ceiling tiles composed of fiberglass or sheetrock typically aren't suspect ACM, but fibrous ceiling tiles are suspect ACM.

Boiler rooms and other building operations rooms can often be a cacophony of piping and wiring. Basically, a building's piping is designed to convey heat or heating materials, water, or electricity. The inspector often has the difficult task to identify the different piping systems in a building and be able to distinguish between them.

Heating pipes—Pipes in a building may convey steam to heat the building, or natural gas or petroleum to supply the boiler with combustible fuel to heat the building. Steam pipes are always insulated, both to prevent heat loss and to protect building occupants. Natural gas pipes leading to a boiler are usually not insulated, but fuel oil piping can be, especially if the oil has been pre-heated, as is the case for #6 fuel oil or bunker oil.

Water pipes—Each water pipe has one of three purposes: to convey heated water to baseboard radiators or stand-alone radiators for heating purposes; to convey drinking water to facility bathrooms, kitchens, and drinking fountains; or to convey waste water from the building to a sewer or septic system. Except for wastewater pipes, water pipes usually are insulated, to prevent the water from cooling off or heating up, as the case may be. Water pipes are typically no more than two inches in diameter, narrower than waste water pipes, which are usually four inches or more in diameter.

Electricity—Electrical conduit is almost never insulated (although the wires inside the conduits always are insulated). Electrical conduit is recognized by its narrow diameter and its connection to outlets and other electrical equipment.

To determine the usage of a pipe, the inspector may have to trace it through the building, across walls and floors, to either of its end points so that its function becomes apparent.

Understanding heating, ventilation, and air conditioning (HVAC) systems is also useful in identifying and classifying suspect ACBM. Older buildings achieved air circulation simply by opening a window and perhaps turning on a floor or ceiling fan. Window-mounted air conditioners soon followed. Modern buildings circulate air through central HVAC systems that utilize an air handler to convey fresh air from outside the building throughout the interior of the building. Suspect ACM in HVAC systems include the insulation on or inside the ductwork (which usually is non-ACM fiberglass, although the adhesive may be ACM) and vibration dampers designed to cushion the air handlers.

15.2 Classifying Suspect Materials

To facilitate grouping of suspect ACBM into categories, the inspector should identify "homogeneous areas," which consist of materials that are alike in color and texture. They also should have been installed at the same time. That way, samples within the same homogeneous area should yield the same analytical results, providing validity to the asbestos survey. Suspect ACBM similar in appearance but found in portions of the building that were constructed or renovated at different times should be considered separate homogeneous areas, unless building records indicate contemporaneous application. If a suspect ACBM is to be sampled, sampling locations must be selected. When in doubt, the inspector should assign suspect materials to separate homogeneous sampling areas.

AHERA defines the following three categories of ACBM:

Surfacing material—This category includes materials that were prepared for application on-site and were applied with a sprayer, a trowel, or some similar piece of equipment. "Spray-on" insulation, typically applied to steel support structures, protects the steel from losing its tensile strength in a fire. Figure 15.2 shows a so-called "popcorn" ceiling texture common in commercial buildings and multi-family apartments for sound mitigation.

Thermal system insulation (TSI)—This category includes materials that are attached to pipes, fittings, boilers, breaching, tanks, ducts, or other components to prevent heat loss or gain or water condensation. Typically, these materials are manufactured off-site and cut to size onsite. The so-called aircell piping insulation shown in Figure 15.3 is always ACM; the asbestos was placed inside the honeycomb structure to provide insulation to an otherwise poorly-insulated cardboard structure (Figure 15.4).

Miscellaneous ACBM—This category includes all building materials not included in the first two categories. Miscellaneous materials inside a building include floor tiles, ceiling tiles (see Figure 15.5), various adhesives, and joint compound applied to wallboard. There may be many types of miscellaneous materials in a building that are not building components, such as theatre curtains and lab table tops. Because they are not building materials, they are not required to be covered in an AHERA survey.

FIGURE 15.2
Spray-on "popcorn" ceiling texture. (Courtesy of GZA GeoEnvironmental, Inc.)

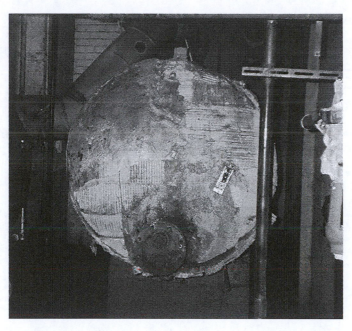

FIGURE 15.3
Suspect ACM in the form of insulation wrap around a hot water tank (Courtesy of GZA GeoEnvironmental, Inc). It may be composed of asbestos, fiberglass, or some other synthetic material, and needs to be investigated as to its composition.

FIGURE 15.4
Suspect ACM in the form of aircell piping insulation. (Courtesy of GZA GeoEnvironmental, Inc.)

FIGURE 15.5
Suspect ACM in the form of lay-in ceiling tiles. (Courtesy of GZA GeoEnvironmental, Inc.)

Roofs are often complex systems consisting of many layers and materials, many of which may contain asbestos, especially tars that are used as sealants around chimneys and other roof openings, and felt papers which underlie shingles and other resilient roof coverings, and are designed to prevent moisture intrusion. Exterior wall shingles and corrugated Transite™ panels (see Figure 15.6), are cementitious, prefabricated products that were applied to the exterior of buildings because of their durability.

FIGURE 15.6
Corrugated Transite™ panels on a building exterior. (Courtesy of GZA GeoEnvironmental, Inc.)

Suspect ACBM can also be distinguished by their *friability*, which is the ability for the material to be pulverized using hand pressure. The AHERA requires the inspector to physically touch the suspect material to determine its friability.

15.3 Functional Spaces

The AHERA developed the concept of a *functional space* to simplify the inspection and the grouping of homogeneous materials. Functional spaces are spatially distinct units within a building. An example of a functional space is a classroom, which is enclosed and separate from the rest of the building. A boiler room also its own functional space. Pipe chases, air shafts, elevator shafts, and return air plenums are also separate functional spaces, even though they are unoccupied.

The AHERA provide the investigator considerable latitude in defining functional spaces. There doesn't need to be a correlation between the functional spaces and homogeneous materials—one homogeneous material may be found in several functional spaces. A long corridor could be divided into separate functional spaces if doing so is useful to identify important distinctions in the conditions and/or disturbance of ACBM.

15.4 Performing the Asbestos Survey

The objective of an asbestos survey is to identify and locate all ACBM in the subject building. However, not all suspect ACBMs need to be physically sampled for asbestos content. If a building was constructed before 1981, then the inspector can classify a suspect ACBM as a presumed asbestos-containing material (PACM), and managed as ACBM. Therefore, an investigator can classify all suspect ACBM as PACM and not collect any samples for analysis. However, in most cases, such a conservative assumption is not in the interest of the client, and samples are collected and analyzed for asbestos content.

15.4.1 Designing the Survey

Before collecting samples, the inspector should review documentation regarding the construction history of the building, such as renderings, specifications, blueprints, and as-built drawings. They may indicate the locations of suspect ACM, or, in some cases, actually specify the brand and model number of building material to be used, which can then be cross-checked with its manufacturer. These documents can provide good intelligence as to whether a suspect material is ACBM.

The inspector should then visit the site and view as much of the building as is feasible. The inspector should look above suspended ceiling grids in each room, peer into crawl spaces, tunnels, and observe mechanical areas, basements, attics, and air plenums. The ACBM is sometimes found in inaccessible areas. In areas where physical access is impossible (e.g., pipe chases or behind walls), the inspector should try to determine the location and physical condition of the material or make reasonable assumptions.

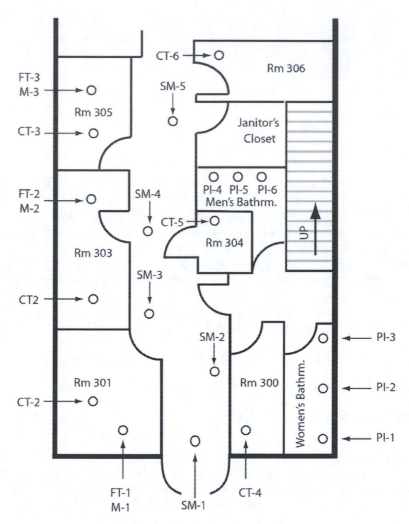

FIGURE 15.7

Scaled drawing of an office building showing the locations of bulk sample collection. In this diagram, the sample names are indicative of the material sampled: "FT" means floor tile; "M" means mastic; "CT" means ceiling tile; "SM" means surfacing material (in this case, "popcorn" ceiling texture); and "PI" means pipe insulation.

For each functional space, the inspector should prepare a scaled diagram that shows all suspect ACBM in the space. Building blueprints or diagrams should be used when available. A typical scaled diagram is presented in Figure 15.7.

15.4.2 Sampling Homogeneous Materials

As per AHERA, at least three bulk samples of TSI must be collected from suspect ACBM in a quantity exceeding 6 linear feet or 6 square feet (one sample is sufficient for smaller quantities). The samples should be randomly distributed—biasing the sample locations could bias the results, especially if the inspector hasn't properly identified the homogeneous area.

For surfacing material, the quantity of the suspect material defines the number of samples to be collected. While the USEPA recommends collecting nine samples from each homogeneous sampling area of surfacing material, AHERA requires the following numbers in its sampling protocol:

Square Footage	Required Samples	Recommended Number
<1,000 ft²	3	9
1,000–5,000 ft²	5	9
>5,000 ft²	7	9

Surfacing materials within a seemingly homogeneous area can vary widely in asbestos content. Therefore, AHERA devised a sample location protocol that forces randomness in the sample locations. It requires the inspector to divide the homogeneous area into a 3 × 3 grid. This can be done by eye—exact measurements are not needed. Samples are then collected from seven of the nine grids (each grid is sampled if the USEPA recommendation is followed). Each sample should be collected from the approximate center of each grid.

AHERA requires two samples of each suspect miscellaneous material. Three bulk samples are recommended, however.

In some instances, especially where demolition or renovation are to occur, it's necessary to identify ACBM in inaccessible areas. In such cases, a *destructive asbestos survey* is conducted, involving the destruction or partial destruction of the obstructing materials. Typical destructive procedures include removing concrete block, cutting holes through plaster or wallboard to access the suspect material, or tearing up floor coverings to search for older, hidden coverings.

15.4.3 Bulk Sampling Procedures

To minimize damage to ACBM and the dangers of possible fiber release, when possible samples should be collected in an unoccupied area. The inspector should wear at least a half-face respirator with disposable high-efficiency particulate filters known as "P100" filters, unless a higher protection factor is necessary or desired. To minimize releasing asbestos fibers during the sampling process, the inspector will wet the surface of the material to be sampled with "amended water" (water with a glycol additive to help penetrate the material) mist from a spray bottle or place a plastic bag around the sampler to capture asbestos fibers generated by the sampling procedure. Sampling is typically accomplished using a cork borer or a knife. The "bulk" sample is then placed into a sample container, which is sealed and labeled with a unique sample identification number. To prevent subsequent release of asbestos fibers from the now-damaged surface, the inspector may apply a sealant or a bridging material such as plaster or other enclosure material such as duct tape to cover the spot where the sample was extracted. The sampling device should be thoroughly cleaned before being used again to prevent cross-contamination.

15.4.4 Sampling Layered Materials

Some materials are manufactured or installed in layers. Plaster typically has two layers that are mixed and installed separately: a brown undercoat, and a white top coat for aesthetics. They must be treated as separate homogeneous areas, although it can be very difficult to do so in the field.

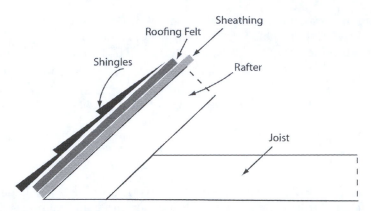

FIGURE 15.8
A schematic diagram of a typical layered roofing system.

Roofs can be complex structures with numerous components. For instance, Figure 15.8 shows a layered roof system consisting of shingles (for durability), a roofing felt (which acts as a vapor barrier), and the rafter (for structural strength), each of which are separate homogenous areas. Often roofing systems will have several different layers of materials. Older buildings may have multiple roofs. Add in roof flashing and other roofing materials, and there can be numerous homogeneous areas of suspect ACM on a single roof.

Another common layered system is wallboard and joint compound, which is applied between wallboard sheets to smooth out the surface. Although they appear to be one material, they are different homogeneous areas and need to be treated as such in an asbestos survey. Separating the joint compound from the wallboard is commonly done in the laboratory.

15.5 Laboratory Analysis of Bulk Samples

The laboratory that will analyze the samples should be a participant in a national or state-sponsored quality assurance program. Accreditation from the American Industrial Hygiene Association (AIHA) is also a desirable factor when choosing a laboratory.

The preliminary analysis of bulk samples should employ polarized light microscopy with dispersion staining (PLM-DS) and stereo-binocular examination. This analytical technique is based on the optical properties of crystalline and non-crystalline substances. It is a quick method for estimating the type and percentage of asbestos in a bulk sample. Figure 15.9 shows how chrysotile fibers appear under a polarizing microscope.

For materials with large quantities of asbestos fibers, PLM-DS analysis will do an adequate job in assessing whether a material is ACBM. For materials with smaller quantities of asbestos fibers, the PLM-DS method has several disadvantages to other methods:

- Because the sample analysis is qualitative, different microscopists can make widely varying estimates of asbestos content for the same sample.
- It has difficulty detecting short fibers, and
- Organic matrixes can hide asbestos fibers in a PLM-DS analysis. Building materials with organic binders include vinyl floor tiles (VFT), adhesives, roofing materials, asphalt shingles, and caulks.

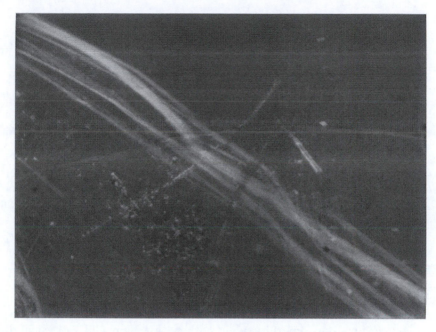

FIGURE 15.9
Asbestos fibers as seen under polarized light microscopy. (Courtesy of Materials and Chemistry Laboratory, Oak Ridge, TN. www.mcl-inc.com.)

For materials that have between 1% and 5% of asbestos using the PLM-DS method, PLM analysis by point counting is recommended and, in some jurisdictions, required. Under the point counting method, the visual field under a microscope is divided into a 10 × 10 grid. The microscopist observes each of the 100 squares in the field of vision and records the presence or absence of asbestos fibers in that square. By tallying the square with asbestos fibers, the microscopist generates a quantitative, more reliable estimate of the percentage of asbestos fibers in the bulk sample.

Because organic matrixes can hide asbestos fibers, the non-friable organically bound (NOB) protocol is recommended and, in some jurisdictions, mandated when PLM has not identified the suspect material as ACBM. Acid is applied to the material to dissolve the organic matrix, and then the material is burned, leaving mainly asbestos residue. If the residue is less than 1% of the original sample, no further analysis is required and the suspect material can be reported as non-ACM. If not, the residue must be analyzed by PLM or by the more accurate yet more expensive transmission electron microscopy (TEM). In a transmission electron microscope, a beam of electrons is transmitted through the suspect material. Asbestos will diffract the electrons at a different rate than the surrounding material, yielding a precise estimate of the type and percentage of asbestos in the material. The TEM is a very reliable identification tool even when the asbestos fibers are very short or thin.

If one or more samples from a homogeneous area contains more than 1% asbestos, then the entire homogenous area is ACBM, even if one or more samples tests negative. In most analyses, the "positive stop method" is employed, in which no more samples from a homogeneous area are analyzed once one of the samples tests positive as ACBM. An example of the use of the "positive stop method" is shown in Table 15.2.

TABLE 15.2

Example of a Portion of a Summary Table of Analytical Results from an Asbestos Survey

Sample No.	Floor	Location	Materials/Items	Analytical Results	
FT-1	3	Room 301	12″ × 12″ floor tiles		NAD
FT-2	3	Room 303			NAD
FT-3	3	Room 305			NAD
SM-1	3	Near front entrance	Popcorn ceiling texture	0.3%	Chrysotile
SM-2	3	Opposite wall of Room 301			NAD
SM-3	3	Near door of Room 301			NAD
SM-4	3	Near door of Room 303			NAD
SM-5	3	Near door of janitor's closet		2.4%	Chrysotile
PI-1	3	Women's bathroom, far wall	Corrugated piping insulation	39.5%	Chrysotile
PI-2	3	Women's bathroom, middle			Positive stop
PI-3	3	Women's bathroom, near door			Positive stop

Note: NAD = No Asbestos Detected.

If there are contradictory results from a group of samples, such as when one of nine samples of surfacing material is positive and the other eight are negative—then the inspector may want to assess whether something went wrong. Are there subtle physical differences between the positive and negative materials? Do they have a different application history? Was there cross-contamination? To answer these questions, sometimes resampling or additional sampling, or analyzing the negative samples using a higher quality analytical method, may be warranted.

15.6 Hazard Assessment

15.6.1 Physical Hazard Assessment

Once the asbestos survey has been completed, a *physical assessment* must be performed. A physical assessment consists of (1) assessing the current condition of the material and (2) assessing the potential for future disturbance. This aspect of the inspection process as is like preparing a map, providing enough knowledge concerning the ACBM to reduce future health hazards posed by these materials.

Functional spaces can be used to provide a physical assessment on a space-by-space basis. Since the condition of a particular homogeneous material may differ in different functional space, the hazard assessment for a particular material may differ from functional space to functional space.

15.6.2 Classifying the Condition of the Asbestos-Containing Building Materials

The potential for building occupants to be exposed to asbestos depends on the condition of the ACBM, the likelihood of disturbance, and the potential for fibers to be transported from their place of origin. The latter issue is not a consideration when performing a hazard assessment under AHERA. The AHERA divides the condition of friable surfacing material and TSI into three categories: good, damaged, or significantly damaged. These categories can be used for miscellaneous ACBM as well.

Good Condition—material with no visible damage or deterioration or showing only limited damage or deterioration.

Damaged—material which has deteriorated or been damaged such that the internal structure (cohesion) of the material is inadequate; which has delaminated such that its bond to the substrate (adhesion) is inadequate; or which, for any other reason, lacks fiber cohesion or adhesion qualities: flaking, blistering, or crumbling; water damage, significant or repeated water strains, scrapes, gouges, mars, or other signs of physical damage.

Significantly Damaged—there has been extensive and severe damage to the ACBM. To decide whether the material is simply damaged or if the damage is significant in nature, the EPA recommends the "10/25 Rule." The 10/25 rule states that a material is significantly damaged if:

- The extent of the damage is roughly 25% of the materials and is localized (see Figure 15.10).
- The extent of the damage is roughly 10% of the material and is evenly distributed throughout the material. This criterion pertains to water stains, gouges, and marks on the ACBM (see Figure 15.11).

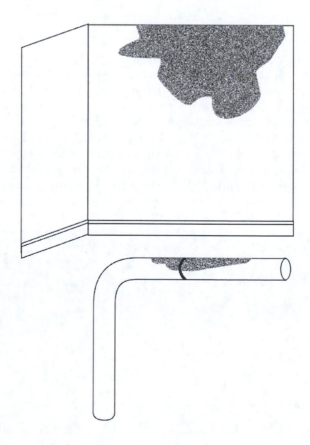

FIGURE 15.10
Example of 25% localized damage that results in a significantly damaged ACM classification. (Courtesy of U.S. Environmental Protection Agency.)

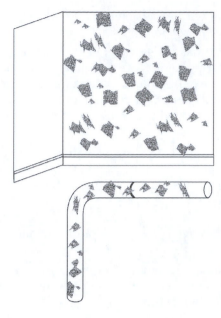

FIGURE 15.11
Example of 10% uniform damage that result in a significantly damaged ACM classification. (Courtesy of U.S. Environmental Protection Agency.)

Accumulation of powder, dust, or debris of similar appearance to the ACBM on surfaces beneath the material can be used as confirmatory evidence. Since the human eye tends to underestimate the percentage of uniform damage in a damaged material, Figures 15.10 and 15.11 can be useful guides for the inspector.

The assessment of miscellaneous materials depends on the type of material to be assessed. For instance, if asbestos-containing ceiling tiles are damaged or significantly damaged, they are unrepairable, but replaceable. Floor tiles and other non-friable miscellaneous materials, on the other hand, require no assessment if undamaged. If non-friable ACM such as floor tiles are damaged, as shown in Figure 15.12, then it would be appropriate for the inspector to recommend removal as the response action, without assigning it a hazard category.

FIGURE 15.12
Significantly damaged floor tiles. (Courtesy of GZA GeoEnvironmental, Inc.)

15.6.3 Classifying the Potential for Disturbance of Asbestos-Containing Building Materials

Once its condition is assessed, the potential for damage or significant damage to the ACBM and PACM also must be assessed. The following are a few types of activities that may release or re-suspend fibers into the air:

- Renovation projects
- Repair and maintenance activities
- Routine cleaning
- Operation of building systems
- Activities of occupants other than service workers
- Deterioration/aging of the ACBM

Exposure factors must also be considered when assessing the potential for disturbance. For example:

- Non-Visible Materials. If ACBM is not immediately visible, the occupants may forget its presence and accidentally expose it in a manner which may result in fiber release.
- Accessibility. The degree to which the material can be physically contacted by a building occupant. The ACBM that is easily accessible can be just as easily disturbed.
- Barriers. Are there barriers to casual contact? If so, are they permanent or temporary? How easily can these barriers be breached?
- Ventilation. Is the material in contact with the building's air stream?
- Air Movement. Air erosion may occur in a return air plenum or fan room, thereby eroding the ACBM.
- Air Plenum. Is the air plenum used for return air? If so, material inside the plenum may be dispersed throughout the building.
- Activity. Heavy activity such as in a gym or machine shop would pose a greater hazard than an area such as a library, or an electrical room with limited access.
- Vibration. Vibrations caused by mechanical equipment, vehicles, loud noises, and other sources could eventually loosen the material.
- Vulnerability of the Occupants. School occupants, and children in general, have a higher potential to disturb ACBM than the occupants of, say, a commercial building.

The difference between potential for damage and potential for significant damage plays an important role in this assessment. The AHERA defines *potential for damage* as follows:

- Friable ACBM is in an area regularly used by building occupants, including maintenance personnel, in the course of their normal activities.
- There are indications and/or there is a reasonable likelihood that the material or its covering will become damaged, deteriorated, or delaminated due to factors such as changes in building use, changes in operations and maintenance practices, changes in occupancy, or recurrent damage.

In addition to these conditions, the existence of major and/or continuing disturbance due to accessibility to building occupants, vibration or air erosion would identify the ACBM as having the *potential for significant damage*.

Changing uses of an area will change the potential for disturbance of its ACBM. Demolition or renovation activities in an area with ACBM obviously change the potential for disturbance. In such cases, the ACBM must be removed prior to the onset of renovation or demolition activities. Other changes in usage of the functional space, such as the conversion of a maintenance room into a classroom, require reclassification of the potential for disturbance. The AHERA requires periodic review of hazard assessments for this reason. It is important to note that unless a material is located behind a permanent enclosure, such as a wall or other air-tight barrier, the potential always exists for the material to be disturbed and damaged.

15.6.4 Hazard Ranking

AHERA provides seven categories by which to assess the current condition of the ACBM and the potential for damage (see Figure 15.13). These categories are used to determine the appropriate *response actions*.

In most cases, ACBM with a 7 hazard ranking will require an active response action, i.e., removal, encapsulation, or enclosure. ACBM with a 1 hazard ranking (no damage with low potential for future damage) can most likely be managed in place under an operations and maintenance (O&M) Plan. The response actions to be taken for materials with hazard ranking from 2 through 6 will require the judgment of the people responsible for the health and safety of the building occupants.

It should be noted that damaged ACBM with a low potential for disturbance is given a higher hazard ranking (4) than undamaged ACBM with a high potential for disturbance (3). This logic falls under the "one in hand, two in the bush" concept. In other words, since damaged ACBM already has the potential to release harmful asbestos fibers into the environment, there is an immediate hazard that needs to be addressed. However, common sense also must enter into the equation. If a library is being converted into a roller rink, then the potential to disturb undamaged ACBM wall plaster is so significant that the risk posed by this material must become a priority.

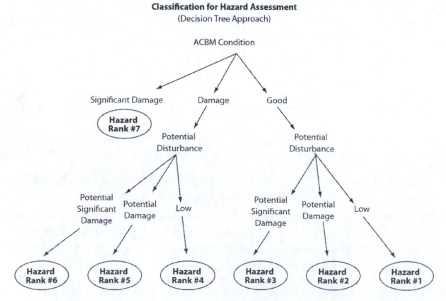

FIGURE 15.13
AHERA hazard potential decision tree.

15.6.5 Air Monitoring for a Hazard Assessment

Hazard assessments for most airborne contaminants involve air sampling (see Chapter 17). Air monitoring is not the preferred method for an asbestos hazard assessment, however, mainly because it only provides a snapshot of conditions at the time of the air monitoring and does not take into account changed or changing conditions. In addition, air monitoring typically doesn't measure conditions at the times of peak disturbance of the asbestos, such as when the area of interest is actively occupied by employees, or when doors open and shut, creating air movement. Lastly, it's more expensive than the preferred method of bulk sampling and hazard assessment. It is the preferred assessment method for asbestos abatement projects, as described later in the chapter.

15.7 Asbestos Abatement

Once the asbestos hazard assessment has been completed, response actions must be chosen if ACM or ACBM were identified to mitigate the asbestos hazard. These methods are known as asbestos abatement.

15.7.1 Types of Asbestos Abatement

There are four types of asbestos abatement: removal, enclosure, encapsulation, and operations and maintenance. Since the most technically involved method of abatement is removal, the rest of this chapter offers far more coverage on asbestos removal than the other three asbestos abatement methods.

For all but minor abatement projects, an asbestos contractor performs the abatement. This contractor typically holds a license that was issued by the state and/or the city in which the project is taking place. The individual workers and the project supervisor are also subject to licensure requirements within their jurisdictions.

15.7.2 Asbestos Abatement Projects by Size

There are three types of asbestos abatement projects under AHERA:

- A large asbestos project involves the removal, encapsulation, enclosure, or disturbance of more than or equal to 160 square feet (SF) of ACBM or more than or equal to 260 linear feet (LF) of ACM.

- A small asbestos project involves the removal, encapsulation, enclosure, or disturbance of between 10 and 160 SF of ACBM, or between 25 and 260 LF of ACBM.

- A minor asbestos project involves the removal, encapsulation, enclosure, or disturbance of less than or equal to 10 SF of ACBM, or less than or equal to 25 LF of ACBM.

OSHA defines four classes of asbestos work, two that involve asbestos removal. *Class I removal projects* involve the removal of TSI and surfacing ACM (ACBM is an AHERA rather than an OSHA term) and PACM. The OSHA regulations are less stringent for *Class II removal projects*, which involve only the removal of miscellaneous materials.

15.7.3 Designing the Removal Project

Asbestos abatement projects are some of the most highly proscribed activities in the sphere of environmental regulations, with elaborate engineering controls and monitoring protocols implemented to prevent asbestos fibers from leaving the workplace and impacting humans outside of the workplace and protect the asbestos workers within the workplace. The elements of a removal project are described below.

15.7.4 Preparing for Asbestos Removal

Prior to the beginning of a project, the following activities are performed:

- The contractor notifies USEPA, and in some cases the state and local jurisdictions, in writing that an asbestos abatement project is to be performed.

- On the day of mobilization, the contractor posts the appropriate notifications on the building entrance, and warning signs at the entrances to the work area (see Figure 15.14). These signs are in English as well as the primary language spoken by the building inhabitants.

- To prevent the spreading of asbestos fibers throughout the building, the building's HVAC system is shut down and isolated.

15.7.4.1 Preparing the Work Area for the Large Asbestos Project

The preparation of the work area (see Figure 15.15 for the layout of a typical work area) involves the construction of a containment, which utilizes a series of engineering controls designed to prevent asbestos fibers from leaving the work area. Before they are constructed, all movable items are removed from the area that will become part of the containment, which would simplify the movement of workers inside the containment. Carpets, drapes, and other fabrics that can't be reinstalled typically are discarded. In addition, items that can't be moved, such as machinery, light fixtures, blackboards, water fountains, toilets, etc. are wet-wiped or using a vacuum equipped with a high efficiency particulate air (HEPA) filter (colloquially known as "HEPA vacuuming"), and then covered with 6-mil thick polyethylene plastic (colloquially known as "poly") so that they don't become contaminated

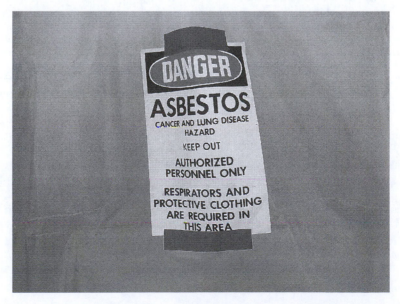

FIGURE 15.14
Asbestos warning sign. (Courtesy of M. Stuckert, Save the 905, Inc., www.RockIsland905.com.)

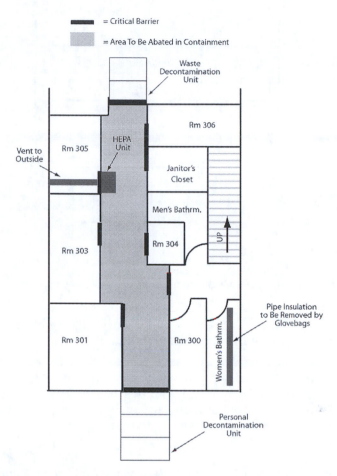

FIGURE 15.15
Layout of a building area about to undergo asbestos removal.

with asbestos fibers during the abatement activities. The work area is now ready for the installation of the engineering controls.

15.7.4.2 Construction of Decontamination Units

Large asbestos projects have personal and waste decontamination units connecting the work area to the rest of the building (see Figure 15.16). The objective of the decontamination units is to provide controlled access to the containment area.

The *personal decontamination unit* allows workers and equipment to move to and from the work area during abatement activities while preventing asbestos from leaving the work area. It consists of the following three rooms:

- *Equipment room, also called the dirty change room.* This room, which adjoins the work area, is a contaminated area where workers exiting the containment leave their equipment, footwear, hardhats, goggles, and other asbestos-contaminated apparel and protective gear.

- *Shower room.* As its name implies, workers exiting the work area shower in the shower room, still wearing respirators since their exposed skin still could be

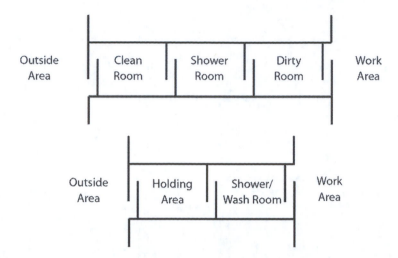

FIGURE 15.16
Personal decontamination unit (top) and a waste decontamination unit (bottom).

contaminated with asbestos fibers. Once the shower water has removed remaining asbestos fibers, the workers are ready to enter the clean room.

- *Clean room.* In the clean room, decontaminated workers can dry their bodies, remove their respirators, dispose of the spent respirator cartridges, and dress. Clean rooms are typically equipped with benches, lockers for clothes and other personal effects, and nails or hooks for hanging respirators. Prior to entry into the work area, workers use the clean room to don their respirator (equipped with new cartridges) and protective clothing.

The second type of decontamination unit, known as the *waste decontamination unit*, is used for the removal of asbestos-contaminated waste from the containment. It has two rooms rather than three, since it lacks an equipment room. Bagged asbestos-contaminated wastes exiting the work area are washed and HEPA vacuumed in the first room, known as the shower/wash room. The decontaminated bags are then placed in the holding area, from which they eventually will be loaded onto a truck for transport to a licensed disposal facility. Personal and waste decontamination units may be combined into one unit if warranted by space limitations outside the work area.

   Decontamination units are typically constructed with wood frames and wood or poly walls and floors. *Airlocks* between rooms to prevent cross-contamination across rooms typically consist of two layers of poly sheeting. Pre-designed decontamination units, usually staged in trailers, are sometimes used on abatement sites, especially for outdoor work.

15.7.4.3 *Critical Barriers*

Construction of the containment is completed with the establishment of *critical barriers* at all other openings into the work area. These openings include windows, doorways, skylights, vents, and air ducts for the HVAC system, and openings in the floors, walls, and ceilings. Critical barriers are composed of two layers of 6-mil poly. It is good practice to secure each layer individually so that the inner layer of poly will remain in place if the outer layer fails.

15.7.4.4 Plasticizing Floors and Walls

To prepare the work area for asbestos removal, the floors and walls are covered fully with protective 6-mil poly (see Figure 15.17). The sheets on the floor should rise at least one foot above the floor and covered with the poly on the walls so that they overlap, preventing creation of a seam through which asbestos fibers could escape. As with the critical barriers, each layer of poly should be installed independently, so the inner layer will remain in place if the outer layer fails. Sometimes, the poly may need to be secured with nails or furring strips, especially if the walls or floors are slippery or the poly is getting too heavy.

15.7.4.5 Electrical Lock-out

Water is typically used to saturate the ACBM prior to removal, creating a humid environment in the work area. To eliminate the potential for a shock hazard, the electrical supply to the work area is de-energized and locked out. This is typically done after the work area has been plasticized, so that workers can work in normal light for as long as possible. Temporary electrical service using *ground-fault interruption* (GFI) protection is used to light the containment and for power equipment.

15.7.4.6 Establishing Negative Pressure

The final engineering control implemented for the work area is the establishment of a *negative air pressure regime*. By creating negative air pressure inside the containment, air will flow from high pressure to low pressure, i.e., from outside the containment into the containment. Thus, to escape a work area, an airborne asbestos fiber inside the containment would have to go against the air flow, which is difficult (though not impossible).

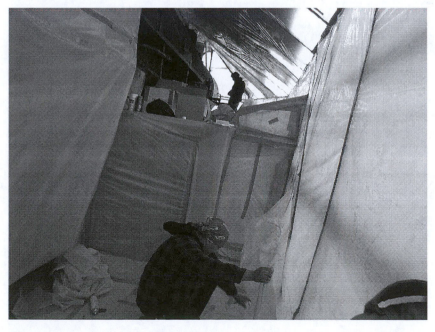

FIGURE 15.17
Preparing a room for asbestos abatement. (Courtesy of M. Stuckert, Save the 905, Inc., www.RockIsland905.com.)

The negative air pressure gradient is maintained with the use of powered exhaust equipment. Units are placed inside the containment that are equipped with HEPA filters, which will capture airborne asbestos fibers prior to exhausting the filtered air outside of the containment (and generally outside the building).

The quantity, size, and the placement of the HEPA units are determined by the size and configuration of the work area. In Figure 15.18, a full containment has been constructed for the removal of the surfacing material (the removal of the piping insulation is discussed later in the chapter).

Regulations require that there be at least one air change every 15 minutes in the work area. The volume of air to be changed every hour in the work area is calculated using the following formula:

$$\text{Volume of Air to be Moved} = \text{Volume of Room} \times \text{\# changes per hour}$$

Assuming that the containment shown in Figure 15.15 is 15′ × 60′ in area and the ceiling is 8′ high, then the volume of air to be moved is calculated as follows:

$$8' \times 15' \times 60' = 7{,}200 \text{ ft}^3 \times 4 \text{ changes per hour} = 28{,}800 \text{ ft}^3/\text{hr}$$

A typical commercially available HEPA-filtered negative pressure generating unit may have an air handling capacity of 1,000–2,000 ft^3 per minute, which is equivalent to 60,000–120,000 ft^3/hr. Therefore, for this space, one HEPA unit operating at 1,000 ft^3/min will provide more than enough air movement to meet regulatory requirements. Note that the quantity of HEPA units required is not based on the quantity, type, or location of the ACM itself.

FIGURE 15.18
HEPA filtration unit outside of a containment. (Courtesy of M. Stuckert, Save the 905, Inc., www.RockIsland905.com.)

HEPA units should be placed to avoid "dead" spaces, or places where no air is flowing. They should also be placed far from the "make-up" air, which is clean air flowing into the containment from the decontamination units, or other auxiliary sources to promote efficient air flow. The HEPA unit shown on Figure 15.18 is situated at the opposite end of the containment from the personal decontamination for that reason. The HEPA unit vent, which is a flexible, plastic duct, extends to a window located in a nearby room through which the filtered air is vented.

Smoke tests may be performed to verify that air from outside the decontamination unit is being drawn into the work area, which would indicate the establishment of a negative air pressure gradient. For quantitative measurements of differential air pressure, a manometer or a magnehelic gauge often is attached to one of the decontamination units, or other locations where such data is desirable. A pressure drop of negative 0.02 and 0.03 inches of water across the barrier is usually indicative of an adequate negative air pressure gradient. Higher pressure differentials may create too much air turbulence for workers to adequately perform their duties in the work area, or can damage the engineering controls in the containment.

15.7.5 Removing the Asbestos-Containing Material

Once the containment is constructed and a negative air pressure environment has been established, the ACM can be removed. Workers enter the clean room of the personal decontamination unit, suit up in protective clothing, don their respirators and confirm the fit, gather their equipment and enter the containment.

15.7.5.1 Wetting the Material

To minimize the creation of airborne asbestos fibers, ACBM is wetted with amended water prior to removal. Dry removal is allowed only in special circumstances and when approved by USEPA, such as when there is an electrical source in the work area that cannot be de-energized. Wetting the material is accomplished using an *airless sprayer*, which operates at a low pressure, thus avoiding the haphazard dislodging of the ACM from the substrate. High pressure wetting, as through a garden hose, may dislodge the ACM before it is adequately wet, resulting in an unacceptable release of asbestos fibers into the workplace.

15.7.5.2 Two-Stage ACM Removal

The ACM removal is accomplished in two stages. The first step is *gross removal*, generally performed using scraping tools for surfacing material and many miscellaneous materials, or a utility knife or other cutting device for thermal system insulation (see Figure 15.19). The removed material should be caught before it falls to the ground, and fallen material should be picked up before it dries up to avoid unnecessary releases of asbestos fibers into the work area. Workers then place the removed ACM in properly labeled waste bags.

The second step, known as *fine cleaning*, involves removing residual ACM that has adhered to the substrate using fine brushes and wet wiping. The HEPA vacuums may also be employed to remove ACM in relatively inaccessible locations in the work area. Fine cleaning should continue until no visible ACM or dust is visible on the abated surfaces.

FIGURE 15.19
Workers removing asbestos inside a containment. (Courtesy of M. Stuckert, Save the 905, Inc., www. RockIsland905.com.)

15.7.5.3 Glovebag Removal

An often-used, specialized procedure for the gross removal of thermal system insulation from horizontal linear features such as piping, involves the use of a *glovebag*. As shown in Figure 15.20, the glovebag is its own containment: a plastic bag designed with two sleeve gloves, a tool pouch, and a small opening that may be used for the insertion of airless sprayers and/or HEPA vacuum nozzles. Tools to be used in removing the ACM from the pipe are placed in the tool pouch, and the slits and the top of the glovebag are taped shut, thus sealing the glovebag to the pipe.

Using glovebags where appropriate can result in huge savings in time and costs, since they avoid the need for the construction of a full containment in the work area.

Before using the glovebag, its integrity can be tested by inserting a smoke tube through the opening in the bag and observe whether smoke leaks out of the bag. If the glovebag is air tight, the worker will insert the nozzle of an airless sprayer into the opening, seal the nozzle opening and wet the material to be removed. The worker can then reach into the bag through the sleeve gloves, access whatever tools are in the tool pouch, and remove the ACM, which falls to the bottom of the bag. Damp rags earlier placed inside the glovebag can be used for the fine cleaning process. Once completed, the tools are returned to the tool pouch, which can then be pulled through the glovebag, tied off and removed intact, eventually to be decontaminated in the personal decontamination unit along with the worker. A HEPA vacuum is then used to remove air from the spent glovebag, which is then twisted and taped off and cut from the pipe. The entire glovebag, with the ACM waste at the bottom, can then be placed into an asbestos waste bag for eventual disposal.

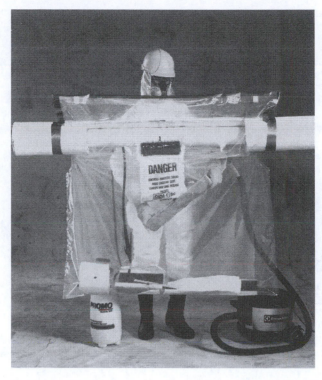

FIGURE 15.20
A glovebag attached to a horizontal pipe. (Courtesy of Grayling Industries, Alpharetta, GA.)

15.7.5.4 Lockdown Encapsulation

Though the substrate may appear to be clean after the fine cleaning has been completed, microscopic asbestos fibers may be present. A *bridging encapsulant*, which is a protective coating or sealant that is designed to prevent residual fibers from becoming airborne, is applied to the abated surface once the surface has been fine cleaned and is dry.

15.7.5.5 Two-Stage Cleaning and Poly Removal

Once the encapsulant has dried and no visible asbestos debris or dust remains, workers remove the first layer of sheeting and place it in asbestos waste bags. This process is then repeated for the remaining layer of poly. With only the critical barriers and decontamination units remaining, a third round of cleaning is then performed. Once the work area passes its third and final inspection, it is ready for clearance air monitoring (see Section 15.7.6.3).

15.7.5.6 Waste Removal

Asbestos-containing wastes are placed in specially marked plastic bags for eventual disposal at a TSCA-licensed landfill (see Figure 15.21). Wastes include the actual ACM removed as well as disposable equipment utilized during the abatement project and materials removed during the preparation of the work area or became cross-contaminated in the course of the project.

FIGURE 15.21
Bagged asbestos waste inside a waste decontamination unit. (Courtesy of M. Stuckert, Save the 905, Inc., www.
RockIsland905.com.)

Asbestos waste bags have warning labels similar to those used on signage for the asbestos project workplace. They also are labeled with the regulatory classification of their waste type. Asbestos wastes are double-bagged to avoid a release of asbestos fibers if the outer bag is damaged. The bags are stored in the holding area of the waste decontamination unit prior to being loaded onto appropriately licensed and labeled trucks for transport to the selected landfill.

15.7.6 Air Monitoring Requirements

Several types of air monitoring are performed during an asbestos abatement project to comply with USEPA and OSHA regulations.

15.7.6.1 Pre-abatement Air Sampling

Air samples are collected inside and outside of the work area after the containment has been constructed and before asbestos removal begins. The objective of the air sampling is to establish pre-abatement, or "background" conditions against which air quality can be judged in the course of the asbestos removal activities.

To collect these pre-abatement air samples, air pumps are set up at various locations inside and outside of the containment. They are equipped with plastic tubing with the sample cassettes at the end of the tubing. Once the required volume of air has been drawn by the air pump, it is shut off and the cassettes are sent to a laboratory, which will analyze the filter inside the cassette for the presence of fibers using either *phase contrast microscopy* (PCM) or TEM. The PCM method has two drawbacks: (1) it cannot distinguish between asbestos and non-asbestos fibers, making it an unreliable measuring tool in dirty areas or areas where fibrous materials, such as clothing, are stored; and (2) it cannot detect fibers less than 5 microns in length or fibers that are less than about 0.25 micrometers in diameter. Analyzing the samples using TEM avoids these limitations but is more expensive than the PCM method.

15.7.6.2 Air Sampling During Removal

During asbestos removal and cleaning activities, three types of air samples are collected:

- Perimeter air samples
- Area air samples
- Personal air samples

Area air samples are collected inside the containment. They are designed to assess the effectiveness of the workers' efforts, such as vacuuming and wetting the asbestos, to minimize airborne asbestos in the work area.

Perimeter air samples are collected around the outside of the containment where asbestos fibers are most likely to escape. Perimeter air monitoring assesses the effectiveness of the engineering controls in preventing asbestos fibers from leaving the work area. On Figure 15.15, perimeter air samples are being collected outside of the two decontamination units, near the HEPA unit exhaust outside the building, and near the critical barrier that was set up to isolate the staircase (in the foyer between Rooms 300 and 304). Together, area air samples and perimeter air samples are known as "daily" air samples.

Area air samples and perimeter air samples usually are analyzed using PCM. The results are compared to the USEPA standard of 0.01 f/cc as well as the results from the background samples. If the air monitoring results suggest a failure of the engineering controls, all work is stopped until the source of the problem is identified and fixed.

Personal air monitoring is required under OSHA and is designed to evaluate worker exposure to airborne asbestos fibers. These measurements can be used to calculate the volume of asbestos fibers a worker is inhaling and determine whether the respirator being worn is adequately protecting the worker.

A personal air monitoring pump is strapped to the worker's waist and attached to a sample cassette by plastic tubing. The cassette is situated near the worker's face to simulate the air that the worker is actually breathing. The worker then conducts daily activities as usual. At the end of the day, the pump is turned off and the cassette is sent to a laboratory for PCM analysis. OSHA regulations require that each job function inside the workplace is monitored. On a typical asbestos removal project, job functions may include workers performing gross removal of ACM, fine removal of ACM, and waste bagging and loadout. In all, at least 25% of the total number of workers in the work area must be monitored. Personal air monitoring is not required if workers are already outfitted in the maximum protective respiratory protection (which is a full-face, supplied air respirator operating in pressure demand mode), or if previous analytical results indicate that workers are being adequately protected and the work being performed inside the work area has not changed. OSHA requires informing workers of the results of the personal air monitoring analyses.

15.7.6.3 Clearance Air Monitoring

At the conclusion of the abatement activities, *clearance air monitoring* is performed to evaluate whether a work area is ready for reoccupancy. In clearance air sampling, worst-case conditions are simulated by using *aggressive sampling techniques*, the assumption being that if airborne asbestos fiber concentrations are acceptable under this condition, fiber concentrations would be acceptable under any other condition.

Prior to the collection of clearance air samples, a leaf blower or a floor fan is activated and pointed towards the abated surfaces to stir up any remaining dust and debris that may contain asbestos. If the resulting analyses meet regulatory standards of 0.01 f/cc, then the area is ready for reoccupancy. Although PCM analyses are often used for the analysis of the clearance air samples, some jurisdictions require the TEM analysis for this most important of air sample analyses. The AHERA requires TEM analysis for clearance air samples in public schools, with a regulatory standard of 70 asbestos structures per square millimeter (s/mm^2).

15.7.7 Small and Minor Asbestos Projects

Small and minor asbestos projects require less preparation and air monitoring than large asbestos projects, although the need for thorough cleaning and attention to worker health and safety doesn't change. For a small asbestos project, a waste decontamination unit is not required. No decontamination units are required for a minor asbestos project. In fact, USEPA requires no more than placing plastic sheeting on the ground, posting signs in the appropriate places, and running yellow caution tape across the entrance to the work space for a small asbestos project. The containment is usually no more than a tent constructed of poly sheeting taped across the surfaces in the work area that require protection.

Only three background, pre-abatement, and clearance air monitoring, inside and outside the containment, are required daily for small asbestos projects. No area or clearance air monitoring is required for minor asbestos projects.

15.7.8 Abatement by Encapsulation or Enclosure

Abatement by encapsulation or enclosure is much less time-consuming than abatement by removal and therefore less expensive. However, these abatement methods are not permanent remedies. The ACBM is still present at the end of the abatement project although the pathway to the receptors (humans) has been removed. These methods are desirable in certain situations, especially where the cost of removal is a deterrent or where removal would result in an unacceptable shutdown of part or all of an operating facility.

For both encapsulation and enclosure, the work area is first equipped with warning signs and established with the use of critical barriers. Loose or hanging ACBM should be removed prior to encapsulation/enclosure. Areas that may be disturbed in the course of the abatement activities are sprayed with amended water and kept damp to reduce airborne asbestos fibers. The work area is then ready for abatement.

Abatement by encapsulation entails treating the whole surface of the ACBM with a *penetrating encapsulant* that binds it tightly to the substrate. Once the encapsulant has dried, clearance air samples are collected using aggressive sampling techniques. The critical barriers can be removed and the area reoccupied once acceptable clearance air monitoring results have been obtained. Only personal and clearance air sampling are conducted.

Abatement by enclosure entails surrounding the ACBM with an airtight enclosure that prevent the emission of asbestos fibers. Such enclosures could be as small as a metal or hard plastic sheathing around piping insulation, or as large as an airtight wood or metal wall around an entire building. Full-building enclosures are desirable when a building contains large amounts of asbestos or is unsafe to enter. The workplace is prepared for abatement as described above. Clearance air samples are collected once the enclosure has been constructed, and the work area is reoccupied once acceptable clearance air monitoring results have been obtained.

15.8 Operations and Maintenance for In-place Asbestos-Containing Material

In school settings, AHERA requires each *local education agency* (LEA) to develop an *operations and maintenance plan* (O&M Plan). The objective of the O&M Plan is to prevent children from being exposed to asbestos hazards remaining in the school. Each LEA assigns a *designated person* to implement the O&M Plan.

An O&M Plan includes scaled drawings showing the locations of the ACBMs, and information regarding its asbestos content and hazard assessment. The O&M Plan also includes:

- Documentation regarding notifications to workers, tenants, and building occupants
- A schedule for periodic inspections of the ACBM by properly trained personnel
- Procedures to be implemented if an asbestos hazard arises
- Training and personal protection requirements for workers who handle ACBM
- Training records for these workers
- Information on the history of ACBM repair and removal

Unlike hazardous substances and petroleum products, asbestos will not go away with time. Therefore, O&M is necessarily a temporary method of abatement, since ACM will eventually degrade and require repair or removal. The O&M Plan is necessarily a "living document," since the condition of the ACBM will change over time.

15.9 Asbestos Surveying and Abatement, and Environmental Consultants

As with most indoor environmental issues, the duties performed by the environmental consultant generally are considered *industrial hygiene*. Unlike most indoor environmental issues, however, the qualifications of personnel providing asbestos-related services are highly prescribed.

An *AHERA inspector* conducts the asbestos survey, the steps of which are described in Section 15.4. The AHERA inspector is not allowed to analyze the samples to avoid a conflict of interest. Instead, a properly certified, independent laboratory performs the analyses. Once the laboratory results are obtained, the AHERA inspector reviews the results and decides whether the adequate information has been obtained or additional sampling is warranted. This includes informing the laboratory whether to use the "positive stop method" for homogeneous areas in which one sample tested positive for asbestos.

The AHERA inspector will prepare a report once the asbestos survey has been completed. The report may include a scaled map showing the sample locations; a table summarizing the results; a section documenting the methodologies used in the survey, including limitations to the survey (such as the identification of inaccessible areas).

The AHERA inspector cannot, however, perform the hazard assessment as per Section 15.6. That task is assigned to an *AHERA management planner*. Since the asbestos survey and the hazard assessment are typically performed at the same time, usually the AHERA inspector and the AHERA management planner are the same person. While it may be advisable to discuss the results of the hazard assessment in the asbestos survey report, the results of the hazard assessment must be put into an AHERA O&M Plan.

AHERA inspectors are required to attend a three-day course approved by the USEPA. The AHERA management planners must attend an additional USEPA-approved two-day course. The courses typically are provided back-to-back, since so many professionals hold both licenses. Annual refresher courses are required to maintain both licenses.

The third title set up under AHERA that is held by environmental consultants is that of *AHERA project designer*, which is the person who designs the response actions to be undertaken to abate the asbestos hazard identified by the AHERA management planner. To obtain this license, the environmental professional must attend a USEPA-approved three-day course, with subsequent annual refresher courses.

Response actions are typically performed by licensed contractors. Asbestos workers and their supervisors also have licensing requirements under AHERA. Workers must attend a USEPA-approved four-day course, with an additional eight-hour course to obtain an AHERA supervisor's license. There are also two-day courses offered to custodial and maintenance workers who are tasked with maintaining ACBM in their buildings as a secondary but not primary job function.

Problems and Exercises

Figure 15.22 shows the basement of a multi-tenant apartment building.

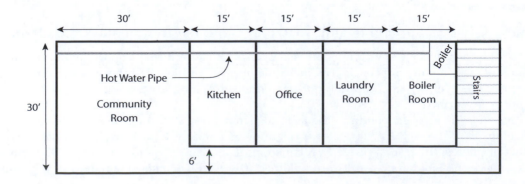

1. The ceilings in all of the rooms in the basement are covered with plaster. How many bulk samples must be collected to assess whether the material is ACBM, as per AHERA? How many bulk samples are recommended under AHERA?

2. The hot water pipe is covered by insulation. How many samples must be collected of the piping insulation to assess whether the material is ACBM, as per AHERA? How many bulk samples are recommended under AHERA?

3. The floor in the kitchen is covered with 12"×12" floor tiles. How many samples must be collected of the floor tiles to assess whether the material is ACBM, as per AHERA?

4. The insulation on the hot water pipe tested positive for asbestos. The community room is being converted into a recreation room, with a ping pong table and exercise equipment. Explain how this change in usage might change the hazard ranking.

5. The plaster on the ceilings throughout the basement tested positive for asbestos. Assuming the ceiling clearance is 8 feet and four air changes per hour are desired, how many HEPA units with an air handling capacity of 1,000 ft^3 per minute are needed in the basement?

Bibliography

Centers for Education and Training, Rutgers University of Medicine and Dentistry (UMDNJ), 1988. Procedures and Practices for Asbestos Control.

Centers for Education and Training, Rutgers University of Medicine and Dentistry of New Jersey (UMDNJ), 1990. Inspecting Buildings for Asbestos-Containing Materials (AHERA Inspector).

Natale, A., and Levins, H., 1984. *Asbestos Removal & Control: An Insider's Guide to the Business.* J. Levins Design, Inc.

New York State Department of Labor, 2004. Industrial Code Rule 56 (12 NYCRR 56).

U.S. Environmental Protection Agency, June 1985. Guidance for Controlling Asbestos-Containing Materials in Buildings. EPA 560/5-85-024. Office of Pesticides and Toxic Substances.

U.S. Environmental Protection Agency, July 1990. Managing Asbestos In Place: a Building Owner's Guide to Operations and Maintenance Programs for Asbestos-Containing Materials. 20T-2003. Office of Pesticides and Toxic Substances.

U.S. Environmental Protection Agency, January 1996. How to Manage Asbestos in School Buildings. Office of Waste and Chemical Management. EPA-910-B-96-001.

U.S. Environmental Protection Agency, May 1999. EPA Asbestos Materials Bans: Clarification.

16

Lead-Based Paint Surveying and Abatement

16.1 Introduction

16.1.1 Lead Hazards

Lead is perhaps the most pervasive of all environmental hazards. It is a hazard in the air, in the subsurface, in surface waters, in drinking water, and, as discussed in this chapter, in paint (see Figure 16.1). Like so many environmental hazards, lead is a significant contaminant because, like asbestos, it has qualities that made it a desirable additive in construction materials, fuels, and a host of other products. Since it is so widespread, it is regulated under almost every major federal environmental statute, including:

- Comprehensive Environmental Response, Compensation, and Liability Act (CERCLA) (Superfund)
- Resource Conservation and Recovery Act
- Toxic Substances Control Act
- Safe Drinking Water Act
- Clean Water Act
- Clean Air Act

The federal Occupational Safety and Health Administration (OSHA) even has its own Lead Standard for Construction (29 CFR 1910.62) to protect workers from lead in the workplace. Lastly, the *Department of Housing and Urban Development* (HUD), which is part of the United States Cabinet, has issued guidelines regarding lead in paint. Although they are guidelines and not regulations, they function as regulations, and profoundly affect the handling and management of lead paint, as described in this chapter.

16.1.2 History of Lead in Paint

Lead had been a component of paint possibly as long as paint itself has existed. Among the reasons for its widespread usage are its abundance in the earth's crust, its malleability, and its durability. Unlike asbestos, which was first used on roofs in the mid-1800s and not inside buildings until the late 1800s/early 1900s, lead use in paint goes back to ancient times.

Although by the 1920s nearly all European countries had banned lead in paint, it was commonly used in the United States until industry voluntarily lowered the

FIGURE 16.1
A can of lead-containing paint.

allowable lead concentration in paint in 1960. Paint that predates 1960 almost always contains lead.

The first official ban of lead in paint in the United States was in 1977, when the Consumer Products Safety Commission (CPSC) banned its use in residences. However, the ban did not pertain to non-residential buildings, such as offices and factories. In 1992, the U.S. Congress passed the *Residential Lead-Based Paint Hazard Reduction Act*, which set the limit of lead in paint at 0.5% by weight, or paint that contains lead at a concentration greater than 1.0 milligrams per square centimeter (mg/cm^2). Paint with a lead content exceeding these limits became known as *lead-based paint* (LBP). Congress delegated the enforcement of this law to HUD. Table 16.1 shows the frequency of occurrence of LBP inside and outside a typical dwelling in the United States based on the age of the building.

16.2 Occupational Safety and Health Administration Lead Standard

The OSHA regulates non-residential lead hazards (29 CFR 1910.62). Since OSHA's mission is to protect workers from workplace hazard, the OSHA lead standard addresses worker exposure to lead. There is no minimum threshold as there is in residential buildings. The mere presence of lead in the workplace triggers the OSHA standard.

TABLE 16.1

The Percentage of All Paint That is Lead-Based, by Year and Component Type

Component Category		Interior	Exterior
Walls/Ceiling/Floor			
	1960–1979	5%	28%
	1940–1959	15%	45%
	Before 1940	11%	80%
Metal Components[a]			
	1960–1979	2%	4%
	1940–1959	6%	8%
	Before 1940	3%	13%
Non-metal Components[b]			
	1960–1979	4%	15%
	1940–1959	9%	39%
	Before 1940	47%	78%
Shelves/Others[c]			
	1960–1979	0%	
	1940–1959	7%	
	Before 1940	68%	
Porches/Others[d]			
	1960–1979		2%
	1940–1959		19%
	Before 1940		13%

Source: HUD, These data are from a limited national survey and may not reflect the presence of lead in paint in a given dwelling or jurisdiction, 1990b and U.S. Department of Housing and Urban Development (HUD), Guidelines for the Evaluation and Control of Lead-Based Paint Hazards in Housing, 1997.

Note: The Information Provided in This Table is Based on a Survey Conducted in 1990.

[a] Includes metal trim, window sills, molding, air/heat vents, radiators, soffit and fascia, columns, and railings.

[b] Includes non-metal trim, window sills, molding, doors, air/heat vents, soffit and fascia, columns, and railings.

[c] Includes shelves, cabinets, fireplace, and closets of both metal and non-metal.

[d] Includes porches, balconies, and stairs of both metal and non-metal.

The main pathway for lead into the workers' bodies is by inhalation, either through inhalation of lead-containing dust or inhalation of lead-containing vapors. And OSHA assumes that all lead on surfaces can become airborne by the generation of either dust or vapors. The OSHA *permissible exposure limit* (PEL) for lead is based on its *time-weighted average* (TWA).

Figure 16.2 is a typical graph of hourly airborne lead readings collected over an 8-hour period (from 8 AM to 4 PM). The readings range from 0.0 to 8.0 micrograms per cubic

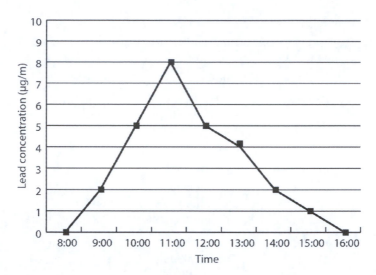

FIGURE 16.2
Hourly measurements used to estimate a time-weighted average.

meter ($\mu g/m^3$), with an average of the nine readings being 3.0 $\mu g/m^3$. This is the TWA of the readings over an 8-hour period. The OSHA PEL for lead in the workplace is 50 $\mu g/m^3$ of air, calculated as an 8-hour TWA, and the action level at which worker monitoring for lead in air is required is 30 $\mu g/m^3$. Some compounds have acute toxicity limits as well as PELs; lead does not, since exposure to high concentrations of lead does not cause acute toxicity in humans. Information on OSHA PELs and acute exposure limits for various compounds and elements is provided in the guide book published by the *National Institute for Occupational Safety and Health* (NIOSH), a division of the federal and Prevention Centers for Disease Control (CDC). The online NIOSH guide is available at http://www.cdc.gov/niosh/npg/default.html.

The three principal categories of work-related exposures to lead are demolition activities, renovation activities, and lead removal activities; the latter of which is discussed later in this chapter. If lead is to be disturbed by worker activity, OSHA requires the performance of an *exposure assessment* to assess potential lead hazards to workers. To conduct an exposure assessment, OSHA requires the collection of a personal air sample for each job classification in each work area for a period of time that represents a full shift. For instance, if an area where lead is known to be present is being demolished, separate tests should be conducted for the worker demolishing the painted concrete block walls and for the arc welder cutting the painted steel beams. The testing should be biased to the surfaces with the highest concentrations of lead in paint, since they presumably would create the worst-case exposure scenario for the worker.

Until the exposure assessment is performed, the employer must assume that the worker is being exposed to lead at a concentration above the PEL and therefore should protect the worker from that exposure. Precautions may include removal of the lead from the workplace prior to the onset of renovation activities, placing the worker in appropriate respiratory protection, or limiting the worker's exposure by decreasing the amount of time over the work shift that the worker disturbs the lead-coated surface.

16.3 United States Department of Housing and Urban Development Guidelines for Lead-Based Paint Inspections

Most of the rest of this chapter is based on the guidelines provided in the HUD publication entitled, "Guidelines for the Evaluation and Control of Lead-Based Paint Hazards in Housing." Although it is not a regulation, it is the *de facto* standard for lead in residential settings. The HUD guidelines apply to residences where children under the age of six reside, as well as non-residential buildings where young children commonly are present, such as community centers and day care centers.

Although lead can become airborne, it does so only in dusty conditions unlikely to be encountered in a residential setting, and it does not easily enter the gaseous phase. The pathway of concern regarding young children is the ingestion pathway. When ingested by young children, lead can disrupt the development of their neurological systems, and young children are prone to ingesting non-foods such as paint chips, especially toddlers and other teething youngsters. As any parent will attest, toddlers tend to sink their newly emerged teeth into anything and everything. A window sill can provide a tempting teething surface to a toddler. In addition, due to their lesser body weight and faster metabolisms, young children have a much lower threshold for lead poisoning than do older children or adults.

16.3.1 Lead-Based Paint and Lead Hazards

By itself, lead in paint does not present a health hazard when maintained properly. However, when it is chipped, blistered, or delaminated, as shown in Figure 16.3, lead paint becomes a health hazard. Under HUD guidelines, a lead hazard is present inside a residence when at least one of the following conditions exist:

- Dust contains lead at a concentration greater than 40 micrograms per square foot ($\mu g/ft^2$) on floors
- Dust contains lead at a concentration greater than 250 $\mu g/ft^2$ on interior window sills
- Dust contains lead at a concentration greater than 400 $\mu g/m^3$ in window troughs and other rough surfaces
- Damaged or deteriorated LBP surface due to contact with another building component
- Chewable LBP surface with teeth marks
- Any other damaged or deteriorated LBP surface

Surface soils can also be sources of lead to young children, since young children commonly will touch the dirt and then put their dirty hands in their mouths, creating a direct ingestion pathway into their bodies.

In addition to industrial sources of lead, soils can become contaminated from the delamination of LBP from exterior surfaces, or the creation of lead-containing dust by the abrasion of LBP on exterior surfaces. Lead can also originate from atmospheric and anthropogenic sources. Any exposed play area with surface soils (i.e., soils within 6 inches of the surface) that contain more than 400 milligrams per kilogram (mg/kg, also called parts per million [ppm]) of lead are considered hazardous to young children.

FIGURE 16.3
Peeling paint on the outside of a window. (Courtesy of GZA GeoEnvironmental, Inc.)

16.3.2 Lead Paint Risk Assessment

To determine if a lead hazard exists in a building where young children are present, a *lead paint risk assessment* is performed. Chapter 5 of the HUD guidelines describes the procedures to use when conducting a lead paint risk assessment, which are summarized here, with some modifications and simplifications.

Lead paint risk assessments are not required in post-1978 buildings or buildings where all LBP has been removed. It is permissible to perform a lead hazard screen for dwellings that are in good condition to determine if a full risk assessment is warranted. However, if the building is in poor condition or was constructed before 1960 and all LBP has not been removed, then a full lead paint risk assessment is warranted. Without a lead paint risk assessment, HUD requires that one assume that all painted surfaces in applicable housing are LBP, and that lead from paint is present in dust inside and outside the building and in the soils surrounding the building.

Performing a lead paint risk assessment entails the following steps:

- Research
- Site walkthrough
- Design the sampling program
- Sampling (dust, soil, and/or paint)
- Evaluation of analytical results
- Control identified lead hazards

16.3.2.1 Data Research

As is typical for an environmental investigation, the first step is research, which entails gathering all available information about the subject matter prior to designing and implementing a sampling program. The environmental consultant, for this task known as the *risk assessor*, should interview the building owner and building manager regarding the history of LBP in the building. Has there been prior testing of LBP? Removal of old paint? Post-1978 renovations or additions to the building? Where are the children's play areas? Have the exposed soils in the play areas been tested for lead content?

Visiting the local health office and interviewing the local health official often yields valuable information in the research phase of the project. Testing of lead levels in blood samples collected from children may be available, as would records of complaints from residents and parents.

The risk assessor should evaluate the information obtained from the research and interviews and compare it with the criteria for lead inspections and removals. The risk assessor should also integrate the documentation with information gathered at the interviews and his/her own personal visual assessment to assess the veracity of the existing documentation.

16.3.2.2 Site Walkthrough

Once the research portion of the lead paint risk assessment is completed, the risk assessor should conduct a walkthrough of the building, especially portions of the building frequented by young children. The risk assessor will look for deteriorating paint surfaces; areas of visible dust accumulation; painted surfaces that are "impact points" or are subject to friction (both of which could eventually lead to paint damage and possible lead release); and painted surfaces that a child may have chewed. Any of these observations would indicate the possible existence of a lead hazard. If the potential paint hazard appears to have been caused by moisture intrusion, then the moisture intrusion must be addressed before addressing the lead hazard itself, since failure to address the source of the lead hazard will cause the lead hazard to return after it has been mitigated.

An inspector (not necessarily the risk assessor) will observe the painted surfaces and classify them as "good" or "poor." Table 16.2 (Table 5.3 in the HUD guidelines) provides a template to assign rankings of the condition of the paint. As noted in the table, the ranking is affected by either the quantity or the percentage of deteriorated paint on a given component.

TABLE 16.2

Categories of Paint Film Quality

Type of Building Component	Total Area of Deteriorated Paint on Each Component		
	Intact	Fair	Poor
Exterior components with large surface areas	Entire surface is in good condition.	<= 10 square feet	>10 square feet
Interior components with large surface areas (walls, ceilings, floors, doors)	Entire surface is in good condition.	<= 2 square feet	>2 square feet
Interior and exterior components with small surface areas (window sills, baseboards, soffits, trim)	Entire surface is in good condition.	<=10% of the total surface area of the component	>10% of the total surface area of the component

16.3.2.3 Designing and Implementing a Sampling Program

Next, the risk assessor will design a sampling plan to quantify the presence of lead caused by LBP. Three different types of testing will be conducted: dust sampling, soil sampling, and paint sampling.

Dust sampling would be conducted if dust was identified in an area that could contain LBP or at one time could have contained LBP. Dust sampling should be biased towards areas where young children are likely to be present, for instance, at the entryway of the building, in the child's principal play area, the child's bedroom, the kitchen, and the bathroom. In multi-family buildings, the risk assessor should collect dust samples from common areas, such as staircases and laundry rooms. Paint on furniture, while definitely a potential hazard to the young child, is generally considered to be the responsibility of the furniture's owner, and is not part of a standard lead paint risk assessment.

Wipe sampling is the preferred method of dust sampling (see Figure 16.4). Typically, dust sampling involves placing a treated, pre-cut piece of gauze on the surface to be tested, allowing dust particles to adhere to the gauze. The gauze is then sent to a laboratory for analysis. Wipe sampling should include the collection of wipe "blanks," which are lead-free wipes, and "spiked samples," which are wipes that are spiked with a known quantity of lead, for quality control purposes.

Usually each wipe sample is collected and analyzed individually, although in certain situations, multiple wipe samples are composited and analyzed as one sample. Compositing is often done to save money, but composite samples often are only useful as a negative screen, since the source of lead on a composite wipe would be ambiguous. Such a detection would only matter if it the lead concentration detected in the composite sample exceeds one of the HUD guidelines for a lead hazard. No more than four wipe samples are allowed to be composited as per HUD guidelines.

FIGURE 16.4
Wipe sampling of a painted surface.

When collecting soil samples for lead content, the same concepts used in dust sampling apply. Sampling should be biased to areas where young children are most likely to play. At least two soil samples must be collected—one from the principal play area, and one from bare soil areas from another location on the property, for quality control purposes. Samples should be collected from the top 1 to 2 inches of soil, the most accessible soil to a child. Composite soil sampling can be performed, although it carries the same limitations as described above for dust sampling.

Drinking water sampling is not required under a lead risk assessment, although it may be desirable to obtain a full picture of the potential lead exposure in a dwelling (see Chapter 14).

When paint and dust testing are warranted in a multi-family dwelling (defined as a dwelling with more than four units), it is unreasonable to expect the risk assessor to obtain access to all of the units. Table 16.3 (which is Table 5.6 in the HUD guidance document) provides HUD's minimum requirements for the number of dwellings to sample in a multi-family building.

Assumptions would then be made about the other units based on the results of the units sampled, based on statistical probabilities. The designated number of units should be selected randomly, except where lead hazards are known to exist or have existed based on prior testing, knowledge of elevated lead in blood levels of children, lead violations noted by a public health inspector, etc. Sampling should be biased to units in which such lead hazards are known to be present or have been present, although these units would not count to the minimum number of dwellings to be targeted since targeted (biased) and random (unbiased) sampling should not be mixed. The inspector can also perform a worst-case inspection, which would target only units where the paint is deteriorated and young children are known to be present. Once the dust and paint samples are collected, they are sent to a laboratory for lead analysis.

16.3.2.4 Assigning Lead-Based Paint Hazard Levels

Table 16.4 (which is Table 5.7 in the HUD guidance document) provides the hazard levels for an LBP risk assessment, both inside and outside the residence. Note that the maximum allowable levels are affected only by the concentration of lead in the paint.

TABLE 16.3

Minimum Number of Targeted Dwellings to Be Sampled among Similar Dwellings (Random Sampling May Require Additional Costs)

No. of Similar Dwellings	No. of Dwellings to Sample
1–4	All
5–20	4 units of 50% (whichever is greater)
21–75	10 units or 20% (whichever is greater)
76–125	17
126–175	19
176–225	20
226–300	21
301–400	22
401–500	23
501+	24 + 1 dwelling for each additional increment of 100 dwellings or less

TABLE 16.4

Hazard Levels for Lead-Based Paint Risk Assessments

Media	Level	
Deteriorated paint (single-surface)	5,000 mg/kg or 1 mg/cm^2	
Deteriorated paint (composite)	5,000 mg/kg or 1 mg/cm^2 divided by the number of sub-samples	
Dust (wipe sampling only)	Risk Assessment	Risk Assessment screen (dwellings in good condition only)
Carpeted floors	100 μg/ft^2	50 μg/ft^2
Hard floors	100 μg/ft^2	50 μg/ft^2
Interior window sills	500 μg/ft^2	250 μg/ft^2
Window troughs	800 μg/ft^2	400 μg/ft^2
Bare soil (dwelling perimeter and yard)	2,000 mg/kg	
Bare soil (small high-contact areas, such as sandboxes and gardens)	400 mg/kg	

mg/cm^2 = milligrams per square centimeter; μg/ft^2 = micrograms per square foot; mg/kg = milligrams per kilogram.

16.4 Designing the Lead Paint Survey

Inspections for the presence of LBP are outlined in Chapter 7 of the HUD guidelines. This section summarizes that chapter. Only a certified LBP inspector can conduct an LBP survey for HUD-financed projects.

The LBP surveys are guided by two concepts: room equivalents and testing combinations. A *room equivalent* is "an identifiable part of a residence." It is analogous to a functional space as defined under AHERA (see Chapter 14). Other than rooms themselves, room equivalents include hallways and stairways, common areas in multi-family buildings such as foyers and entranceways, and exterior areas such as playgrounds and walkways with painted railings. The objective in defining a room equivalent is in identifying building areas with uniform painting histories. For example, a closet is not considered to be a room equivalent because closets usually have the same painting history as the associated room. Exterior walls should be considered the same room equivalent unless there are obvious dissimilarities.

Testing combinations consist of three components: a room equivalent, a building component type, and a *substrate*, which is the material of the painted building component. Examples of testing combinations are provided in Table 16.5. Testing combinations are analogous to homogeneous areas in asbestos survey; they provide the basic unit of the LBP survey. It is important to note that color, the most conspicuous quality of paint, is *not* part of the testing combination,

TABLE 16.5

Examples of Testing Combinations

Room Equivalent	Building Component	Substrate
Master Bedroom	Door	Wood
Master Bedroom	Door	Metal
Kitchen	Wall	Plaster
Garage	Floor	Concrete
Exterior	Siding	Wood
Exterior	Swing set	Metal

since the deeper and older the layers of paint are more likely to be LBP. Therefore, any valid methodology to be used in an LBP investigation must evaluate the lowest layer of paint, that is, the layer of paint on the substrate. As a rule of thumb, testing should be oriented to where the paint is thickest, since the most layers of paint likely will be encountered there.

16.5 Lead Paint Sampling and Analysis

The goal of a lead paint survey is to determine the *lead loading*, also called *area concentration*, on a given portion of a substrate. Lead loading is evaluated by testing the painted surface. This subsection describes the common methods used to test painted surfaces for lead content.

16.5.1 Paint Chip Sampling

A straightforward way to assess the presence of lead in paint is by *paint chip analysis*. In this method, a piece of paint is physically carved off the substrate and analyzed for lead content, either on the premises by a mobile laboratory, or off-site by a fixed-base laboratory. Typically, a fixed-base laboratory provides a higher degree of quality control, although more immediate results will be obtained from using an on-site laboratory. The drawbacks of collecting paint chips for analysis by a fixed-base laboratory are that (1) it doesn't provide immediate results, (2) paint chip sampling damages the sampled surface, (3) the deepest paint layer can be missed; and (4) part of the substrate sometimes detaches with the paint sample, which affects the lead concentration calculations.

16.5.2 Chemical Test Kits

Chemical test kits are simple, inexpensive tools used to assess the presence of lead in paint. These kits, available at most home improvement centers, contain swabs roughly the size of a small cigarette. The swabs contain a reagent that will change color when it comes into contact with lead. To test a painted surface, the paint is scored to the substrate using a sharp instrument. The swab's seal is broken, which releases the reagent. The swab then is rubbed over the incision. The reagent will change color if it comes into contact with lead. The drawbacks of this method are (1) the test is qualitative, which especially is a problem for an LBP survey; (2) failure to score the paint all the way to the substrate will result in failure to evaluate the lowest layer or layers that are most likely to contain lead; and (3) it is prone to false positives when other metals are present.

16.5.3 X-ray Fluorescence Machine

The most reliable and common method of testing for the presence of LBP, and the method recommended by the USEPA, involves the use of an *x-ray fluorescence machine*, generally referred to as an XRF machine. OSHA does not recognize the XRF method as a manner to assess the presence of lead paint in the workplace. In addition, XRF testing should not be performed on paint that is chipped or peeling, since the XRF machine requires an even surface to provide an accurate reading.

The XRF machine contains a radioactive source (typically Cobalt-57 or Cadmium-109) that emits x-rays produced by radioactive decay. When x-rays encounter lead, the lead gives off gamma rays at a certain energy level due to the photoelectric effect. These gamma rays enter an aperture in the machine, when a detector records their presence. The XRF

FIGURE 16.5
Using an XRF machine to test for lead-based paint. (Courtesy of Bay Area Lead Detectors Blog, http://www.bayarealeaddetectors.com/blog/.)

machine then converts the amount of gamma rays detected into the concentration of lead in the tested surface.

To operate an XRF machine, the machine operator places the aperture at the end of the machine against the surface to be tested (see Figure 16.5). When the machine operator presses the trigger, the chamber that contains the radioactive source is opened, and the x-rays impinge on the paint surface being tested. The time that the aperture is opened depends upon the machine. Some instruments automatically adjust for source decay, i.e., when the radioactive source begins to wear out.

XRF machines, because they contain a radioactive source, must be registered with the Nuclear Regulatory Commission. Their transportation across state lines is regulated, as are their usage and their disposal.

16.5.3.1 Machine Calibration

Because each machine will provide inaccurate readings if it is not used or maintained properly, the machine operator must follow the manufacturer's instructions on the usage of the machine. Integral to the operation of all XRF machines (and all quantitative analysis equipment), is calibration.

Machine calibration is designed to ensure that the lead readings obtained are precise, that is, reflecting the actual lead concentration in the paint tested. Calibration must be

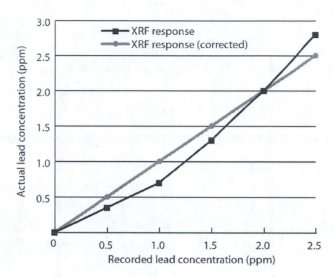

FIGURE 16.6
A typical five-point calibration curve.

performed the first time a machine is used that day, after the machine has been transported (since vibrations and jostling during transit can affect the machine's operations), and at selected landmarks in the course of the testing day (for instance, every 4 hours, or after a certain number of tests are run).

Ideally, a calibration check will verify the machine's precision. The most important reading to ensure for accuracy is at the bright line between LBP and non-LBP, i.e., 1.0 mg/cm². If a surface contains lead at a concentration of 12.0 mg/cm² and the XRF machine records the lead concentration at 8.0 mg/cm², it is less of a concern than if a surface contains lead at a concentration of 1.2 mg/cm² and the XRF machine records the lead concentration at 0.8 mg/cm². Figure 16.6 shows the results of a typical five-point calibration curve and correction.

16.5.3.2 X-ray Fluorescence Testing Protocols

Once the XRF machine is calibrated, the survey begins. Testing protocol calls for at least one "shot" per testing combination except for walls in room equivalents. If, in the judgment of the machine operator, a surface is sufficiently large that there might be variations in its painting history, then it should be tested more than once. The XRF machine stores the results of each test, which then can be downloaded onto a computer.

Keeping track of each testing combination tested can be a challenge. Figure 16.7 shows the shot locations for the walls of a portion of the third floor of a fictitious school. Each shot can be defined by its building component type, the room equivalent, and the cardinal direction of the shot. For instance, the shot along the north wall of Room 303 would be labeled "W303N," the shot of the wall in the hallway just outside the men's bathroom (the southern part of the hallway) as "WHALLS," and so on.

16.5.3.3 Data Interpretation

Interpretation of the results of the XRF testing is guided by the *XRF performance characteristic sheet* provided by the manufacturer. One of the items specified on the XRF performance

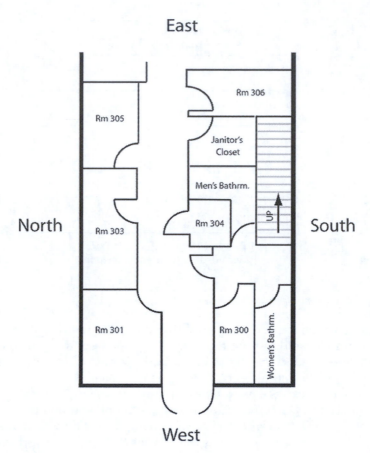

FIGURE 16.7
A diagram of an area about to undergo an LBP survey.

characteristic sheet is the "inconclusive range" of the machine. For instance, the sheet for one popular machine indicates that its inconclusive range is between 0.8 and 1.2 mg/cm^2. This means that all measurements falling into that range are statistically uncertain and cannot be used to determine the presence of absence of LBP on that testing combination. In such cases, the investigator can either (1) retest the testing combination; (2) confirm the results by collecting a paint chip sample and sending it to a laboratory for analysis; or (3) assume that the paint is LBP. The investigator cannot assume the paint is non-LBP based on an inconclusive test.

The XRF performance characteristic sheet also provides information on how to calculate a *substrate correction*. Some substrates may interfere with the XRF measurements by either contributing x-rays or inhibiting them from entering the machine recording chamber. The XRF performance characteristic sheet will tell the user how to perform a substrate correction.

For example, let's say that a reading from a testing combination is 1.1 mg/cm^2. However, the XRF performance characteristic sheet indicates that one must subtract 0.5 mg/cm^2 for that substrate. Therefore, what first appeared to be a reading indicating the presence of LBP would now be 0.6 mg/cm^2, which indicates the absence of LBP.

To be certain that the appropriate substrate correction is being applied, HUD advises that in a single-family residence the investigator obtain three readings on bare substrate from two different locations. The substrate correction would be the average of the six readings.

16.6 Data Documentation

Once the field work has been completed and the testing results are in, the inspector will prepare a survey report. The report should contain all of the relevant information about the methods used in the investigation, the results of the investigation, and the qualifications of the investigator. As with all environmental investigation reports, this report should answer the where, when, why, who, and how questions. The qualifications and certifications of the investigator should be documented, and all test results should be reported, even test results that were inconclusive or otherwise unusable in the data interpretation. Labeling of the testing combinations should be sufficiently obvious that someone who had never been to the facility could determine the locations of the LBP identified by the investigation.

Many jurisdictions have mandatory reporting requirements for LBP. All mandatory reporting should be made in accordance with the regulations, and the report should document that the required notifications were made.

16.7 The Housing and Urban Development Guidelines for Lead-Based Paint Hazard Abatement

As with asbestos, a lead paint hazard can be abated either by enclosure, encapsulation, or removal (removal is the most invasive and usually the most expensive; HUD suggests avoiding this abatement method when possible). In accordance with its Lead Renovation, Repair, and Painting Rule (RRP Rule), the USEPA requires that LBP abatement be performed by USEPA-licensed contractors. Excepting this licensure requirement, LBP abatement is governed mainly by guidelines, although some states have their own regulations regarding LBP abatement. This section describes the three abatement methods.

16.7.1 Abatement by Enclosure

An *enclosure* is a "dust tight" barrier over the surface containing the LBP. The HUD guidelines indicate that the barrier must be rigid and durable, able to last without replacement for up to 20 years. In the case of contaminated soils, the barrier could be a synthetic rubber surface commonly used on modern playgrounds or some similar impenetrable barrier. For exterior walls, it could be vinyl or aluminum siding; for interior walls, paneling or sheetrock. As with all methods of abatement, a certified risk assessor must certify the integrity of the enclosure.

16.7.2 Abatement by Encapsulation

Encapsulation is the process by which the LBP-covered surface is coated with either a liquid-applied coating or an adhesive material, such as wallpaper. The primary difference between enclosure and encapsulation is that for enclosure the barrier is rigid, typically applied with screws, nails, or other such materials, whereas with encapsulation, the material adheres to the surface without the aid of such materials. As with enclosures, the encapsulant should be designed to last for up to 20 years.

Encapsulation is not appropriate on highly deteriorated surfaces, where the encapsulant would delaminate along with the paint it is designed to isolate, or high friction surfaces, where the longevity of the encapsulant is likely to be compromised. Paints can be used for encapsulation only if they are designed to withstand hazards associated with lead paint, such as resistance to chewing. The encapsulant manufacturer typically specifies application methods, appropriateness for various substrates, and testing procedures to verify the effectiveness of the encapsulation activities. In all cases, starting with a clean surface is key to the effectiveness of the encapsulation activities.

16.7.3 Abatement by Removal

Removal of lead paint differs from asbestos removal mainly because lead is not considered to be an airborne contaminant, or a hazard to older children and adults. Procedures for LBP removal are far less prescribed than asbestos removal. The HUD guidelines typically are used in conjunction with OSHA regulations governing the worker safety.

As with any environmental investigations or remediation, the first and most important rule is to "do no harm." Disturbance of a hazardous material such as lead-based paint releases lead into the air where it can be inhaled by workers, residents, and children. Most precautions taken in a lead paint removal project are designed to prevent harmful releases from occurring.

The LBP removal follows the steps of other types of remediation: design, implementation, and verification. The design portion of the lead paint removal deals mainly with protection of residents and workers, establishing procedures for paint removal, and establishing protocols for verification of the LBP removal process.

16.7.3.1 Worker Protection

OSHA requires that workers wear proper *personal protective equipment* (PPE) to protect them from lead inhalation. An initial exposure assessment will allow the contractor to determine the requisite level of respiratory protection to be worn by the workers. In most cases, paint removal workers will wear half-face air purifying respirators equipped with HEPA filters.

Workers also wear protective suits, not to protect their skin, since lead cannot penetrate the dermal barrier, but to prevent lead-containing dust from accumulating on their clothes and contaminate portions of the building outside of the work area. The protective suits are removed as workers leave the work area and are placed in bags for eventual proper disposal.

16.7.3.2 Protecting the Residents

To protect residents, the area where paint removal activities occur must be unoccupied. Warning signs are posted, and openings leading outside the work area are isolated or protected. The level of work area protection is not nearly as elaborate and proscribed as for asbestos abatement, mainly because lead dust particles are not as aerodynamic as asbestos fibers and therefore unable to travel as far or remain in the air as long as lead dust. In addition, lead inhalation does not present the same threat to workers as does the inhalation of asbestos fibers, so the degree of protection needed for populations other than young children is less involved.

16.7.3.3 Preparing the Work Area

In general, the work area is prepared similarly to an asbestos project. The level of protection to be implemented for a given lead paint removal project will depend upon the following factors:

- The size of the surface(s) to be abated
- The type of hazard control methods to be used
- The extent of existing contamination
- Building layout
- Vacancy status of the dwelling

Prior to all work area preparation, dust and severely deteriorated paint should be removed (see Figure 16.8), since they are likely to be disturbed and possibly released outside the work area during the work area preparation activities.

There are four levels of work area protection defined by HUD:

Level 1 protection pertains to removal actions of less than 2 square feet of LBP. For a Level 1 job, warning signs are placed outside the abatement area. One sheet of 6-mil polyethylene sheeting ("poly") is placed in the abatement area and extending 5 feet beyond the abatement area. Vents are sealed within 5 feet of the area, and furniture is covered within 5 feet of the area.

FIGURE 16.8
Paint dust fallen from a wall in a residence. (Courtesy of GZA GeoEnvironmental, Inc.)

Level 2 protection is used when there is to be 2–10 square feet of LBP disturbance per room, or there is to be an interim control of a lead hazard while a permanent solution is being designed. These cases call for two layers of poly in the work area with an airlock on the doorway to the work area, as described in Chapter 15. All vents inside the room are to be sealed with poly and all furniture removed from the room before removing the LBP.

Level 3 protection is used for a Level 1 or Level 2 project that will occur over multiple days. For these projects, the residents are allowed to return to the dwelling at the end of each work day even though the lead abatement activities are still in progress. The rooms where lead abatement is occurring must be firmly secured, and warning signs must be present at the main entranceway to the dwelling. The top layer of poly, which will be contaminated with lead dust and removed LBP, must be removed at the end of the work day. The guidelines also suggest the collection of dust samples outside of the area on an as-needed basis to verify that lead-containing dust hasn't escaped from the work areas.

Level 4 protection is used for large projects, i.e., those projects in which greater than 10 square feet of surface will be disturbed. For such jobs, there must be warning signs on the building exterior and airlocks on all doors.

When removing paint from exterior surfaces, HUD recommends avoiding windy days, to prevent the airborne spread of lead-containing dust or particulates in the course of the abatement activities.

Lead removal can also pertain to the removal of lead-contaminated soils. In these cases, a temporary fence or barrier tape is placed 20 feet around the surfaces or soils to be abatement to restrict residents and others from the work area. Whether residents can return to their dwelling at the end of the workday will depend on the size and duration of the project. One layer of 6-mil poly is placed on the ground, extending to 10 feet beyond the surfaces or soils to be abated. Movable items within 20 feet of the surfaces or soils to be abated are removed from the area, and entryways to the building within 20 feet of the surfaces or soils to be abated are closed off. Warning signs are placed around the perimeter of the work area.

16.7.3.4 Lead Paint Removal Procedures

There are several approved methods for the removal of lead paint. The simplest method is building component replacement, e.g., taking off the door with lead paint and replacing it with a new door painted with lead-free paint. Removing paint from a building component may involve scraping tools, heat guns and paint strippers, although a paint stripper may pose an additional hazard to abatement workers if it contains compounds that are inhalation hazards to the workers. However, HUD prohibits any removal method that will create airborne lead-containing dust or gases, such as sanding, dry removal, or burning.

Gross removal of paint is followed by HEPA vacuuming and wet washing. Vacuums must be supplied with HEPA filters since lead dust can be sufficiently small that they may not be trapped by conventional vacuum filters. For all lead removal methods, negative air pressure is not required because of the poor aerodynamics of lead dust. Simple water may be used in the wet washing process, although sometimes detergents are added to the water to improve cleaning.

For Levels 1, 2, and 3 area preparations, one round of HEPA vacuuming and wet washing within 5 feet of the work area is sufficient. For Level 4 area preparation, a "three-pass" cleaning process is used. In the first pass, workers will HEPA vacuum the surface and wet wipe. The HEPA vacuum removes most of the dust and dirt, and the wet wash helps to dislodge the remaining dust and dirt from the surface. These steps are followed by a second round of HEPA vacuuming. A HEPA spray-cleaning vacuum can be used as an alternative to the three-pass cleaning system.

Once the LBP has been removed from the surface, the contaminated poly is placed in waste bags and the workers and their removal equipment are removed from the work area. At least one hour after completion of the removal activities, an independent third party inspects the work area. The one-hour time interval allows for any dust to settle out of the air. If the inspector determines that the lead removal has been successful, then clearance sampling can begin. If there is evidence that lead remains in the work area, then the entire cycle of preparation, removal, cleaning, and inspection must be repeated.

16.7.3.5 Clearance Wipe Sampling

Unlike asbestos removal, clearance sampling involves wipe sampling rather than air sampling. It should be noted that XRF testing is not allowed for clearance sampling because of its difficulty in accessing the presence of dust on a surface. HUD recommends the following protocol for collection of discrete clearance wipe samples:

- Collect wipe samples from window wells, window sills, and floors.
- Wipe samples may be discrete, individual samples or composite samples.
- Samples should be biased towards areas where a lead hazard was present, or to high traffic areas.
- Two dust samples from at least four rooms in the dwelling should be collected and tested, if the paint removal project covered the entire house. These samples should be collected from the window wells, the window sills, and the floors. In common areas, there should be at least one wipe sample for every 2,000 square feet of floor.
- A wipe sample should also be collected within 10 feet of each airlock, as applicable, to determine the effectiveness of the containment system.

If the inspector chooses to composite the clearance wipe samples, then three composite samples should be collected for every batch of four rooms. Only wipe samples collected from the same building components can be composited. Compositing must be performed in the field, not in the laboratory. The same rules (and limitations) apply for compositing clearance samples as investigation samples.

The following guidance values are used to determine whether the work area can be cleared for reoccupancy.

- Floors: 40 µg lead/ft^2
- Window sills: 250 µg lead/ft^2
- Window troughs and other rough surfaces: 400 µg lead/ft^2

If one or more clearance samples fail to meet the above standards, then the entire cleaning and testing process needs to be repeated.

Once finished, all porous surfaces should be coated with a sealant. Surfaces should be completely dry before the sealant is applied.

There are reporting requirements for various jurisdictions, and reoccupancy of the abated areas may require regulatory approval and possibly a sign-off by a licensed professional.

16.8 Lead Paint and Environmental Consultants

Many of the activities described in this chapter require special licensing from either the USEPA or the various state governments, and all of the activities require direct field experience under the tutelage of a licensed practitioner to ensure the proper performance of the activity. Regulations require that a USEPA-licensed risk assessor perform lead paint risk assessments. HUD requires that only a certified LBP inspector can conduct an LBP survey.

While there is no national license for operating XRF machines, the manufacturers typically recommend that the operator go through formal training prior to using the machine. Consultants can design and conduct small-scale operations and maintenance activities, such as the repair or removal of small quantities of LBP, although many facilities train their maintenance staffs to perform these activities.

Problems and Exercises

1. The XRF measurements of a testing combination yielded readings of 0.26, 0.49, and 1.97 mg/cm^3. Assuming no substrate correction, what can you conclude about the presence of LBP for this testing combination?

2. A composite paint chip sample contains lead at a concentration of 0.80 mg/cm^3. What can you conclude about the presence of LBP in the testing combination?

3. One of seven XRF readings collected for a testing combination was above 1.00 mg/cm^3, with the substrate correction applied. What possible reasons are there for this discrepancy?

4. What are the dangers in getting a false positive from an XRF reading? From getting a false negative?

Bibliography

Dartmouth Toxic Metals Research Program. www.dartmouth.edu/~toxmetal/.

Getting the Lead Out: A Lead Prevention Timetable, April 1997. *The Cooperator.* http://cooperator.com/article/getting-the-lead-out/full.

U.S. Department of Health and Human Services, October 1991. Centers for Disease Control. Preventing Lead Poisoning in Young Children.

U.S. Department of Housing and Urban Development (HUD), 1997. Guidelines for the Evaluation and Control of Lead-Based Paint Hazards in Housing.

17

Indoor Air Quality Investigation and Mitigation

17.1 Introduction

Indoor air quality is a broad topic that incorporates many topics discussed elsewhere in this book, such as asbestos, lead paint, and mold, as well as many of the chemical pollutants discussed in Chapter 4. Because it intersects with so many other environmental problems, *indoor air quality* (IAQ) can be a very difficult problem to diagnose. The IAQ investigations and remediations are interdisciplinary, bringing in toxicology, architecture, and many engineering disciplines. This chapter focuses on the contributions that can be made by the environmental consultant to the identification and remediation of indoor air quality problems.

It should be noted that many of the contaminants that create IAQ problems are also regulated under the Clean Air Act. However, the Clean Air Act focuses on regional and area pollution and air pollution sources. It does not apply to indoor air problems, which are far more local in nature and impact.

17.1.1 History of Indoor Air Problems

Until the mid-1940s, commercial buildings in the United States typically had windows that could be opened manually by the building occupants. Air conditioning, if available, was supplied by window-mounted units. Post-war commercial buildings typically were constructed with exterior walls that were permanently sealed, and central heating, ventilation, and air conditioning (HVAC) systems became the norm. The trend towards fully sealed commercial buildings greatly accelerated in the 1970s, when the energy crisis increased the desire for energy efficient buildings with minimum air leakage through rickety windows. Individuals no longer had direct control over their breathing air conditions.

Indoor air quality as a health issue gained national prominence in 1976 with the outbreak of what became known as *Legionnaire's Disease* at an American Legion convention in Philadelphia, Pennsylvania. The *Legionellosis* bacterium had bred in the cooling tower of the hotel's air conditioning system, through which it spread through the entire building. This event prompted a call for regulating climate control systems.

With the rise in indoor air quality complaints, by the early 1990s the term *sick building syndrome* came into vogue. This term referred to buildings that were inherently defective, resulting in conditions ranging from uncomfortable to downright toxic to its inhabitants. Its appearance was enhanced by the increasing trend to use building materials that were chemically treated or that emitted chemicals when new, especially *volatile organic compounds* (VOCs), which are described in Chapter 4. Indoor air quality, as an environmental problem, was here to stay.

17.1.2 Health Effects of Indoor Air Pollutants

Health-related effects from an indoor air pollutant can result from a single exposure, multiple exposures, or prolonged, constant exposures. Symptoms typically include irritation to the eyes, nose, and throat, headaches, dizziness, and fatigue. Long-term exposure to indoor contaminants can lead to more serious illnesses, including respiratory diseases, heart disease, and cancer.

As with mold, indoor air pollutants more often affect susceptible populations, including children, the elderly, and people suffering from asthma, allergies, and immuno-compromised illnesses (as well as contact lens wearers). This is buttressed by the fact that many indoor contaminants such as mold (see Chapter 19) lack established dose-response relationship between exposure and health effects, which makes it difficult to connect the symptoms with the contaminant. Because of the amount of time the typical office worker spends indoors, and the sealed nature of the modern office building (as opposed to the common residence), office workers are particularly vulnerable to IAQ problems.

17.1.3 Indoor Air Investigation Triggers

The IAQ investigations are ordinarily triggered by real or perceived health impacts or odor complaints by a resident or a building worker. The IAQ investigations may also be triggered by a property transaction or financing, in which case they are performed as supplements to a Phase I Environmental Site Assessment, or as part of a risk assessment, often for a lending institution or insurance company.

17.2 Sources of Indoor Air Pollution

It is helpful to conceptualize indoor air pollutants as originating either inside the building or outside the building and migrating into the building. They all share inhalation as their primary pathway. Some of the main causes of indoor air problems are described below.

17.2.1 Poor Air Flow

Improperly maintained or improperly operated ventilation systems are a prime source of indoor air quality problems. Such systems can affect the entire building or portions of a building.

The air flow rates recommended by the American Society of Heating, Refrigerating and Air-Conditioning Engineers, Inc. (ASHRAE) are 15 cubic feet per meter (cfm) per person in an industrial setting and 20 cfm per person in an office setting. The generally accepted range for relative humidity is 30%–50%, although it may vary based on the time of year, as discussed below.

Poor air flow can cause a buildup of carbon dioxide from human exhalation. ASHRAE recommends are that indoor CO_2 concentrations not exceed 700 parts per million (ppm) above background concentrations. As of this writing, the generally accepted outdoor concentration of CO_2 at sea level in the United States is 385 ppm. This means that an indoor contribution of 1,085 ppm, or 700 ppm above background, is needed for the indoor air to exceed the ASHRAE allowable CO_2 concentration of 700 ppm.

The range of indoor temperatures and relative humidity recommended ASHRAE varies by the time of year and the type of indoor environment. However, even within those ranges, some people will complain about the temperature and humidity of the indoor air. For that reason, ASHRAE recommends using an 80% "occupant acceptability," meaning that one should assume that 10% of all people in a given environment, at any given time, will claim that the temperature and/or relative humidity are too low (10%) or too high (another 10%), even with all conventional parameters being within the established range of acceptability.

17.2.2 Combustion Products

Many indoor air pollutants are the byproducts of incomplete combustion of heating fuel. Sources include the building's heating system (if it burns fossil fuel), outdoor maintenance equipment, such as lawn mowers, snow and leaf blowers, and motor vehicles, especially in a residence with an attached garage. The byproducts of incomplete combustion include carbon monoxide (CO), nitrogen oxide (NO), nitrogen dioxide (NO_2), various volatile and semi-volatile organic gases, and particulates.

Carbon monoxide is a colorless, odorless gas that deprives the brain of oxygen, which can lead to nausea, unconsciousness, and death. OSHA's one-hour exposure limit for CO in the workplace is 50 ppm, although the National Institute for Occupational Safety and Health (NIOSH) recommends a limit of 35 ppm. The American Conference of Governmental Industrial Hygienists (ACGIH) recommends an 8-hour time-weighted average (TWA) limit of 25 ppm for CO exposure.

Exposure to NO and NO_2 can irritate the respiratory tract, although high levels of exposure, unlikely in an indoor air setting, can cause more severe and permanent damage. There are no standards for NO and NO_2 in indoor air.

17.2.3 Dust and Particulates

Particulate emissions from incomplete combustion can cause respiratory irritation. Dust and particulates can contain contaminants, such as heavy metals (including lead) and asbestos, which can create their own set of problems. Particulates are usually described by their average diameter. For instance, *PM_{10} particles* (generally referred to as PM_{10}) are less than 10 microns in diameter. The USEPA recommended limit for PM_{10} in indoor air is 150 micrograms per cubic meter ($\mu g/m^3$) for a 24-hour TWA.

17.2.4 Ozone

Ozone is a critical, useful component of the earth's upper atmosphere. When down in the breathing zone, however, it is a respiratory irritant that can cause breathing problems because it is able to react with many chemicals found indoors and create chemicals that are toxic to humans. Organic compounds formed by the reactions between ozone and many common indoor pollutants include organic compounds that can be more odorous, irritating, and/or toxic than the pollutants they derive from. Research has shown that indoor ozone levels in the breathable air should not exceed 20 parts per billion (ppb).

17.2.5 Volatile Organic Compounds

The VOCs are emitted from literally thousands of solid or liquid products and byproducts. They may have short- and long-term adverse health effects, depending on the chemical.

The VOCs are widely used as ingredients in household products, such as paints and lacquers, paint strippers, cleaning supplies, waxes, mothballs, and air fresheners. Chemicals typically found in garages and maintenance areas, such as gasoline and other fuels, automotive products, and industrial cleaners, can contain VOCs. The VOCs are in office products such as inks, correction fluids, and office equipment such as copiers and printers. They also can be combustion byproducts, or byproducts of chemical reactions.

Another source of VOCs common to both offices and residences is clothing that has been dry cleaned using perchloroethene, or "perc" (see Chapter 4). Not all perc is removed from the clothing in the drying process, so dry cleaned clothing in a closet or on a person is a potential source of VOCs.

The VOCs may also originate as off-gases from building materials. Formaldehyde can off-gas from carpeting. Most carpets being sold today are advertised as "low VOC" carpets; resins from pressed wood products, whose usage has greatly increased in recent years; *urea formaldehyde foam insulation* (UFFI), which was used as insulation in many homes in the 1970s; and paints and varnishes.

17.2.6 Bioaerosols

Biological hazards typically are *bioaerosols*, which are biological agents that remain suspended in air. They include mold (discussed in Chapter 19), viruses, pollen, dust mites, and animal hair and dander, all of which can produce a variety of health-related effects in susceptible populations. Legionella, discussed at the beginning of the chapter, is a bioaerosol that grows in slow-moving or still, warm water, such as water found in evaporative cooling towers or showerheads, thus making it a hazard to commercial as well as noncommercial buildings.

17.2.7 Tobacco Smoke

Environmental tobacco smoke (ETS) is caused by smoking inside the building or smoking near the ventilation intake of a building. It is becoming rarer due to smoking bans established in most non-residential buildings. In large quantities, second-hand smoke can be as cancer-causing to the bystander as the smoker. From an IAQ perspective, it can cause respiratory irritation, and can react with ambient ozone to form deadly carbon monoxide.

17.2.8 Pesticides

Pesticides, either from application inside the building, storage inside the building, migrating inside of the building from an outdoor application, can affect indoor quality. Of particular concern is the indoor application of pesticides meant to be used outdoors, since they are likely to persist indoors and be a constant threat to the health of the building occupants.

17.2.9 Subsurface Contamination

Vapor intrusion, an indoor air quality issue that is triggered by vapors emanating from soils or groundwater that are contaminated with chemicals, usually VOCs, is discussed in Chapter 10.

17.3 Heating, Ventilation, and Air-Conditioning Systems

Since the building's HVAC system is such a crucial component to the whole indoor air issue, it pays to have a working understanding of how the system works to perform a proper IAQ study.

The HVAC system includes all heating, cooling, and ventilation equipment serving a building. The heating and cooling systems generally run independently of each other, although they and the ventilation system, use a central component known as the *air handler* (see Figure 17.1). The air handler brings in outdoor air which often is mixed with air that has circulated through the building, heats or cools the air as needed, sends it into the building, and then retrieves it for re-conditioning and reuse.

Outdoor air enters the air handler through the air intake, which is usually designed to allow in a minimum amount of outdoor air during extreme outside temperature conditions. Outdoor air and recirculated air are mixed in the air handler, and then sent through heating and cooling coils to regulate the temperature of the air delivered to the building.

Air intakes are equipped with air filters that generally are designed to keep out particulates. The efficiency of the air filter is measured by its *minimum efficiency reporting value* (MERV), a value designed by ASHRAE. Table 17.1 shows the ranges of MERVs and the particulate sizes captured by an air filter with a given MERV.

FIGURE 17.1
A typical industrial air handling unit.

TABLE 17.1

Minimum Efficiency Reporting Values (MERVs)
for Air Filters

MERV Rating	Minimum Particle Size
1–4	>10.0
5–8	10.0–3.0 μm
9–12	3.0–1.0 μm
13–16	1.0–0.3 μm
17–20	<0.3 μm

μm = *micrometer.*

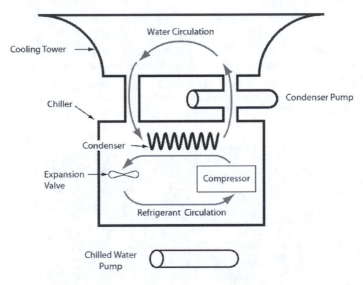

FIGURE 17.2
A schematic diagram of a commercial air conditioning system.

Figure 17.2 is a schematic diagram of the air conditioning ("chiller") portion of an HVAC system. A compressor compresses the air, thus heating it. A condenser removes moisture from the air, after which it goes through an expansion valve, which cools the air. The cooled air is then blown into the air distribution system by a supply fan. Air is usually distributed in commercial buildings through ducts located above suspended ceilings and behind walls. The air circulates through the user space, and then is pulled through a ceiling-mounted return into ducts and returned back to the air handler (see Figure 17.3). In some cases, the space above the dropped ceiling itself is used for returning air to the air handler. This is done as a cost-saving device, but if this air picks up materials or growths that have fallen onto the dropped ceiling, it would become contaminated and become a pathway to spread contaminants throughout the building.

Two portions of the air conditioning system can spawn microbial and bacterial growth if not properly maintained. The condensate created by the condenser typically collects in the drain pan under the cooling coil and exits the air handler via a deep seal trap. If not allowed to dry periodically, or allowed to build up cellulosic materials, drain pans can

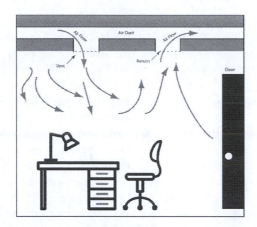

FIGURE 17.3
A supply and return air system in an office.

be a breeding ground for bioaerosols. The same holds true for cooling towers, which, as its name implies, are designed to remove waste heat from a building, typically from an industrial building with manufacturing processes that generate heat (see Figure 17.2).

Some buildings are kept under slightly positive air pressure relative to the outdoors to reduce infiltration of outside air, which could result in moisture and high humidity inside the building. For that reason, air flow from the air handler is sometimes greater than the air flow back to the air handler.

17.4 Performing the Indoor Air Quality Investigation

More than any other type of investigation discussed in this book, the type of IAQ investigation depends greatly on the nature of the air quality problem. Therefore, the investigator should gain a thorough knowledge of the nature of the IAQ problem and be able to place the problem in a physical and temporal context, or the investigation is doomed to failure.

Investigations typically fall into one of three categories: epidemiological, environmental, or engineering. An *epidemiological IAQ assessment* will emphasize the health-related aspects of the triggering event and will seek the cause or causes of the outbreak. An *environmental IAQ assessment* will emphasize the source of the odor complaints, or some other triggering event relating to the storage or usage of chemicals in and around the building. The *engineering IAQ assessment* will emphasize air flow issues and seek their remedy. The methods used to perform the steps discussed below will depend on the given situation, and which category of investigation is being conducted.

17.4.1 Building Reconnaissance

An initial walkthrough of the problem area provides information about the four basic factors influencing indoor air quality: occupants, contaminant sources, pollutant pathways, and the air circulatory system. It is helpful to perform the walkthrough to coincide with the time of the day that the complaint occurred or conduct multiple walkthroughs throughout the day if the timing of the problem may be an issue.

The investigator should assess quality of air in the building, the presence of objectionable odors. Is the air stale, too dry, or too damp? Is it too hot or too cold inside the building, or are there large temperature variations depending on the time of day or where you are inside the building? Is outdoor air entering the building through uncontrolled pathways, such as leaking windows, doors, or gaps in the exterior construction?

The quality of the breathable area should be noted, particularly in the area of the complaint or complaints. If the problem is widespread, the air circulation system could be the source of the problem, especially as a potential pathway for contaminants. Investigating the air distribution system should begin with the components of the HVAC system(s) that serve the complaint area and surrounding rooms but may need to expand if connections to other areas are discovered. The investigator should look for evidence of deterioration, corrosion, water damage, or excessive dust or debris within the system, and determine if the air distribution system has any blocked vents, inoperative fans, or clogged filters. Where health-related issues are involved, the investigator should check the drain pans for a buildup of water that can allow bacteria and mold to breed.

The investigator should check for pollution sources near the air intakes, including neighboring properties. The investigator should observe patterns of traffic, construction activity, and other potential sources in the neighborhood of the building and inquire about outdoor ambient air problems in the area.

Most commercial buildings have ample available information on the HVAC system's construction and its maintenance. The investigator should review available construction and operating records and collect detailed information on the HVAC system and possible pollutant pathways. Safety data sheets (SDSs), which must be readily available at the facility as per OSHA, provide information regarding the chemicals used in the HVAC system and other building equipment.

Identifying chemicals used by the building occupants can lead the inspector to potential sources of odors/chemicals, be it chemicals used by employees, chemicals worn by employees, or materials that are new and might be off-gassing. It is important to identify equipment that use chemicals, including office equipment, learn about their usage patterns, and identify items that are not equipped with local exhaust.

The combustion equipment used by the building would be a prime suspect if combustion byproducts are the contaminants of concern. Areas with water damage should be identified, as well as the locations of new furnishings that could be emitting VOCs or other gases. The investigator should evaluate the quality of housekeeping practices and look for deteriorated or moldy building components and equipment.

17.4.2 Interviews

A critical piece of the IAQ assessment is the interview phase. The investigator must confirm the health-related issue, which could range from one mildly ill person to a building-wide epidemic, by interviewing, at a minimum, the people who have lodged the complaints. The investigator should encourage the complainants to describe their symptoms as precisely as possible and compared with the medical histories of the affected persons, so correlations can be made.

Since the information provided by the interviewees can be biased and conflicting, the investigator should be skeptical about the information gathered during the interview. In many cases, people believe that they are suffering from indoor air pollutants, when in fact they are suffering from a problem not related to the indoor air. Checking whether other people in the work area are suffering from similar symptoms may help the inspector connect the dots. Workplace stresses can manifest themselves as health complaints; the investigator should be aware of pending or recent layoffs, pay cuts, collective bargaining negotiations, etc. in the workplace.

Interviews with maintenance personnel at the building also help the investigator develop a history of the building as it relates to indoor air issues. Personnel could be asked if there have been any recent fires, roof leaks, or smoke damage inside the building. Recent modifications, maintenance, or problems with the building's HVAC system are important to know. Such information can be obtained from building management as well as the local building and health officials.

If it appears that the problem could be due to elevated carbon monoxide levels or the presence of Legionella, the situation should be treated is an emergency and the investigator should leave the premises along with the other building occupants.

17.4.3 Diagnosing the Problem

When diagnosing the indoor air problem, the investigator should consider the location and timing of the complaints. The investigator must look for patterns in the timing of the complaints, in the symptoms, and in the location of the complaints. Did the onset of the symptoms coincide with an indoor pesticide application? Are they cyclic, for instance, are they present in the morning, and gradually taper off in the afternoon? Has there been a change in the room use or the number of occupants in the room? Has the room been renovated, say, with the installation of new carpeting?

If the complaints are widespread, with no spatial pattern, then the building's HVAC system is a prime suspect. It is also possible that contaminated outdoor air is finding its way inside the building, or some objectionable cleaning material being used in the building. Localized problems tend to have a localized source: some chemical in the area, or maybe a problem with the HVAC's air vent or return.

If the complaints lack consistency, that is, spatially related but with different symptoms, the investigator must ask: could these diverse symptoms have the same cause? Are the symptoms continuous? Are they present outside of the workplace? Do they start when the individuals arrive at work, or are they still present when the individuals have been away from the office for an extended period of time?

If a single individual has the problem, then the problem may have nothing to do with the building. If a localized problem can't be identified, then the individual may be mistaking a personal problem for a problem related to indoor air quality.

17.5 Air Measurement Methods

Air measurements are usually performed as part of an engineering IAQ assessment. If poor air flow is initially suspected as the cause of the IAQ problems, the investigator can go right to measuring the air.

17.5.1 Air Measurement Devices

Air is generally measured using direct reading instruments, i.e., instruments specifically designed for collecting one type or a few limited types of data. For instance, carbon dioxide

FIGURE 17.4
Collecting multiple gas readings using a direct reading instrument.

is measured using a CO_2 monitor. The direct reading instrument shown in Figure 17.4 can measure CO_2, temperature, humidity, and carbon monoxide simultaneously, making it and meters like it useful tools when collecting air measurements. As with all quantitative measuring instruments, it should be properly calibrated before being used.

Chemical smoke tubes can be used to determine pollutant pathways in the study area, between sections of the building, as well as from outside the building. A micromanometer (or equivalent) can measure the pressure gradient between areas. Switching air handlers or exhaust fans on and off, opening and closing doors, and simulating the range of operating conditions in other ways can reveal the different ways that airborne contaminants move within the building.

Scented oils, whose smell can be tracked throughout the study area, are sometimes used to determine pollutant pathways. However, the nose can quickly become desensitized to odors, making this method unreliable. A tracer gas such as sulfur hexafluoride (SF_6) can provide qualitative as well as quantitative information on air flow patterns inside a building.

17.5.2 Locations of Air Measurements

The investigator should collect indoor air measurements in portions of the building where the occupants are known to spend their time and that are most representative of occupant conditions in that part of the building. If representative areas cannot be identified, then measurements should be collected in the middle of the room or zone. If temperature is the issue, measurements should be collected in locations where the most extreme values of temperature are likely to occur, such as near doors and windows, and corners away from and close to heating or cooling equipment. If temperature stratification is suspected, the investigator should record temperatures from multiple heights in the room. Absolute humidity only needs to be measured at one location in a building with an HVAC system, since humidity within the area of coverage usually homogenizes in a short period of time.

17.6 Air Sampling Methods

The decision to collect air samples for laboratory analysis should be made with caution. In general, the more focused the sampling effort, the more effective it will be, since air sampling results sometimes raise more questions than they answer. Before collecting air samples, the investigator should have a clear understanding of what substances should be tested for, where and what time during the day the tests should occur, and what decision criteria will be employed when interpreting the results.

17.6.1 Air Sampling Locations

Once the chemicals of interest have been selected, the investigator must decide where and when to collect the samples. Samples may be biased towards "worst case" conditions, such as measurements during periods of maximum equipment emissions, minimum ventilation, or disturbance of contaminated surfaces. Worst case sample results can be very helpful in characterizing maximum concentrations to which occupants are exposed and identifying sources for corrective measures.

Often it is advisable to obtain samples during average or typical conditions as a basis of comparison. It may, however, be difficult to know what conditions are typical. Exposures to some pollutants may vary dramatically as building conditions change. Devices that allow continuous measurements of key variables can be helpful.

17.6.2 Air Sampling Methods

The most frequently used method for testing for a variety of pollutants involves the use of *summa canisters*. Summa canisters, which are also discussed in the vapor intrusion section of Chapter 10, are stainless steel chambers that are devoid of air when received from the laboratory. In the field, they are fitted with a regulator that can be adjusted to allow a given air flow into the canister. When the canister is full, the regulator is closed and the canister is shipped to a laboratory for analysis. The air inside the canister can be analyzed for VOCs, pesticides, polychlorinated biphenyls (PCBs), polycyclic aromatic hydrocarbons (PAHs), dioxin, and many other classes of chemicals using other USEPA-approved analytical methods.

Testing for Legionella typically involves the collection of water samples or surface samples when the bacterium's presence is suspected, which are generally places where warm water is present. Samples are usually incubated and analyzed in the same manner as mold samples (see Chapter 19).

One popular method for testing for a specific compound is the use of *Draeger tubes*. Draeger (also spelled Dräger) tubes are glass tubes that are filled with a chemical that will react with the compound of interest. They are connected to a pump, which is designed to draw air into the attached tube at a set pumping rate. If the reactant is exposed to the compound of interest, it changes color. The tube is graduated, so that the quantity of the reactant that changes color can be correlated with the concentration of the compound of interest in the air sample. Short-term tubes are designed for tests lasting 10 seconds to 15 minutes. The less-common long-term tubes are designed to provide TWAs over a 1- to 8-hour period.

17.7 Indoor Air Mitigation

Once the indoor air problem has been identified, the investigator should determine the extent of the problem. This could entail running additional tests for additional contaminants, in other parts of the building, or at different times of the day or even different times of the year.

After the nature and extent of the problem has been determined, a mitigation plan is developed. Mitigation methods generally involve:

- Improved ventilation;
- Air cleaning;
- Pathway constriction or elimination; or
- Source reduction or elimination.

Better air cleaning could be achieved by replacing or fixing broken or damaged existing equipment. Particulate control could be enhanced by installing electrostatic precipitators and particulate filters.

Fixing ventilation problems can, however, be as complicated as taking down walls, rerouting ductwork, or redesigning the HVAC system. Complicated or extensive mitigation efforts usually invoke the services of a professional engineer experienced in mitigation if indoor air quality problems.

Removing the sources of the indoor air problems, where feasible, is the simplest way of solving an indoor air quality problem caused by chemicals or biological agents. Improved housekeeping procedures can solve a myriad of indoor air quality problems. Vacuuming carpeting and fabric-covered furniture regularly can remove animal dander, dust mites, and other annoying and allergenic particulates from the workspace or residence. Removing dust from hard surfaces, toys, and the like is also recommended. Pollutant reduction practices include improved chemical storage and handling procedures, changing to less toxic chemicals, and workplace rules limiting chemicals in personal care products, such as hairsprays. Techniques for controlling the pathways for pollutants may include keeping motor vehicles away from air intakes and ensuring that building heating systems that use fossil fuels are properly adjusted to ensure more complete combustion and can adequately ventilate to the outdoors.

Remediation for Legionella outbreaks in commercial buildings often include flushing the source with hot water (>160°F; 70°C), sterilizing standing water in evaporative cooling basins, replacing shower heads, and in some cases flushing the source with heavy metal salts, which will kill the Legionella.

17.8 Indoor Air Quality and Environmental Consultants

As with asbestos, lead-based paint, and mold, indoor air quality typically is the purview of industrial hygienists, or professional engineers in situations where the remedy involves altering the building or the configuration of its ventilation system. The industrial hygienist typically conducts all aspects of the investigation, assisted by epidemiologists and engineers as needed. Contract laboratories usually perform the sample analyses. Mitigation plans are designed by industrial hygienists or professional engineers, depending on the type of mitigation involved. Outside contractors typically implement the mitigation plans.

Problems and Exercises

1. High levels of carbon monoxide are measured on the third floor of an office building. List the possible reasons for the high measurement, assuming that the measuring instrument was not malfunctioning.

2. Some, but not all workers in an office have complained about the same health symptoms. The health symptoms are consistent with poor indoor air quality.

3. Workers in a building complain about the quality of the indoor air in the building only in early morning hours. What could be causing this indoor air problem?

4. A flood recently impacted the first floor of the building. What possible indoor air quality issues could arise from the flooding?

Bibliography

American Society of Heating, Refrigerating and Air-Conditioning Engineers, Inc., 2004. *ASHRAE Standard 55-2013—Thermal Environmental Conditions for Human Occupancy*, ASHRAE.

American Society of Heating, Refrigerating and Air-Conditioning Engineers, Inc., 2007. *ANSI/ASHRAE Standard 62.1—Ventilation for Acceptable Indoor Air Quality*, ASHRAE.

Bower, J., 2001. *The Healthy House*. The Healthy House Institute.

Dräger, Safety AG & Co. KGaA, 2008. *Dräger-Tubes & CMS Handbook*, 15th ed. Dräger, Safety AG & Co. KGaA.

Interstate Technology & Regulatory Council, January 2007. Vapor Intrusion Pathway: A Practical Guideline. www.itrcweb.org/Team/Public?teamID=22.

New Jersey Department of Environmental Protection, January 2018. Vapor Intrusion Guidance, Version 4.1.

Salthammer, T., Ed., 1999. *Organic Indoor Air Pollutants—Occurrence, Measurement, Evaluation*, Wiley-VCH.

Spengler, J.D., and Samet, J.M., 1991. *Indoor Air Pollution: A Health Perspective*. Johns Hopkins University Press, Baltimore, MD.

Spengler, J.D., Samet, J.M., and McCarthy, J.F., 2001. *Indoor Air Quality Handbook*. McGraw–Hill, New York.

Tichenor, B., 1996. *Characterizing Sources of Indoor Air Pollution and Related Sink Effects. ASTM STP 1287.* ASTM, West Conshohocken, PA.

U.S. Environmental Protection Agency and the United States Consumer Product Safety Commission, April 1995. The Inside Story: A Guide to Indoor Air Quality. EPA Document #402-K-93-007.

U.S. Environmental Protection Agency, July 2000. Home Buyer's and Seller's Guide to Radon. 402-K-00-008.

U.S. Environmental Protection Agency, November 2002. Draft OSWER Guidance for Evaluating the Vapor Intrusion to Indoor Air Pathway from Groundwater and Soils (Subsurface Vapor Intrusion Guidance). Office of Solid Waste and Emergency Response. EPA-530-D-02-004.

U.S. Environmental Protection Agency et al., January 2005. IAQ Tools for Schools. EPA Document #402-K-95-001 (Third Edition).

U.S. Environmental Protection Agency, 2010. Review of the Draft 2002 Subsurface Vapor Intrusion Guidance. Office of Solid Waste and Emergency Response.

18

Radon Investigation and Mitigation

Radon intrusion into a building is a special subset of indoor air problems. This chapter focuses on radon, its origins and effects, the steps in a radon investigation, and the design and construction of a radon mitigation system.

18.1 What Is Radon?

Unlike many if not most of the environmental pollutants discussed in this book, radon, atomic number 86, has no beneficial uses. It is a "noble gas," which means that it does not react with other elements. Unlike the other noble gases, however, radon is *radioactive*, emitting an alpha particle within what is known as the "uranium decay series" (see Figure 18.1). That alpha particle, when inhaled, can damage the delicate tissue deep inside the lungs, which increases the likelihood of lung cancer. Radon is the second leading cause of lung cancer in the United States after cigarette smoking (National Cancer Institute, 2018).

Although people might think that uranium is only found in areas where it is mined, there are traces of uranium in rock formations throughout the United States. Buildings constructed on top of rock formations that contain trace concentrations of uranium have the potential to develop a radon problem. The United States Environmental Protection Agency (USEPA) estimates that one in 15 homes in the United States have elevated concentrations of radon.

What makes radon a problem specific to building interiors is its mobility as a gas. As shown on Figure 18.2, radon atoms can work their way into a building through cracks in the foundation or walls, gaps between construction joints or suspended floors and openings that are part of the building structure, such as floor and basement wall penetrations designed to connect the building to outside sewer, electric, telephone, and natural gas utilities. Radon can even enter a building entrained in drinking water. Entry of radon into the building often is facilitated by the lower vapor pressure inside the building compared to the underlying soil or rock or the exterior barometric pressure, creating a pressure gradient that facilitates radon's intrusion into the building. Once inside the building, radon concentrations can increase, especially if there is little exchange with the outside air, which usually occurs when the house is sealed in times of extreme temperatures (summer or winter), or in portions of buildings with poor ventilation, such as in the basement of a residence.

Radon can't be smelled, tasted, or seen, making it an insidious threat to building occupants. The USEPA has set an action limit for radon inside buildings at 4.0 *picocuries per liter of air* (pC/l). A picocurie is an extremely small number, equal to about the decay of two radioactive atoms per minute. This concentration forms a bright line by which radon hazards inside a building are assessed.

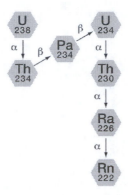

FIGURE 18.1
The Uranium-238 decay series leading to the formation of radon.

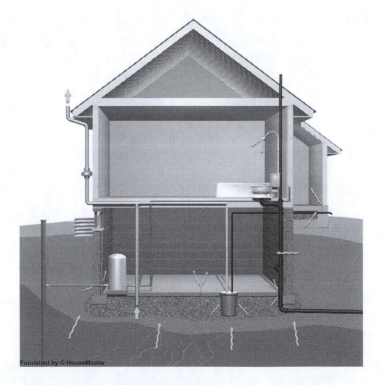

FIGURE 18.2
A diagram showing various paths through which radon enters a building. (Courtesy of HouseMaster Home Inspections. Housemaster® is a registered trademark of DBR Franchising, LLC.)

Millions of homes in this country have been tested for radon. Because of this large database, and because radon is typically related to the geology of the underlying rock and soil, states and territories have developed databases of radon potential by county and in many states by municipality. Following the USEPA protocol, areas are rated as Zone or Tier 1 (high radon potential), Zone or Tier 2 (moderate radon potential) or Zone or Tier 3 (low radon potential). Figure 18.3 shows the radon potential for counties in the United States.

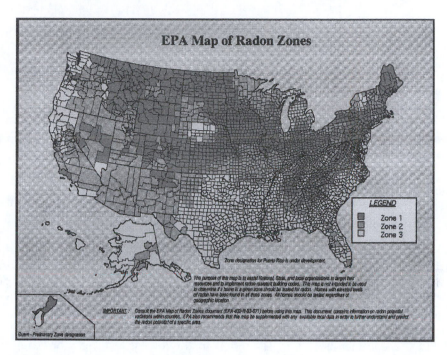

FIGURE 18.3
A map of radon potential zones in the United States. (Courtesy of U.S. Environmental Protection Agency, http://www.epa.gov/radon/images/zonemapcolor_800.jpg.)

These tier designations never should be interpreted as meaning that a radon hazard is or is not present inside a building. Tier ranking is best used in due diligence situations as a guide as to whether a radon survey is warranted. A Tier 1 ranking, the highest risk category, is more likely to trigger a radon survey in a due diligence scenario than would a Tier 3 ranking. In all cases, the only way to determine whether a radon hazard is present is to test for the presence of radon inside the building.

18.2 Radon Investigations

Radon can be tested using active devices or passive devices. *Active radon testing devices*, which can run on batteries or alternating current, continuously measure and record the amount of radon or its decay products in the air. These devices can provide time-specific information on the radon condition inside a building. They are far more expensive than passive screening devices and require a qualified operator to run and maintain the equipment. Active radon testing devices are generally used in industrial settings where radioactive elements are being stored or handled.

Far more common are so-called *passive radon testing devices*. They are called passive because they do not require a power source to operate. Typical passive devices include charcoal canisters, alpha-track detectors, charcoal liquid scintillation devices, and electret ion chamber detectors (see Figure 18.4). These devices are widely available on line and

FIGURE 18.4
A passive radon measuring device. (Courtesy of From the National Institute of Health, Bethesda, MD.)

at retail stores catering to do-it-yourself homeowners. They typically are small enough to fit into the palm of a hand and are black in color so that they can be placed discreetly in various portions of a building. Within the canisters are chambers that are designed to either collect radioactive particles or record the change in ionization to the collection media caused by the radioactive particles. They are then packaged and sent to a laboratory that will measure the radon level implied by the canister contents.

The USEPA recommends testing for radon in the lowest level of a building "suitable for occupancy." This is because the lowest level of a building has the maximum contact with the underlying soil and rock formations and is closest to the penetrations or cracks through which the radon would enter the building, hence the area most likely to contain elevated concentrations of radon. The level needs to be "suitable for occupancy" because the radon limit of 4.0 pC/l is based on risk data that assumes that a minimum residence time in that area for residents to be exposed to the threshold dosage of radon.

Radon testing should be biased to simulate worst-case occupancy conditions at the level being tested. Worst-case conditions should be simulated by sealing the room by closing the doors and windows and disabling all air exchangers such as air conditioners at least 12 hours before conducting the test. That said, the test should be conducted under normal heating & cooling conditions; the investigator will have to balance these priorities in buildings located in hot climates. The canisters should be placed in areas that are not drafty, and placed in or near breathing zone, to simulate exposure conditions for the building occupant. They are typically kept in place for a minimum of 48 hours; the canister manufacturer will specify the recommended residence time for its product.

Weather conditions in the course of the test should be noted. Low barometric pressure, high wind conditions, and storms could affect the test results.

It is often tempting to place the canisters in areas where people are unlikely to reside, such as a boiler or laundry room, because of the decreased likelihood for interference, tampering, or foot traffic. Such canister placement is only acceptable if the test in that room will yield similar results to a test that would be run in a room suitable for occupancy. If a typical building resident spends no more than an average of two hours per week in the basement of a house, say, doing the laundry, the test will not adequately simulate actual resident exposure to radon gas.

In most scenarios, multiple radon canisters are employed within a building to avoid a poor test due to a defective or damaged canister, or due to interference or tampering by a building occupant. Canisters should also be placed in more than one room at the desired building level so that spurious results from unusual building or geologic conditions could be identified by the person interpreting the data. It is also advisable to collect radon information from the next level up in the building, so that if there are exceedances of radon levels, they can be fully understood before implementing a remedy.

Large buildings with multiple levels and multiple rooms on each level pose a challenge to the investigator, since it may be impractical or costly to perform a radon survey in every room suitable for occupancy. In such situations, a site visit prior to the deployment of the radon canisters may be useful. For instance, if certain rooms have more penetrations or more foundation cracks than other rooms, the worst-case scenario would be biasing the testing towards those rooms. If those rooms pass the radon test, then the investigator would be justified in assuming that rooms with less propensity for the presence of radon gas would pass as well. If there are known geological variations beneath the building, however, testing should be done in rooms overlying all known geological strata, since that can affect the quantity of source material for the radon and therefore the amount of radon available to enter the room.

While there is a bright line of 4.0 pC/l for assessing a radon hazard, the USEPA does recognize that natural variations and test variations can affect the results of a radon test. The USEPA's "Home Buyer's and Seller's Guide to Radon" indicates that a result of 4.1 pC/l has a 50% chance of actually being under 4.0 pC/l. If the recorded radon concentrations are close to 4.0 pC/l, it may be advisable to rerun the tests to ensure that the radon hazard has been adequately assessed.

18.3 Radon Mitigation

The easiest and cheapest way to mitigate a radon hazard in a building is to install a radon mitigation system inside the building during construction. For new construction, the existence of a radon hazard could be known from data collected in a prior building on the property, or from nearby buildings, if the geological formations are the same. A radon mitigation system also could be installed prophylactically if the building is located in a Tier 1 or even a Tier 2 area, or if the building owner just wants to be conservative.

A typical radon mitigation system is shown on Figure 18.5. A sub-slab gas-permeable layer [A] is placed below the building foundation, which acts as a low-pressure sump for the collection of gases beneath the building slab. It is covered with plastic sheeting [B], and utility penetrations are sealed and caulked [C], thus preventing uncontrolled entry of gases into the building. A vertical vent pipe [D] is installed to channel gases from beneath the building slab to the roof, where a junction box [E] which operates a small fan vents the gases safely into the atmosphere. As for drinking water, a breathable air *point-of-entry treatment system*, similar to the POET system described in Chapter 14, will control the entry of radon into the building via that pathway.

Figure 18.6 is a diagram of a radon mitigation system installed in an existing building. Because the sub-foundation engineering controls cannot be retrofitted in an existing building, multiple penetrations beneath the slab are warranted. Existing buildings may have sub-slab barriers that cannot be seen. Sub-slab barriers, varying construction practices, possibly created by multiple rounds of construction or renovation, will affect air flow beneath the slab and therefore affect the effectiveness of a radon mitigation system. Alternatively, French drains, floor trenches, or other slab penetrations can impact the effectiveness a radon mitigation system. In such cases, pressure testing is warranted in

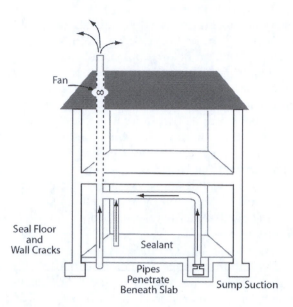

FIGURE 18.5
Components of a passive radon mitigation system. (Courtesy of U.S. Environmental Protection Agency, http://www.epa.gov/radon/pubs/rrnc-tri.html.)

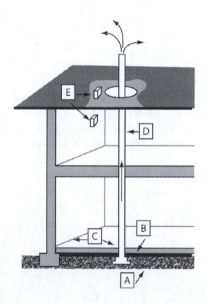

FIGURE 18.6
Diagram of a post-construction radon mitigation system. (Courtesy of U.S. Environmental Protection Agency.)

the lowest level of the building. Pressure tests in the lowest level of the building will demonstrate the extent of the radius of influence a vacuum will have in pulling vapors from beneath the slab, which will dictate the number of penetrations needed for an effective radon mitigation system. This process is similar to the process of vapor intrusion mitigation, described in Chapter 10.

All unwanted pathways will be sealed to prevent uncontrolled entry of gases into the building.

The vertical pipes are then connected to the main riser that vents the vapors safely to the roof of the building. The diameter of the pipes will depend upon the desired air flow. The route of the pipes through a multi-story building can be tricky. Ideally, they can be routed through unoccupied areas, but sometimes they must go through occupied areas. Typically, sheetrock pipe chases are designed to hide the pipes for aesthetic reasons. For the same reason, piping rarely is routed along the exterior wall of the building, although this would be less costly. Lastly, the fan is installed. The air moving capacity of the fan will depend upon the desired air flow.

18.4 Radon Investigation and Mitigation, and Environmental Consultants

Environmental consultants may conduct passive radon testing, although certain jurisdictions may require the consultant to be licensed to perform the testing. As with vapor mitigation (see Chapter 10), the design and construction of a radon mitigation system is in the purview of the environmental consultant. Typically, the design plans must be signed by a professional engineer, and construction of the radon mitigation system is overseen by an engineer working under the P.E.

Problems and Exercises

1. Radon canister testing requires a specific range of temperatures and relative humidity in the area being tested. Explain why conditions outside of the specified range of temperatures and relative humidity could affect the test results.

2. Describe two methods by which potential radon concentrations on an undeveloped property could be assessed.

3. Many of the other elements in the uranium decay series also are radioactive. Why aren't they regulated in the same manner as radon?

4. An elevated reading of radon is measured in one of several radon samples inside a building. Explain the conditions under which a building-wide radon mitigation system would be installed.

Bibliography

Bell, William. Massachusetts Department of Public Health, 1992. Radon: What You Don't Know Can Hurt You!

National Cancer Institute, 2018. https://www.cancer.gov/about-cancer/causes-prevention/risk/substances/radon/radon-fact-sheet.

Oregon Department of Health Services, 2016. Radon Protection Services. www.oregon.gov/oha/ERD/Pages/January-Radon-Action-Month.aspx.

U.S. Environmental Protection Agency, 1992. Indoor Radon and Radon Decay Product Measurement Device Protocols. USEPA Office of Air and Radiation.

U.S. Environmental Protection Agency, 2009. A Citizen's Guide to Radon. http://www.epa.gov/radon/pubs/citguide.html. Retrieved 2008–06–26.

U.S. Environmental Protection Agency, 2012. A Citizen's Guide to Radon: The Guide to Protecting Yourself and Your Family From Radon. EPA402/K-12/002.

US. Environmental Protection Agency. www.epa.gov/rpdweb00/understand/chain.html.

19

Mold Investigation and Mitigation

19.1 Introduction

Mold is the most common of biological hazards. Part of the kingdom known as fungi, over 100,000 varieties of mold have been identified, and scientists estimate that another 100,000 varieties are yet to be identified. Molds can survive in the most hazardous of environments in the form of spores, which are shells that protect the mold from harm until they find an environment in which they can thrive, at which point they shed their shells and begin to live. Mold typically will grow in circular patterns known as *colonies* (see Figure 19.1).

19.1.1 Health Impacts from Mold

Because mold has been around humans for as long as there have been humans, our species has adapted to being around mold, and is immune from most of its health hazards. In fact, many molds, such as the mold from which penicillin is extracted, are beneficial to humans.

It is the molds that produce *mycotoxins* that create the health issues. There are over 100 molds (a minute fraction of all molds) that are known to produce mycotoxins, which are compounds produced by molds that are toxic to humans and animals. Mycotoxins enter the human body by inhalation, and then generally attach the fat-rich organs in the human body, such as the liver or the kidneys. Temperature, pH, as well as the presence of other microbial agents, such as fungi, bacteria, and viruses, can affect the ability of a mold to produce mycotoxins.

The health effects of mold do not follow a dose-response relationship (see Chapter 8), which is how risks from chemicals and radionuclides are generally assessed. Instead, health effects caused by mycotoxins vary significantly by individual susceptibility. Populations most impacted by mold include the elderly, people with allergies or asthma, or people who are immuno-compromised. Adverse health effects can be acute, due to high, long-term exposure, or chronic, due to long-term exposure. The lack of a dose-response health impact relationship has prevented the scientific and regulatory communities from developing permissible exposure limits and maximum permissible concentrations for mycotoxins.

19.1.2 Types of Toxic Molds

The best known toxic mold is *Stachybotrys chartarum (atra)*, or black mold. Exposure to the any of the 16 species of *Stachybotrys* can cause allergic rhinitis, skin irritation, sinus infection, pink eye, fatigue, and, in rare instances, brain damage and memory loss. There is evidence linking it to acute pulmonary hemosiderosis (bleeding in the lungs) in infants. *Stachybotrys* is not a hardy mold, and usually fares poorly competing with other molds for food and water. Consequently, the levels of *Stachybotrys* commonly found inside a building

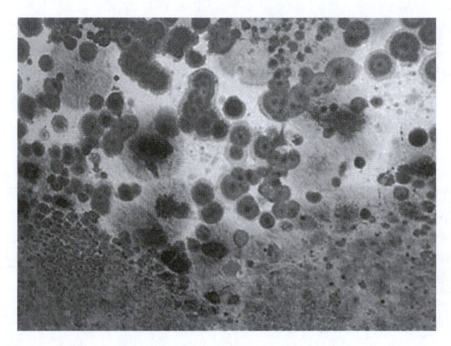

FIGURE 19.1
Mold growth as seen through a microscope. (Courtesy of Robert Kolodin.)

tend to be low compared with other molds, although that does not diminish its potential for harm. There are no known beneficial uses of *Stachybotrys*.

Aspergillus (sp.) is typically found in soil and originates from decomposing plant matter, household dust, building materials, and ornamental plants. There are 160 species of *Aspergillus*, 16 of which are toxic. Exposure to toxic forms of *Aspergillus* can cause sinus and local infections in vulnerable populations. These infections can be fatal in immune-compromised people. Certain species of *Aspergillus* produce the cancer-causing mycotoxin known as alfatoxin.

Penecillium (sp.) is commonly found in soil, food, cellulose, and grain. Some of the estimated 2,000 species of *Penecillium* are well-known for their medicinal value. Exposure to toxic forms of *Penecillium* can lead to skin, lung, and gut infections in immuno-compromised people.

Other, less common forms of toxic mold include some species of *Fusarium*, *Trichoderma*, and *Memnoniella*.

19.1.3 Conditions Conducive to Mold Growth

Molds live comfortably in temperatures ranging from 40°F to 100°F (4°C to 37°C), the same temperature range as is found inside buildings. As with most living things, molds need oxygen, a nutrient source, and water to grow and amplify. Foods for molds, in general, are items with cellulosic content, such as wood products, paper products, and organic floor and wall coverings, and other common building materials.

Moisture usually derives from *water intrusion* or from humid conditions. Water intrusion can be caused by flooding or water leaking into the building or from its piping. Figure 19.2 shows a water-damaged ceiling tile. Relative humidity above 60% (the norm is 30% to 50%) inside a building also is conducive to mold growth.

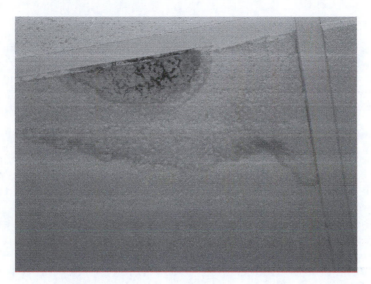

FIGURE 19.2
Mold growth on a water-damaged ceiling tile. (Courtesy of GZA GeoEnvironmental, Inc.)

19.2 Conducting a Mold Survey

There are several reasons for conducting a *mold survey*, which involves the identification of mold and conditions conducive to mold growth. Among the reasons to conduct a mold survey is in response to a health-related issue, reports of unpleasant musty or moldy odors, or the direct observation of mold. Another significant trigger for a mold survey is in support of a property transaction.

In 2006, ASTM International issued a standard designed to guide practitioners conducting mold surveys. However, the standard, known as E2418-06, was withdrawn in 2015, with no replacement in the works. Therefore, the guidelines to mold surveys provided in this chapter follow generally accepted practices rather than a published standard.[1]

The goal of a mold survey is to identify mold and conditions that are conducive to mold growth, such as deferred maintenance of a commercial building's air heating, ventilation, and air conditioning (HVAC) systems, and moisture conditions. The scope of the mold survey needs to be understood by all parties prior to the onset of field activities. Surveys triggered by a health-related issue usually will be more comprehensive than a survey triggered by, say, a property transaction, in which it would be acceptable to observe "representative" offices in an office building, or "representative" apartments in a multi-tenant building, and then extrapolate the possible presence of mold in the uninspected portions of the building.

The three major steps in a mold survey are (1) a walk-through survey; (2) document reviews; and (3) interviews.

---

[1] In 2015, because ASTM Standard E2418-06 had not been updated in nine years, it was withdrawn as per ASTM regulations governing technical committees.

19.2.1 Walk-through Reconnaissance

During the walk-through reconnaissance, the investigator looks for evidence of mold growth and water intrusion, the latter being in many ways more important to identify than the mold growth itself. The most important areas to investigate are those where mold was reportedly seen or smelled, or where water intrusion is suspected or possible. Such areas should include places where water is known to be present, including bathrooms and kitchens; portions of the building near water bodies or septic systems; and areas with evidence of water damage or musty odors, especially basements. In these areas, the inspector will look for mold growth on cellulose-containing surfaces, which include cardboard, paper, wallboard and ceiling tiles (both wallboard and ceiling tiles have paper surfaces front and back, with minerals and mineral fibers within). A thorough mold survey should also include an inspection of the HVAC system.

Some building materials, such as drywall covered with vinyl wallpaper, carpet, or wood paneling, may act as vapor barriers, trapping intrusive moisture underneath their surfaces and thereby providing a moist environment where mold can grow. Therefore, the investigator also should observe "hidden areas" such as crawl spaces, utility tunnels, and the tops of ceiling tiles. Several types of meters are available to aid the inspector in identifying building materials that have experienced water intrusion.

Moisture meters are used to record the amount of moisture in a building component. *Borescopes* are used to physically look into spaces not otherwise visible to the inspector, such as the interstitial space between wallboard units and crawl spaces. Access to the interstitial space between wallboard units can be obtained by removing lighting switch plates or electrical outlet plates. Borescopes are sometimes used in conjunction with wall check sampling devices, which enable the inspector to collect bulk samples of suspect mold growth. Remote moisture meters are also used to measure moisture behind walls (see Figure 19.3).

Infrared thermography (IT) measures the thermal condition of walls and floors by producing an image of infrared light (which is invisible to the human eye) that objects emit. The image depicts variations in temperature which need to be verified with a moisture meter whether the variations in temperature are related to an excursion in the amount of moisture in the object. An IT recording device converts the variations in infrared light into spectra observable by the human eye. By interpreting the image recorded, known as thermogram, or thermograph, the investigator can identify surfaces that have experienced water intrusion.

19.2.2 Document Review and Interviews

As with many types of environmental investigations, reviewing relevant documents and conducting interviews with knowledgeable parties enables the investigator to build a conceptual model of mold and water intrusion conditions inside the building. The investigation should record known mold-related health issues among building occupants, past instances of water intrusion, and building maintenance issues which might have resulted in water intrusion that would favor the growth of mold. When health issues are involved, these two steps are critical, and should be performed before the walk-through survey is conducted, so the walk-through survey can be biased towards the areas of concern.

Documents available from the building owner, building tenant, or the local health department will provide important information pertaining to the mold survey, especially when a health-related incident has occurred. Useful documents may include prior mold

FIGURE 19.3
Using a moisture meter to measure moisture behind a wall. (Courtesy of GZA GeoEnvironmental, Inc.)

and water intrusion surveys; indoor air quality reports; health department inspections, violations, etc., which may be related to mold conditions; and engineering reports, especially HVAC maintenance and repair reports, structural engineering reports or property condition assessments, which may touch on water intrusion issues.

Whenever possible and as applicable, the people with mold-related symptoms should be interviewed. Such interviews can be tricky, since allergic symptoms can appear for a variety of causes in addition to the presence of mold. Assessing the timing of events and the location of things are critical. Did the symptoms appear before or after water intruded into the basement? Are there other factors that could be contributing to the health issue? People with knowledge of the building and its systems should be interviewed as well.

19.3 Mold Sampling and Analysis

Unlike many other types of environmental investigations, mold can be observed; mold sampling is not required and often not even needed. Once mold is observed and its visible extent is delineated, the investigator can go right to the remediation step without the time and expense of collecting samples. Often, however, there may be reasons to collect mold samples even when the presence of mold is obvious, such as part of the medical evaluation process or a litigation.

19.3.1 Bulk Sampling Methods

Sampling methods for mold fall into two general categories—bulk sampling and air sampling. There are four types of *bulk sampling methods.*

- *Gross sample collection* involves using a hand tool to scrape or cut a piece of the suspect material, which is placed in an airtight bag and sent to a laboratory for analysis.
- *Contact, or tape sampling* involves pressing a sticky material, such as tape, onto the surface suspected of containing mold. This sampling method, like gross sample collection, requires visual confirmation of the presence of mold to be effective (see Figure 19.4).
- *Wipe sampling* involves applying a gauze pad or equivalent onto a measured portion of the surface suspected of containing mold.
- *Microvac dust vacuum sampling* uses a micro-vacuum to collect dust samples for laboratory analysis. This equipment used in this process is similar to air sampling, as described below. It is the only bulk sampling method that does not require visual confirmation of the presence of mold. This sampling method is typically employed when it is unfeasible to make visual confirmation due to existing conditions, such as when obstructions prevent visual inspection for mold.

Two types of analysis can be performed on bulk samples: *viable analysis* and *non-viable analysis.* Viable analysis looks for molds that are living or have the potential to live and are incubated in a laboratory prior to analysis. Non-viable analysis looks for all molds and mold-related structures, such as spores, that is, molds that have died as well as those that are living or have the potential to live. Samples to be analyzed for viable molds must

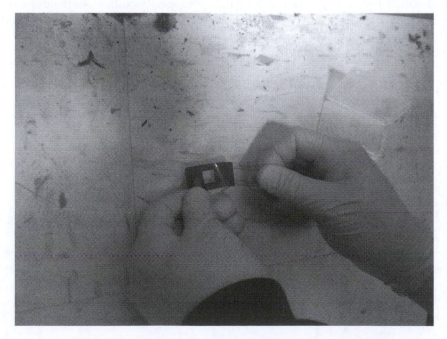

FIGURE 19.4
Using tape to collect a bulk mold sample.

TABLE 19.1

Total Mold Analytical Results

Location	Results (Count/gram)	Primary Fungal Spores
Header Over Door, Family Room	78,200	*Cladosporium, Amerospores*
North Wall of Family Room	2,300,000	*Cladosporium*
Bulk Insulation at Western Roof	13,835	*Cladosporium*
Bulk Insulation in Attic	9,384	*Cladosporium, Nigrospora*

TABLE 19.2

Viable Mold Analytical Results

Location	CFU/m	Primary Viable Molds
New Kitchen Area	966	*Cladosporium, Penicillium, Fusarium*
Basement Crawl Space	530	*Penicillium*
Basement Office Area	2,097	*Penicillium, Cladosporium*
Master Bedroom	141	*Aspergillus, Penicillium*
Outside Background	3,588	*Cladosporium*

incubate for a period of time with the assistance of a growth media, such as an agar culture, to give the spores an opportunity to germinate and form colonies.

Table 19.1 provides a summary of non-viable analyses of bulk samples collected inside a house. The results, shown in *mold counts per gram*, indicate that mold is present at the four sampling locations, which is typical, but is far more prevalent on the north wall of the family room. Since there are no standards for the presence of mold, all four detections can be considered significant; their significance is left to professional judgment. In the case presented in Table, 19.1, the molds detected are not associated with mycotoxins.

Table 19.2 shows the analytical results for viable mold analyses, measured in *colony forming units per meter (CFUs/m)*. The CFUs are used since each mold spore can grow into a colony, forming the characteristic circular mold growth pattern shown in Figure 19.1. The quantities are generally less than the quantities shown in Table 19.1, and do not suggest a huge infestation of mold. On the other hand, some of the mold geneses shown in the table are associated with mycotoxins. Mold geneses could include many species, so that *speciation* would be warranted to identify which species within the genus are present in the sample, thus aiding the identification of molds associated with mycotoxins.

19.3.2 Air Sampling for Mold

Because mold spores are ubiquitous, air sampling for mold spores cannot conclusively distinguish between the normal background presence of spores and the presence of mold spores due to excessive mold growth. Also, as with any type of air sampling, mold air samples provide only a snapshot of the moment in time in which the sampling occurred. They do not take into account mold variations by season, by time of day, or by temperature or relative humidity. Due to these limitations, air sampling is not part of a routine mold assessment. On the other hand, air sampling for mold can be useful to assist in locating mold that couldn't be visually located or bulk sampled, in support of a health-related investigation or a mold-related litigation, or to assess the effectiveness of mold remediation efforts.

Selecting the air sampling locations must be decided on a case-by-case basis. At a minimum, at least two samples should be collected indoors: one in the area of interest and one near the area of interest, to provide information on the potential spread of mold through the building. Because mold is in the ambient air, a background sample should be collected from an area unaffected by mold growth, typically an outdoor location on the property.

To collect an air sample for non-viable mold analysis, an air pump is attached to an impact cassette (Figure 19.5), which is equipped with a sticky surface that will capture the mold, mold spore, or other bioaerosol when drawn into the cassette. The collection method calls for running the air pump at 15 liters per minute (L/min) for up to ten minutes. A wall check sampling device, connected to an air pump, can enable the collection of air samples from within wall cavities. The cassette is sent to a laboratory, where the sticky surface is removed, stained, and analyzed under a conventional optical microscope. As with bulk samples, air samples can be analyzed either for viable mold or non-viable mold. The laboratory will normalize the analytical results to a standard based on the air pump flow rate and time used to collect the sample.

To collect an air sample for viable mold analysis involves *liquid impingement* or *gravitational sampling* techniques. Under the liquid impingement technique, an air pump pulls bioaerosols into a liquid collection medium, which is then placed into an agar plate. The gravitational technique involves using an air pump to send bioaerosols impinging through an impactor designed to hold an agar growth medium. The sample collection method calls for running the pump at 28.3 L/min for up to ten minutes. The agar plate is then sent to a laboratory to incubate for a few days prior to analysis for viable mold.

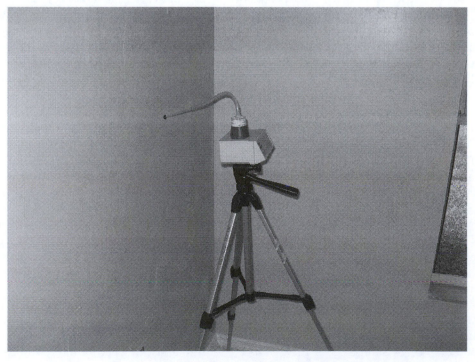

FIGURE 19.5
Collecting an air sample from behind a sheetrock wall. (Courtesy of Robert Kolodin.)

TABLE 19.3

Air Sampling Results for Total Mold

Sample Location	Room 303		Room 305		Outdoor	
Spore Types	Raw Count	Count/m³	Raw Count	Count/m³	Raw Count	Count/m³
Alternaria	0	0	0	0	2	88
Cladosporium	3	132	4	176	25	1100
Penicillium	2	88	2	88	8	352
Stachybotrys	2	88	0	0	0	0

For all air sampling methods, sample analysis should follow analytical methods recommended by the American Industrial Hygiene Association (AIHA), the American Conference of Governmental Industrial Hygienists (ACGIH), or other professional guidelines.

19.3.3 Interpretation of Air Sampling Results

An example of mold air sampling results is provided in Table 19.3. In this table, the analytical results are normalized to counts per cubic meter of air. Of significance in Table 19.3 is the detection of *Stachybotrys* in Room 303, as well as *Penicillium*, some species of which can produce mycotoxins. It should be noted that only two actual *Stachybotrys* mold or mold spores were detected in Room 303. This is a small number, which does not differ statistically from a reading of, say, one or four, and therefore can be misinterpreted. However, the presence of any *Stachybotrys* is a concern, especially since no *Stachybotrys* was detected in the outdoor baseline sample.

19.4 Mold Remediation

As stated earlier, there are no regulations regarding mold remediation, only guidelines. The most widely used guidelines are described in this section of the chapter.

19.4.1 Water Intrusion Mitigation

The first step in any mold remediation activity is addressing the water intrusion problem that contributed to the mold growth. There are several practical steps that can be taken to prevent water intrusion inside the building, including the following:

- Fix leaky plumbing and leaks in the building envelope as soon as possible.
- Look for condensation and wet spots and mitigate the sources of the moisture.
- Maintain low indoor humidity, below 60% relative humidity (RH), ideally 30%–50%, if possible.
- Keep HVAC drip pans clean, flowing properly, and unobstructed.
- Vent moisture-generating appliances, such as dryers, to the outside where possible.
- Don't let foundations stay wet. Provide drainage and slope the ground away from the foundation.

19.4.2 Worker Protection during Mold Remediation

Acute exposure to mold, known as *organic dust toxic syndrome* (*ODTS*), can affect workers performing mold remediation. A more chronic mold-related illness is *hypersensitivity pneumonitis* (*HP*), which results from repeated exposures to the same toxic mold and therefore a concern to mold remediation workers. However, the lack of a dose-response health relationship makes it difficult to establish reliable guidelines to protect workers from exposure to mold. OSHA recommends that remediation workers use respiratory protection and protective suits when performing mold removal (see Figure 19.6). Such personnel protective equipment are discussed in Chapter 15.

19.4.3 Mold Remediation Methods

In most cases, the best method for remediating mold-impacted building components is removal and replacement. Simply scrubbing with water detergent or a damp rag will remove mold from hard surfaces, such as resilient flooring. Wet vacuums should be used to remove excess moisture from porous building components, carpeting, and furniture. The final cleanup should include drying the remediated area, then using a high efficiency particulate air (HEPA) vacuum to remove remaining debris. Mold-contaminated materials should be placed in plastic bags or sheets to reduce the potential for recontamination. As with asbestos, an encapsulant can be utilized post-removal to lock in the remediated surface, or as a stand-along remediation method, especially for situations where removal is impractical, such as for structural components.

FIGURE 19.6
Mold removal worker wearing a protective suit and respiratory protection. (Courtesy of GZA Geo Environmental, Inc.)

Using *biocides* such as chlorine bleach on the mold-impacted areas will kill the mold; however, dead mold can still be toxic, since the mycotoxins may still be present, lying in wait within their spores to wreak havoc. Therefore, it should only be used in limited circumstances, most importantly when a building's structural components, which cannot be safely or inexpensively removed, are affected.

19.4.4 U.S. Occupational Safety and Health Administration Guidelines for Workplace Preparation

There is some disagreement in the scientific community whether projects should be defined by their size, in lieu of meaningful dose-response data. Nevertheless, OSHA and USEPA have issued guidelines for mold remediation.

The OSHA guidelines identify four different project levels, based on the square footage of mold-impacted building components to be remediated, and establishes a separate project level for the remediation of HVAC systems in commercial buildings. The USEPA defines three different project levels. Many of the materials and procedures described here are intentionally similar to materials and procedures used in asbestos abatement, as described in Chapter 15.

A *Level I remediation project* (labeled a "small" project by the USEPA) involves removal of less than 10 square feet (SF) of impacted building components. Level I remediation can be performed by maintenance personnel who are trained to use respiratory protection and wearing half-face respirators (minimum). The work area should be unoccupied, with no containment necessary. Standard dust suppression methods are used, and materials that cannot be adequately cleaned are placed in airtight, 6-mil polyethylene ("poly") bags and sent off-site as ordinary waste. Nearby surfaces should be cleaned with a damp cloth or mop and a detergent solution, then allowed to dry. The project is completed when the remediated area is dry and visibly free of contamination and debris. No clearance testing is required for a mold remediation project.

A *Level II remediation project* covers the remediation of 10 to 30 SF of impacted building components (the USEPA labels as a "medium" project remediation of 10 to 100 SF of impacted building components). The USEPA guidelines define three levels of mold remediation projects: small (<10 SF), medium (10 to 100 SF), and large (>100 SF). A Level II project differs from a Level I project in that it requires the placement of secured plastic sheeting over surfaces that could become contaminated in the course of the mold remediation activities. For medium-sized projects, the USEPA recommends the use of a "limited" containment, which consists of an enclosure consisting of one layer of 6-mil poly around the work area and sealing off of air vents, doors, and other openings to outside the work area, and the establishment of negative pressure within the limited containment.

A *Level III remediation project* covers the remediation of 30 to 100 SF of contiguous impacted building components. A Level III remediation project differs from a Level II project in that it requires the use of workers experienced in mold remediation and requires the establishment of a "limited" containment as described above.

A *Level IV remediation project* (a "large" project, as defined by USEPA), entails remediation of more than 100 SF of mold-impacted materials. On Level IV projects, workers should wear full-face respirators, protective clothing, and gloves. The work area should be completely isolated prior to the onset of remediation activities and placed under negative air pressure with the usage of an exhaust fan equipped with a HEPA filter (see Figure 19.7). Engineering controls such as double-sheeted airlocks and a decontamination unit should be utilized to prevent cross-contamination of nearby areas (see Chapter 15 for a detailed

FIGURE 19.7
Entrance to a fully contained mold abatement work area. (Courtesy of GZA GeoEnvironmental, Inc.)

description of a decontamination unit). The project manager should consider vacating the surrounding areas during the mold abatement activities, especially people who are prone to mold-related illnesses.

19.4.5 Heating, Ventilation, and Air Conditioning System Remediation

HVAC systems impacted by mold present a special problem, since the mold will spread throughout a building as long as the HVAC system is contaminated. Remediation of a contaminated HVAC system involves shutting down the source of the water into the system and removal of cellulosic materials inside the HVAC system, whose presence indicates a malfunctioning or poorly maintained system). Biocides can be used as recommended by manufacturer, especially in cooling coils, condensation pans, and other HVAC components that typically hold moisture.

19.5 Mold Surveying and Remediation, and Environmental Consultants

There are no nationally recognized certifications in the field of mold (except for certified industrial hygienists), although certain states now have no certification requirements for the person or persons conducting any portions of the mold survey. Excepting those states, mold surveys typically are performed by professionals with the same type of training provided to asbestos and lead-based paint workers, but who lack certifications specific to mold. Environmental consultants typically don't get involved in routine mold remediation activities, since there are no air monitoring requirements, at least at the federal level. Consultants may be asked to inspect the remediated area to ensure that no mold hazard or water intrusion issue remains.

Problems and Exercises

1. What evidence exists that there is no across-the-board dose-response relationship between mold exposure and health effects?

2. High levels of toxic molds are measured in an indoor air sample. However, there is no visual or olfactory evidence of mold or evidence of water intrusion in the area. List three possible reasons that high levels of toxic molds were measured in the sample.

3. Evaluate the pros and cons of non-viable analysis versus viable analysis of bulk samples for the presence of mold.

4. Discuss the advantages and disadvantages of collecting air samples for mold using a high flow rate over a short period of time versus using a low flow rate over a long period of time.

Bibliography

American College of Occupational and Environmental Medicine, 2002. Adverse Human Health Effects Associated with Molds in the Indoor Environment.

American Industrial Health Association, 2004. AIHA Guidelines On Assessment, Remediation and Post-Remediation Verification of Fungi in Buildings.

ASTM International, 2006. Standard Guide for Readily Observable Mold and Conditions Conducive to Mold in Commercial Buildings: Baseline Survey Process E2418–06 (withdrawn).

Canada Health Department, 1995. Guide to Recognition and Management of Mold.

GZA GeoEnvironmental, Inc., 2003. Mold Identification, Recognition, Measurement, Toxicity and Abatement (unpublished).

National Clearinghouse for Worker Safety and Health Training, 2005. Guidelines for the Protection and Training of Workers Engaged in Maintenance and Remediation Work Associated with Mold.

New York City Department of Health and Mental Hygiene, 2008. Guidelines on Assessment and Remediation of Fungi in Indoor Environments.

Occupational Safety and Health Administration, 2003. A Brief Guide to Mold in the Workplace. OSHA Safety and Health Information Bulletin (SHIB).

U.S. Environmental Protection Agency, Office of Air and Radiation, Indoor Environments Division, March 2001, reprinted in September 2008. Mold Remediation in Schools and Commercial Buildings. EPA 402-K-01–001.

Appendix A: List of Abbreviations

AAI	All Appropriate Inquiry(ies)
ABIH	American Board of Industrial Hygiene
ACBM	asbestos-containing building materials
ACGIH	American Conference of Governmental Industrial Hygienists
ACM	asbestos-containing material
ACOE	Army Corps of Engineers
AE	acid extractable compounds
AHERA	Asbestos Hazard Emergency Response Act
AIHA	American Industrial Hygiene Association
AIPG	American Institute of Professional Geologists
AOC	area of concern
APR	air purifying respirator
ARARs	applicable or relevant and appropriate requirements
AS	air sparging
ASHARA	Asbestos School Hazard Abatement Reauthorization Act
ASHRAE	American Society of Heating, Refrigerating and Air-Conditioning Engineers, Inc.
AST	above-ground storage tank
ASTM	American Society for Testing and Materials
ATSDR	Agency for Toxic Substances and Disease Registry
AULs	activity and use limitations
BAF	bioaccumulative factors
BAT	best available technology
BN	base-neutral compounds
BOD	biochemical oxygen demand
BSAF	sediment/soil-to-biota bioaccumulation factors
BTEX	benzene, toluene, ethylbenzene, and xylenes
CAA	Clean Air Act
CADD	computer-aided drawing and design
CATEX (or CE)	categorical exclusion
CDC	Centers for Disease Control and Prevention
CEQ	Council on Environmental Quality
CERCLA	Comprehensive Environmental Response, Compensation, and Liability Act
CESQGs	conditionally exempt small-quantity generators
CEU	continuing education unit
CFR	Code of Federal Regulations
CFS	cubic feet per second
CFU	colony-forming unit
CHMM	Certified Hazardous Materials Manager
CIH	Certified Industrial Hygienist
CNRP	Certified Natural Resources Professional
COD	chemical oxygen demand
CORRACTS	Corrective Action database

CP	Certified Professional
CPSC	Consumer Products Safety Commission
CREC	controlled recognized environmental condition
CWA	Clean Water Act
DAP	diffuse anthropogenic pollution
dBA	a-weighted noise decibels
DCE	dichloroethene
DDT	dichlorodiphenyltrichloroethane
DEHP	diethylhexyl phthalate
DEIS	draft environmental impact statement
DNAPL	dense non-aqueous phase liquid
DO	dissolved oxygen
DQO	data quality objectives
DRO	diesel range organics
EA	environmental assessment
ECSM	ecological conceptual site model
EIS	Environmental Impact Statement
EP	environmental professional
EPCRA	Emergency Planning and Community Right-To-Know Act
ESA	Endangered Species Act
ESA	environmental site assessment
ETS	environmental tobacco smoke
FEIS	final environmental impact statement
FEMA	Federal Emergency Management Agency
FID	flame ionization detector
FIFRA	Federal Insecticide, Fungicide, and Rodenticide Act
FOIA	Freedom of Information Act
FONSI	finding of no significant impact
FS	feasibility study
GAC	granular activated carbon
GFI	ground-fault interruption
GHG	greenhouse gas
GIS	geographic information system
GPR	ground penetrating radar
GRO	gasoline range organics
HAAS	haloacetic acid
HAP	hazardous air pollutant
HASP	health and safety plan
HAZWOPER	hazardous waste operations and emergency response
HDPE	high-density polyethylene
HEPA	high efficiency particulate air
HMI	hazardous materials inventory
HP	hypersensitivity pneumonitis
HREC	historical recognized environmental condition
HRS	hazard ranking system
HUD	U.S. Department of Housing and Urban Development
HVAC	heating, ventilation, and air conditioning
IAQ	indoor air quality
IASL	indoor air screening level

IDIQ	indefinite delivery/indefinite quantity
IHMM	Institute of Hazardous Materials Management
IRIS	Integrated Risk Information System
IRM	interim remedial measures
ISCO	in situ chemical oxidation
ISCR	in situ chemical reduction
ISO	International Standards Organization
IT	infrared thermography
LBP	lead-based paint
LCSP	lead content sampling plan
LEA	local education agency
LEDPA	least environmentally damaging practicable alternative
LEED	Leadership in Energy and Environmental Design
LEPC	local emergency planning committee
LF	linear feet
LNAPL	light non-aqueous phase liquid
LQG	large quantity generator
LSP	Licensed Site Professional
LSRP	Licensed Site Remediation Professional
LUST	leaking underground storage tank
MCL	maximum contaminant level
MCLG	maximum contaminant level goals
MERV	minimum efficiency reporting value
MIP	membrane interface probe
MNA	monitored natural attenuation
MSL	mean sea level
MSW	municipal solid waste
MTBE	methyl tertiary-butyl ether
NAAQS	National Ambient Air Quality Standards
NAPL	non-aqueous phase liquid
NCP	National Contingency Plan
NEPA	National Environmental Policy Act
NESHAP	National Emission Standards for Hazardous Air Pollutants
NGO	non-governmental organization
NIOSH	National Institute for Occupational Safety and Health
NOAEL	no observed adverse effect level
NOB	non-organically bound
NOI	notice of intent
NPDES	National Pollution Discharge Elimination System
NPDWR	national primary drinking water regulation
NPL	National Priorities List
NPS	non-point source
NREP	National Registry of Environmental Professionals
NSDWR	National Secondary Drinking Water Regulations
NTCHS	National Technical Committee for Hydric Soils
NTNCWS	non-transient, non-community water system
NTU	nephelometric turbidity units
NWI	National Wetlands Inventory
O&M	operations and maintenance

ODTS	organic dust toxic syndrome
ORP	oxygen reduction potential
OSHA	U.S. Occupational Safety and Health Administration
PACM	presumed asbestos-containing materials
PAH (also PNA)	polycyclic aromatic hydrocarbons (also known as polynuclear aromatics)
PCBs	polychlorinated biphenyls
PCE	perchloroethene (also perchloroethylene, tetrachloroethene, tetrachloroethylene, perc)
PCM	phase contrast microscopy
PDB	passive diffusion bag
PE	Professional Engineer
PEL	permissible exposure limit
PFAS	per- and polyfluoroalkyl substances
PFOA	perfluorooctanoic acid
PFOS	perfluorooctane sulfonate
PG	Professional Geologist
PID	photoionization detector
PLM	polarized light microscopy
POET	point of entry treatment
POPs	persistent organic pollutants
POTW	publicly owned treatment works
POU	point of usage
PPE	personal protective equipment
PPRTV	provisional peer reviewed toxicity values
PRP	potentially responsible party
PSD	prevention of significant deterioration
PVC	polyvinyl chloride
PWS	Professional Wetland Scientist
QA	quality assurance
QAPP	quality assurance project plan
QC	quality control
RCRA	Resource Conservation and Recovery Act
REA	Registered Environmental Assessor
REC	recognized environmental condition
RFA RCRA	facility assessment report
RFB	request for bid
RfC	reference concentration
RfD	reference dose
RFI	RCRA facility investigation
RFP	request for proposal
RI	remedial investigation
RME	reasonable maximum exposure
ROD	record of decision
RQD	rock quality designation
SAP	sampling and analysis plan
SAR	supplied air respirator
SARA	Superfund Amendments and Reauthorization Act
SDS	safety data sheet

SDWA	Safe Drinking Water Act
SEIS	supplemental environmental impact statement
SF	square feet
SHPO	state historic preservation officer
SHWS	state hazardous waste site
SITE	Superfund Innovative Technology Evaluation
SOCs	synthetic organic contaminants
SPCC	spill control and countermeasures
SQG	small quantity generator
SSDS	sub-slab depressurization system
SVE	soil vapor extraction
SVOCs	semi-volatile organic compounds
SWL	solid waste landfill
SWMU	solid waste management unit
SWPPP	storm water pollution prevention plan
SWS	Society of Wetland Scientists
T&M	time and materials
TAL	target analyte list
TCCD	2,3,7,8-tetrachlorodibenzo-p-dioxin
TCE	trichloroethene
TCL	target compound list
TCLP	toxicity characteristic leachate procedure
TEM	transmission electron microscopy
TICs	tentatively identified compounds
TIF	tax increment financing
TICs	tentatively identified compounds
TMDL	total maximum daily load
TOC	total organic carbon
TPH	total petroleum hydrocarbons
TRI	toxic release inventory
TRV	toxicity reference values
TSCA	Toxic Substances Control Act
TSDF	treatment, storage, and disposal facility
TSI	thermal system insulation
TTHM	trihalomethanes
TWA	time weighted average
UCL	upper confidence limit
UFFI	urea formaldehyde foam insulation
USACOE	U.S. Army Corps of Engineers
USEPA	U.S. Environmental Protection Agency
USFWS	U.S. Fish and Wildlife Service
USGS	US Geological Survey
UST	underground storage tank
VEW	vapor extraction well
VISL	vapor intrusion screening level
VOCs	volatile organic compounds
VSI	visual site inspection
XRF	x-ray fluorescence
ZVI	zero-valent iron

Appendix B: State Environmental Departments

State	State Environmental Agency	Acronym	Web Address
Alabama	Department of Environmental Management	ADEM	www.adem.state.al.us
Alaska	Department of Environmental Conservation	ADEC	www.dec.state.ak.us/
Arizona	Department of Environmental Quality	ADEQ	www.azdeq.gov/
Arkansas	Department of Environmental Quality	ADEQ	www.adeq.state.ar.us/
California	California Environmental Protection Agency	Cal EPA	www.calepa.ca.gov/
Colorado	Department of Public Health and Environment	CDPHE	www.cdphe.state.co.us/
Connecticut	Department of Environmental Protection	DEP	www.ct.gov/dep/site/default.asp
Delaware	Department of Natural Resources and Environmental Control	DNREC	www.dnrec.delaware.gov
Florida	Department of Environmental Protection	DEP	www.dep.state.fl.us/
Georgia	Department of Natural Resources	GADNR	www.gadnr.org/
Hawaii	Department of Health	HDOH	www.hawaii.gov/health/
Idaho	Department of Environmental Quality	DEQ	www.deq.idaho.gov/
Illinois	Illinois Environmental Protection Agency	IEPA	www.epa.state.il.us/
Indiana	Department of Environmental Management	IDEM	www.in.gov/idem/
Iowa	Department of Natural Resources	DNR	www.iowadnr.gov/
Kansas	Department of Health and Environment	KDHE	www.kdheks.gov/environment
Kentucky	Department for Environmental Protection	DEP	www.dep.ky.gov
Louisiana	Department of Environmental Quality	DEQ	www.deq.louisiana.gov/portal/
Maine	Department for Environmental Protection	DEP	www.maine.gov/dep
Maryland	Maryland Department of the Environment	MDE	www.mde.state.md.us
Massachusetts	Massachusetts Dept of Environmental Protection	MassDEP	www.mass.gov/dep/
Michigan	Department of Environmental Quality	DEQ	www.michigan.gov/deq
Minnesota	Minnesota Pollution Control Agency	MPCA	www.pca.state.mn.us/
Mississippi	Department of Environmental Quality	MDEQ	www.deq.state.ms.us/
Missouri	Department of Natural Resources	DNR	www.dnr.mo.gov/env/index.html
Montana	Department of Environmental Quality	DEQ	www.deq.mt.gov/default.mcpx
Nebraska	Nebraska Department of Environmental Quality	NDEQ	www.deq.state.ne.us/
Nevada	Nevada Division of Environmental Protection	NDEP	www.ndep.nv.gov/
New Hampshire	Department of Environmental Services	DES	www.des.nh.gov/
New Jersey	Department of Environmental Protection	NJDEP	www.state.nj.us/dep/
New Mexico	New Mexico Environmental Department	NMED	www.nmenv.state.nm.us/
New York	Department of Environmental Conservation	NYSDEC	www.dec.ny.gov/
North Carolina	Department of Environment and Natural Resources	NCDENR	http://portal.ncdenr.org/web/guest
North Dakota	Department of Health—Environmental Division	DoH	www.ndhealth.gov/EHS/
Ohio	Ohio Environmental Protection Agency	Ohio EPA	www.epa.state.oh.us/
Oklahoma	Department of Environmental Quality	DEQ	www.deq.state.ok.us/
Oregon	Department of Environmental Quality	DEQ	www.oregon.gov/DEQ/
Pennsylvania	Department of Environmental Protection	DEP	www.depweb.state.pa.us/
Rhode Island	Department of Environmental Management	DEM	www.dem.ri.gov/

(Continued)

State	State Environmental Agency	Acronym	Web Address
South Carolina	Department of Health and Environmental Control	DHEC	www.scdhec.gov/
South Dakota	Department of Environment and Natural Resources	DENR	www.denr.sd.gov/
Tennessee	Department of Environment and Conservation	TDEC	www.state.tn.us/environment/
Texas	Texas Commission on Environmental Quality	TCEQ	www.tceq.state.tx.us/
Utah	Department of Environmental Quality	DEQ	www.deq.utah.gov/
Vermont	Department of Environmental Conservation	DEC	www.anr.state.vt.us/dec/dec.htm
Virginia	Department of Environmental Quality	DEQ	www.deq.state.va.us/
Washington	Washington State Department of Ecology	WA DOE	www.ecy.wa.gov/ecyhome.html
West Virginia	Department of Environmental Protection	DEP	www.dep.wv.gov/Pages/default.aspx
Wisconsin	Department of Natural Resources	WDNR	www.dnr.wi.gov/index.asp
Wyoming	Department of Environmental Quality	DEQ	http://deq.state.wy.us/
Washington DC	District Department of the Environment	DDOE	www.ddoe.dc.gov/ddoe/site/

Index

A

AAI, *see* All Appropriate Inquiry(ies) (AAI)

above ground storage tanks (ASTs), 97, 100–101, 112

acetone, 50, 53, 150–151, 202, 225

acid extractables (AEs), 58

acid rain, 29

ACM, *see* asbestos-containing material (ACM)

activity and use limitations (AULs), 95, 170, 192, 242–243

adjoining property, 104–106, 114–116, 118

advection, 87, 134, 170

AEs, *see* acid extractables (AEs)

Agency for Toxic Substances and Disease Registry (ATSDR), 181

AHERA, *see* Asbestos Hazard Emergency Response Act (AHERA)

AIPG, *see* American Institute of Professional Geologists (AIPG)

aircell piping insulation, 304–305

air pathway, 124

air purifying respirator (APR), 140, 348

air purifying unit, 227

air sparging (AS), 187, 201, 212

air stripping, 196–197

alfatoxin, 376

aliphatics, 43, 51

alkanes, 49, 51

alkenes, 49, 52, 205

alkynes, 53

All Appropriate Inquiry(ies) (AAI), 91–92, 105, 120

alpha particle, 63–64, 367

American Conference of Governmental Industrial Hygienists (ACGIH), 310, 383

American Industrial Hygiene Association (AIHA), 310, 383

American Institute of Professional Geologists (AIPG), 9

American Society of Heating, Refrigerating and Air-Conditioning Engineers, Inc. (ASHRAE), 355, 357

Applicable or Relevant and Appropriate Requirements (ARARs), 39, 185–186

APR, *see* air purifying respirator (APR)

aquiclude, 85

aquifer, 60, 84–86, 115, 121–122, 125, 167, 171–173, 192–194, 196–198, 201, 204–205, 249, 274, 288–289

analysis, 182–184

aquitards, 85, 134, 169

ARARs, *see* Applicable or Relevant and Appropriate Requirements (ARARs)

aroclor, 59–60

aromatics, 50, 53–54

artesian condition, 85

asbestos abatement

airless sprayer, 323–324

air samples

aggressive air sampling, 328

area air samples, 327

clearance air samples, 325, 328–329

perimeter air samples, 327

personal air samples, 327

critical barriers, 320–321, 325, 327–328

decontamination units, 319–320, 323–328

encapsulant, 325, 328

encapsulation, 317, 325, 328

enclosure, 317, 328–329

glovebag removal, 324–325

ground-fault interruption, 321

negative air pressure, 321–322

operations and maintenance for in-place asbestos-containing material, 315–317, 329

personal decontamination unit, 319–320

phase contrast microscopy (PCM), 326–328

small and minor asbestos projects, 317, 328

transmission electron microscopy (TEM), 326

waste decontamination unit, 320, 326, 329

waste removal, 326

asbestos-containing building material (ACBM), 301, 303–309

asbestos-containing material (ACM), 301, 307–309, 312, 315–317, 328–330

health problems related to asbestos

asbestosis, 301

lung cancer, 301

mesothelioma, 301

regulatory history, 301

types of asbestiforms, 299

Asbestos Hazard Emergency Response Act (AHERA), 32, 342